BIOTECHNOLOGY IN AGRICULTURE, INDUSTRY AND MEDICINE SERIES

# TECHNOLOGIES AND MANAGEMENT FOR SUSTAINABLE BIOSYSTEMS

# BIOTECHNOLOGY IN AGRICULTURE, INDUSTRY AND MEDICINE SERIES

**Biotechnology and Industry**
*G.E. Zaikov (Editor)*
2007. ISBN: 1-59454-116-7

**Biotechnology in Agriculture and the Food Industry**
*G.E. Zaikov (Editor)*
2004. ISBN: 1-59454-119-1

**Agricultural Biotechnology: An Economic Perspective**
*Margriet F. Caswell, Keith O. Fuglie, and Cassandra A. Klotz*
2003. ISBN: 1-59033-624-0

**Biotechnology, Biodegradation, Water and Foodstuffs**
*G.E. Zaikov and Larisa Petrivna Krylova (Editors)*
2009. ISBN: 978-1-60692-097-8

**Governing Risk in the 21st Century: Lessons from the World of Biotechnology**
*Peter W.B. Phillips (Editor)*
2006. ISBN: 1-59454-818-8

**Research Progress in Biotechnology**
*G.E. Zaikov (Editor)*
2008. ISBN: 978-1-60456-000-8

**Biotechnology and Bioengineering**
*William G. Flynne (Editor)*
2008. ISBN: 978-1-60456-067-1

**Biotechnology: Research, Technology and Applications**
*Felix W. Richter (Editor)*
2008. ISBN: 978-1-60456-901-8

**Biotechnology: Research, Technology and Applications (Online Book)**
*Felix W. Richter (Editor)*
2008. ISBN: 978-1-60876-369-6

**Industrial Biotechnology**
*Shara L. Aranoff, Daniel R. Pearson, Deanna Tanner Okun, Irving A. Williamson, Dean A. Pinkert, Robert A. Rogowsky, and Karen Laney-Cummings*
2009. ISBN: 978-1-60692-256-9

**Industrial Biotechnology and the U.S. Chemical and Biofuel Industries**
*James R. Thomas*
2009. ISBN: 978-1-60741-899-3

**Biosensors: Properties, Materials and Applications**
*Rafael Comeaux and Pablo Novotny (Editors)*
2009. ISBN: 978-1-60741-617-3

**Industrial Biotechnology: Patenting Trends and Innovation**
*Katherine Linton, Philip Stone, Jeremy Wise, Alexander Bamiagis, Shannon Gaffney, Elizabeth Nesbitt, Matthew Potts, Robert Feinberg, Laura Polly, Sharon Greenfield, Monica Reed, Wanda Tolson, and Karen Laney-Cummings*
2009. ISBN: 978-1-60741-032-4

**Biochemical Engineering**
*Fabian E. Dumont and Jack A. Sacco*
2009. ISBN: 978-1-60741-257-1

**Perspectives on Lipase Enzyme Technology**
*J. Geraldine Sandana Mala and Satoru Takeuchi*
2009. ISBN: 978-1-61741-977-8

**Technologies and Management for Sustainable Biosystems**
*Jaya Nair, Christine Furedy; Chanakya Hoysala and Horst Doelle (Editors)*
2009. ISBN: 978-1-60876-104-3

BIOTECHNOLOGY IN AGRICULTURE, INDUSTRY AND MEDICINE SERIES

# TECHNOLOGIES AND MANAGEMENT FOR SUSTAINABLE BIOSYSTEMS

JAYA NAIR,
CHRISTINE FUREDY,
CHANAKYA HOYSALA
AND
HORST DOELLE
EDITORS

Nova Science Publishers, Inc.
*New York*

For permission to use material from this book please contact us:
Telephone 631-231-7269; Fax 631-231-8175
Web Site: http://www.novapublishers.com



**LIBRARY OF CONGRESS CATALOGING-IN-PUBLICATION DATA**

Technologies and management for sustainable biosystems / Jaya Nair ... [et al.].
p. cm.
Includes index.
ISBN 978-1-60876-104-3 (hardcover)
1. Water reuse. 2. Salvage (Waste, etc.) 3. Sustainable development. 4. Biological systems. I. Nair, Jaya.
TD429.T43 2009
628.1'68--dc22

2009027062

*Published by Nova Science Publishers, Inc. ✢ New York*

# CONTENTS

# PREFACE

Over the past centuries, biosystems relied on natural processes for the management of natural resources. The provision of water, food, energy and shelter throughout the world were integrated with biosystems and organic wastes were managed sustainably. Increased urbanization, the green revolution and its dependence on mechanical technologies, coupled with lifestyle changes and market-driven economic centralism, is consuming resources in an unsustainable way. The reassessment and refinement of the old approaches to biosystems would allow the development of decentralised technologies promising real commercial gains without sacrificing a sustainable approach. Such systems, however, require greater efficiency than natural systems to achieve the desired outcomes, and so the management of new technologies is critical.

This book is a compilation of papers presented at the first independent international conference of the International Organisation of Biotechnology and Bioengineering (IOBB) conducted on July, 6-9, 2008 at Fremantle, Western Australia. It was organised by the Environmental Technology Centre, Murdoch University, Western Australia under the title "Technologies and Strategic Management of Sustainable Biosystems". Papers were presented in the areas of integrated biosystems for waste treatment and reuse, nutrient management, health and environmental issues with biosystems, community participation, governance, education and training and case studies. Some of the papers were published in a special issue of the international journal *Bioresource Technology* while those published here were selected by a team of international reviewers from the remaining presentations.

The chapters in this book will be of interest to academics, land developers, policy makers and community workers in understanding various options for using biosystems for the sustainable management of resources. They will be a teaching and reference source for students, researchers and community developers to understand the scope for onsite biological systems for the management of water, waste, energy and food production.

The book has five sections:

| | |
|---|---|
| Section One: | Wastewater treatment and reuse |
| Section Two: | Integrated biosystems |
| Section Three: | Constructed wetlands for wastewater treatment |
| Section Four: | Solid waste management |
| Section Five: | Community governance |

This book covers a wide range of topics spread throughout these 5 sections (with many chapters in each) such as wastewater treatment (7), integrated biosystems (6), constructed wetlands (2), solid waste management (4) and community governance (5).

This conference attracted delegates from the Asia-Pacific region where the integrated biosystems approach has been practised for centuries and many have been up-scaled to meet modern challenges. We acknowledge the support of AusAid in providing travel funding to several delegates from developing countries and thank all the sponsors of the conference, in particular Biolytix Pty Ltd., Environmental Science at Murdoch University, Mindarie Regional Council and LotteryWest (Western Australia).

We thank all the reviewers for their valuable feedback on the manuscripts. We are also grateful to IOBB and all the members of the organising committee of IOBB2008 for helping to organize a very successful conference – IOBB2008.

Dr. Jaya Nair
Dr Christine Furedy
Dr Chanakya Hoysala
Dr Horst Doelle

# SECTION ONE: WASTE WATER TREATMENT

In: Technologies and Management for Sustainable Biosystems ISBN: 978-1-60876-104-3
Editors: J. Nair, C. Furedy, C. Hoysala et al. 

*Chapter 1*

# BIOSYSTEMS FOR GROWING A NON-FOOD BIOFUEL CROP (ARUNDO DONAX) WITH SALINE WASTEWATER

***Tapas K. Biswas*** [*] ***and Chris M.J. Williams***
South Australian Research & Development Institute,
GPO Box 397, Adelaide, SA 5001, Australia

## ABSTRACT

There are no shortages of poor quality water and lands in Australia. Ongoing drought in many regions has renewed interest in alternate uses of saline wastewater rather than disposal to evaporation basins. This paper reports on the underutilised resources of wastewater and saline land to grow a new second generation biofuel crop, namely *Arundo donax* L. (giant reed). *Arundo donax* (Adx) is a perennial, rhizomatous grass that has been grown in every mainland state of Australia for over 150 years. Adx together with other cellulose feed stocks could form the basis of a new biofuel and/or pulp/paper industry for Australia. An integrated biosystem such as "Serial Biological Concentration (SBC)" offers the ability to create a number of financial opportunities whilst concentrating the saline waste water stream to a manageable mass and at the same time growing biomass/biofuel crops.

**Keywords:** Biofuel, *Arundo donax,* wastewater, saline biosystem, drainage, giant reed.

## INTRODUCTION

More than 50 % of the cropped land in Australia is affected by soil acidity, sodicity and salinity problems with an estimated annual impact to the agriculture of A$2,559 million [NLWRA, 2002]. Globally, available fresh water is becoming scarcer as is arable land. Increasing conversion of many food crops into bioethanol has greatly increased the price for

[*] Corresponding author. Tel: +61 08 8303 9730; Fax: +61 08 8303 9473 Email:tapas.biswas@sa.gov.au

food crops such as corn, sugarcane, soybean and canola. For example, in the USA at present, about 30% of corn seed is used for ethanol production. As a consequence, there has been an acute shortage for staple food in many countries.

Introduction of high-yielding, non-food biomass crops to support the change to renewable energy policy is inevitable. To lessen the burden of this food shortage, there is a growing need for sustainable biosystems to use marginal land and waste waters for second generation biofuel or pulp/paper crops [Williams et al. 2007]. *Arundo donax* L. (Adx), commonly known as giant reed, has many potential uses as feedstock for biofuel, pulp/paper or fodder production [Lewandowski et al. 2003; Spafford, 1941; Williams et al., 2008a; Williams et al. 2006]. It is a perennial rhizomatous grass that has persisted for over 150 years in Australia [Jessop et al. 2006].

The Serial Biological Concentration (SBC) system [Blackwell et al. 2005] is one such integrated biosystem, which takes saline water through a series of potentially productive cells or stages and enables communities to generate returns from wastewater flows by growing salt tolerant crops such as Adx rather than discharging to evaporation basins.

Adequate drainage flow (also called 'leaching fraction') is a key to creating a sustainable SBC system. It requires a simple design, which is robust and inexpensive. The SBC system can be installed on existing saline lands, maintaining sustainability of the SBC whilst reclaiming the saline landscape if irrigated with good quality water. Salt can be prevented from reaching river systems using a stand alone viable series of filter cells such as, an agricultural production system including growing a biofuel crop (Adx), a range of saline aquacultures, production of a range of pure salts, and in some cases generating electricity by using solar pond technology.

Giant reed is invasive in riparian systems subject to torrential flooding in many regions of the world [Bell, 1997]. However the lack of fertile seed production limits the establishment of new, distant colonies to those arising from the spread of stem and rhizome fragments [Williams et al. 2008a]. The reputation of Adx as an invasive species in certain riparian systems has discouraged research to develop its full utilisation as a commercial crop. Based on appropriate selection of planting sites (eg. no plantations in riparian zones subject to flooding) and crop hygiene (eg. use of buffer areas, covered transport), Adx could be grown with a manageable level of risk [Czako and Marton, unpublished data; Williams et al. 2008a, b].

This paper describes available saline soil and water resources for an integrated biosystem SBC growing *Arundo donax* (Adx), which has the potential to form the basis of new biofuel and/or pulp paper industries in Australia.

## Saline Water and Soil Resources in Australia

Australia has large reserves of saline (> 5000 mg/L of total soluble salts) ground water with 3,434 GL/year of sustainable groundwater [Ball et al., 2001] unsuitable for drinking or irrigation of traditional crops (figure 1). In addition, urban and peri-urban sewage wastewater produced annually is 1824 GL, of which only 156 GL is reused [Boland et al. 2006].

Along the lower part of the Murray River there are about 124 Salt Interception Scheme (SIS) bores that yield over 10 GL/year of highly saline water, mostly 20 to 50 dS/m, by intercepting the ground water flow before it can enter the river. Similarly, there are also large quantities of saline wastewaters produced from mining and intensive primary industries (abattoirs, wool scouring, and wineries).

Saline soils in Australia and South Australia are estimated to cover 2.6 and 1.4 million ha, respectively [DWLBC, 2008]. In many situations large areas of saline, marginal soils exist adjacent to the highly saline water resources. For example, at Bolivar, South Australian water agencies could treat sewage to a low quality (e.g. class D, high salinity) and use or sell such wastewaters to grow salt tolerant biofuel crops on nearby saline, marginal lands. This would be preferable to discharging to the sea or to evaporation basins. Biofuel crops could be grown as rainfed crops on thousands of hectares of saline soils with moderate rainfall (>400 mm) in the south east of South Australia.

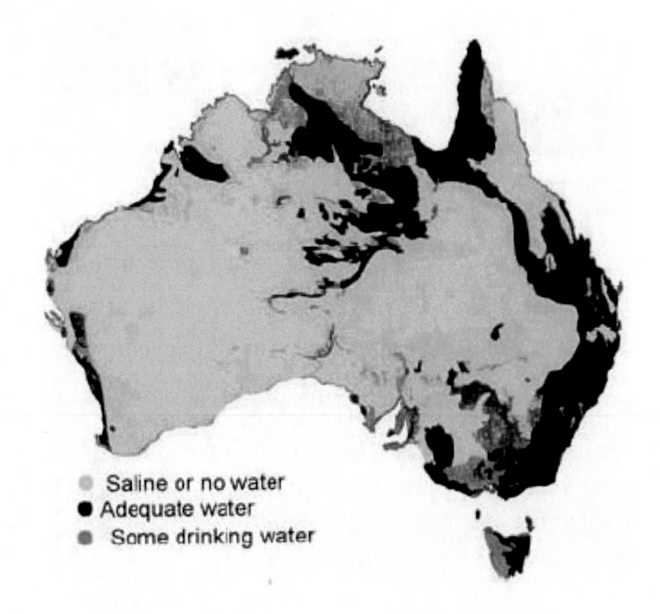


Figure 1. Saline water resources in Australia [source NLWRA, 2008].

## SERIAL BIOLOGICAL CONCENTRATION (SBC) BIOSYSTEM FOR GROWING ADX WITH SALINE WASTEWATER

The SBC biosystem, as shown in figure 2, replicates three stages of a land filter system (cell) to produce drainage waters having different salinities at the end of each filtration event due to 33% leaching fraction, which is often possible to maintain. Therefore, by having a sequence of filter cells, as described by Biswas et al. (2002) and Jayawardane et al. (2001) the SBC biosystem allows the growth of salt tolerant crops like Adx and at the same time achieves both a reduction of drainage volumes and maximises the financial returns from the crops produced. The filter cell uses a system of flood irrigation and subsurface agricultural drains to process sewage effluent by stripping out nutrients, and mitigating pathogens, suspended solids and Biological Oxygen Demand (BOD) using a combination of volatilisation, oxidation, reduction (i.e. denitrification), soil adsorption and plant uptake processes. This sequential agricultural process is terminated when the drainage water is too salty to sustain economic crop production.

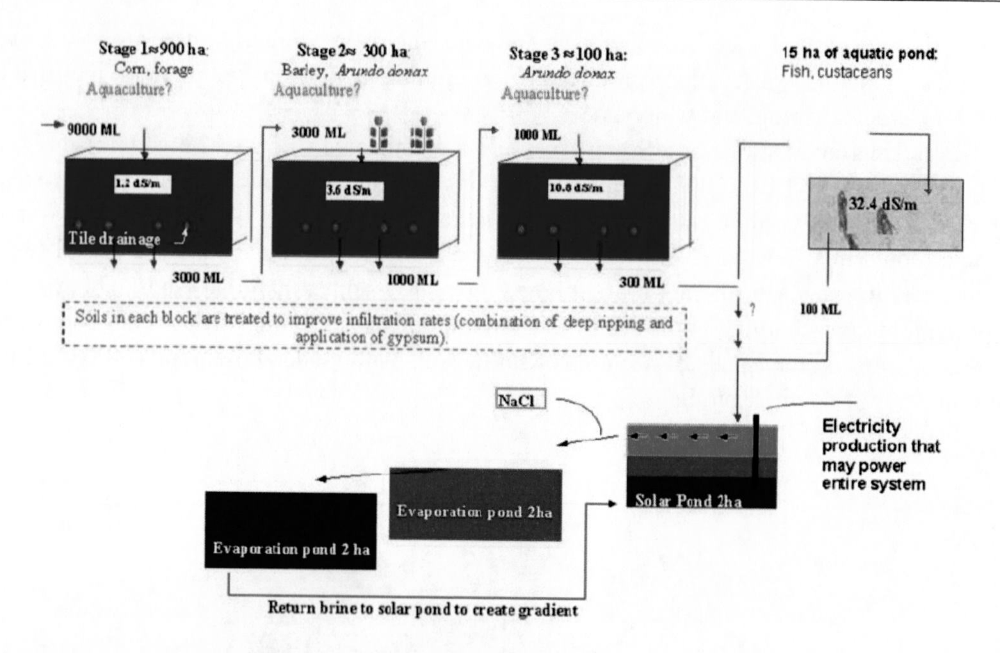


Figure 2. Schematic representation of possible layout, flows and concentrations of salt in SBC biosystem [modified after Blackwell et al. 2000].

The filter cells operation can be planned on a fortnightly cycle, where irrigation is applied over 2 days, followed by a 1-2 day post-irrigation equilibration period and an 8-10 day pumping period. During the pumping period wastewater slowly passes through soil 0.9 to 1.2 m deep agricultural drains and then to a collection sump for use at the next cell. The pump is shut down for 2 days to reach equilibrium before receiving the next vent of irrigation. The cycle is then repeated. During this regulated flow of effluent, nutrients are adsorbed on the surface of soil particles or taken up by the crop and weeds.

The final 3 components of the SBC system consist of production of a range of aquatic species in ponds of varying salinities, salt gradient solar ponds to produce energy and evaporation basins to produce pure salts for industrial use and/or stockpiling for disposal. It is estimated that a well planned SBC system can generate enough electricity through the solar ponds to run all the pumps for irrigation and drainage required for running that SBC biosystem.

## Biomass Yields, Carbon Sequestration and Nutrient Uptake by Adx and Other Crops

In an experiment in Australia, as described by Williams et al. (2008a), Adx which received 21 ML/ha/year of winery wastewater, the Loveday rootstock at the Barmera site produced the highest biomass yields of 45.2 t/ha of dry tops including 35.9 t/ha of dry, bare stems in the first year (table 1). These high biomass yields were similar to those reported at the highest yield sites in Spain [Lewandowski et al. 2003] and at Roseworthy, South Australia [Williams et al. 2006]. The latter studies were conducted on arable land with good quality irrigation water. The biomass yields of of Adx of 45.2 t/ha/year equated to 4 times the dry matter yield/ha/year of irrigated blue gum (*Eucalptus globolus*) stands [Paul and Williams, 2006; Williams et al. 2008a].

The exceptional yields (45 dry t/ha/year) of the irrigated Loveday rootstock grown at Barmera on marginal land far exceeded those from traditional biomass crops grown on arable land with *ad lib* irrigation. For example, Biswas et al. (2002) reported 15 t/ha of total tops dry matter per season for forage sorghum grown on arable land and irrigated *ad lib* with secondary treated sewage near Griffith, New South Wales.

Photosynthesis by Adx was the main mechanism for the large amount of organic carbon (20.6 dry t/ha) sequestered in the dry tops in the first year at the Barmera site for the Loveday rootstock (table 1). Adx crops could qualify for carbon (C) credit programs if introduced in Australia (annual sequestration of carbon). Uptake of nitrogen (N), phosphorus (P) and potassium (K) by the plant tops of Adx from the Loveday rootstock was 528, 22 and 664 kg/ha, respectively, during the first year of growth (table 1).

**Table 1. Biomass yield and nutrient removal by *Arundo donax* (Adx) in South Australia**

| Adx Rootstock/ land type | Biomass yield (t/ha/year) | | | Nutrients removed in Adx tops (kg/ha/year) | | | |
|---|---|---|---|---|---|---|---|
| | Leaf | Stem | Tops | Org C[1] | N | P | K |
| Barmera (saline soil) irrigated | | | | | | | |
| Loveday | 9.3 | 35.9 | 45.2 | 20577 | 528 | 22 | 664 |
| Roseworthy (arable land) | | | | | | | |
| Irrigated | 10.3 | 40.7 | 51.0 | 22200 | 773 | 40 | 832 |
| Dryland[2] | 4.0 | 11.4 | 15.4 | 6500 | 282 | 17 | 331 |

[1] Org C: Organic Carbon

[2] Dryland grown at Roseworthy, South Australia with 461 mm rainfall.

In the SBC system [Blackwell et al. 2000] reported growing different high value crops at the different salinities. Yields of a selection of these crops over a range of salinities for a period of two years (1998-99) are given in figure 3. Yields reported at 12% moisture content showed that yield levels were maintained up to stage 2, which was irrigated with 3.6 dS/m water, but declined to about half in stage 3 when irrigated with 10.8 dS/m water except for sunflower which performed well even with the highest EC water.

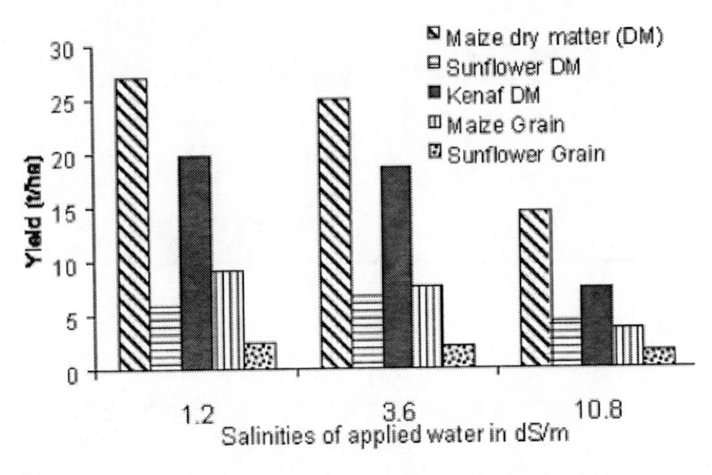


Figure 3. Crop yields at different stages in SBC [Blackwell et al., 2000].

## SECOND GENERATION BIOFUEL CROPS TO USE SALINE LANDS AND WATERS

First generation biofuel is made from the edible organs of food crops. Second generation biofuel is processed from specialist non-food cellulosic biomass crops and agriculture and forestry wastes. Demand for food crops (cereal grains, oilseeds, sugarcane) grown on arable lands to produce first generation biofuel in America and elsewhere is outstripping supply. This has led to rapidly increasing global food prices, very low world grain reserves and increased risks of pending famine.

Second generation biofuel could reduce oil well-to-vehicle wheels carbon dioxide production by up to 90% [Shell, 2007] compared to fossil fuels. Such biofuel may be one of the most cost-effective methods of renewable, low-carbon energy for vehicle transport [Shell, 2007]. Although costs of production are comparable for grain ethanol and cellulosic biofuel, the current estimated higher capital costs of factories processing cellulose may be a short term barrier to their commercialisation [Wright and Brown, 2007]. However, new technologies for biomass-to-liquid (BTL) fuel are advancing rapidly for second generation biofuel. For example, a biofuel company recently processed (in laboratory scale tests in Norway), samples of Australian grown, giant reed (*Arundo donax* L.) into BTL ethanol. *Arundo donax* produced 299 L of ethanol /t of stem dry matter biomass, in a process time of less than 24 hours (Williams et al. 2008a). This equates to 11,000 L/ha of bioethanol (figure 4) from Adx from 40 t/ha of dry tops /year, compared to 4,400 L/ha from corn kernels (*Zea mays*), 4,600 L/ha from switch grass (*Panicum virgatum*) and 8,800 L/ha from sugarcane (*Saccharum officinarum*), the latter three crops from Bourne (2007).

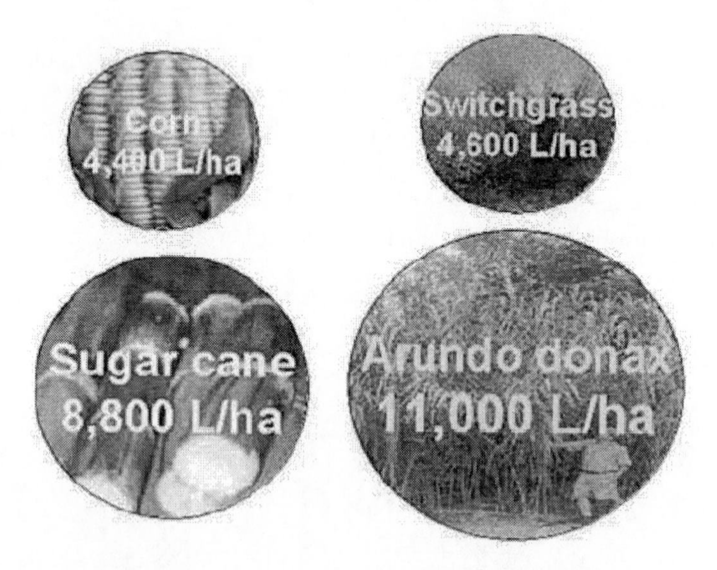


Figure 4. Comparative potential bioethanol production (L/ha/year) from food/non-food crops. Modified from (Bourne, 2007).

The calorific value of 19 MJ/kg of dry Adx biomass [Williams et al. 2008b], this is equivalent to 1 tonne of dry Adx biomass combustion generating 19 GJ or 5320 kWh (electricity for 266 homes for 1 day).

## BIOSYSTEM SOIL SALINITY DYNAMICS AND THE SALT TOLERANCE OF ADX

Periodic measurement of soil water salinity using a SoluSAMPLER$^{TM}$, a suction tube device designed by [Biswas, 2006], and expressed as EC of soil water (referred hereafter as ECswe) were conducted within the Loveday stand. The ECswe readings increased from 5 dS/m in January to 18 - 50 dS/m by April, 2007 (figure 5). In comparison, salinity of ocean water is approximately 55 dS/m. Winter rainfall, which was associated with salt dilution and leaching, reduced ECswe to 17-33 dS/m by the end of June, 2007.

Accordingly, Adx Loveday rootstock was exposed to ECswe of 18-50 dS/m, on average half that of ocean water, from February to June 2007. This coincided with the time of maximum growth rate and very high biomass yields of Adx at Barmera (table 1). According to Flowers et al. (1986), halophytes are those plants that can complete their normal annual life cycle under conditions of over 150 mM (15 dS/m) rootzone salinity. We classed Adx as a halophyte due to its tolerance to prolonged periods of extreme salinity [Williams et al. 2008]. Ratios of K:Na in plant parts were calculated to further assess the salt tolerance of Adx. For Adx Loveday rootstock, K:Na ratios for leaf, stem and rhizomes of 62:1, 15:1, and 8:1, respectively were calculated. The high K:Na ratios of Adx leaves and stems were likely associated with sodium exclusion through a preference for potassium. This is a known mechanism for salt tolerance in halophytes [Flowers et al. 1986].

Blackwell et al. (2005) reported the changes in soil profile salinity over five years in a SBC Stage 1 filter cell (figure 6). The figure shows that a saline soil was progressively ameliorated due to leaching of the excess salts through provision of a subsurface drainage system and irrigation with 1.2 dS/m water. The soil salinity changes were very marked during leaching in the first two years, and very gradual in the subsequent years. The field soil salinities appeared to be gradually shifting towards the predicted soil salinity profile, using the model in FAO Irrigation and Drainage Paper No. 29 [Ayers and Westcott, 1985] for irrigation water EC of 1.2 dS/m

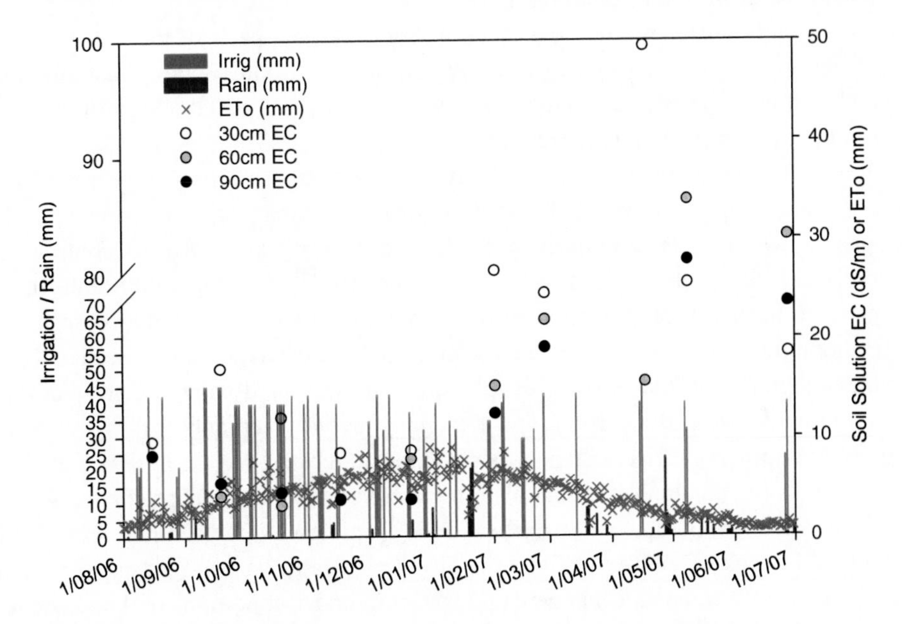


Figure 5. Changes in soil solution EC for first year growth of Adx at Barmera, SA irrigated with winery wastewater.

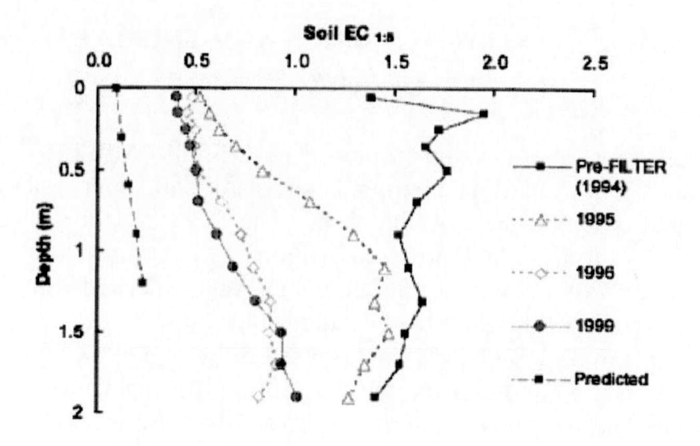


Figure 6. Changes in soil salinity of SBC stage 1 (after Blackwell et al. 2005)

## WEED RISK ASSESSMENT AND BIOSECURITY

A major deterrent to the use of Adx as a biomass crop is its invasive properties in certain riparian ecosystems, especially areas subject to torrential flooding, with Adx occurring in all inhabited continents [Randall, 2002]. Adx is likely to be invasive only when planted in these riparian and floodplain systems, particularly those subject to torrential flooding such as those that arise from annual snow melts in California. Outside such aquatic systems, there is limited potential for rhizome and stem fragments to be broken off, spread and regenerate. Unlike some invasive bamboos, Adx rhizomes have a limited lateral spreading growth habit, with plants forming well defined clumps [Lewandowski et al. 2003]. Most significantly, no fertile seed production has been reported for Adx in the USA [DiTomaso and Healy, 2007] or Australia. Williams et al. (2008a) found no fertile seeds when 400 seeds/stand were tested from five wild stands in South Australia.

Adx is currently a declared noxious weed in South Africa, parts of the USA (California, Texas and Nevada) and in the Sydney basin region of New South Wales, Australia. Despite being widely grown in Australia for over 150 years, it has not posed a significant threat in comparison to other grass or aquatic/riparian weeds.

The most effective treatment to control unwanted Adx plants reported by Bell (1997), in the USA is foliar application of a 1-3% solution of glyphosate (with a non-ionic surfactant) applied post flowering and pre-dormancy. This can give up to 100% control in one application [Bell, 1997]. The SARDI research team on Adx, is working with South Australian weed experts, to further develop a strategy from which the potential 'weed risk' from current and future plantings of Adx can be readily managed. The bases of this strategy are: (a) appropriate geographical placements of plantings in the landscape, avoid plantings in floodplains (1 in 50 year flood risk), and especially avoid riverbanks subject to torrential flooding. A preliminary buffer distance of 500m is suggested for Australia; and (b) crop hygiene protocols to contain Adx at the planting sites (secure transport protocols, soil buffer strips of 50 m width from adjacent properties) and (c) implement a weed risk levy (Williams et al. 2008b). The lack of viable seed and the absence of rapid lateral root expansion means that Adx is likely to be a manageable new biomass crop under conditions outlined above (e.g. terrestrial, non flood plain ecosystems).

## Benefit-Cost Analysis of Adx Grown for Industry

Benefit:cost analyses were carried out on two potential Adx growing enterprises (Black and Williams, unpub report). Based on the results shown in table 2, these two enterprises were profitable in Australia.

Secondly, benefit:cost analyses were conducted for these enterprises based on levying the enterprises at 50 c/t dry matter to conduct surveys, control any escaped Adx that might establish on roadside shoulders as a result of transporting Adx tops and report to the public [Black and Williams, unpublished report]. Based on results in table 2 and the fact that Adx does not produce viable seed together with the restriction that Adx should not be planted on floodplains near riparian systems, we concluded that the benefits of the Adx enterprises at Barmera and South East of South Australia far exceeded the costs of survey and control of any volunteer Adx stands.

**Table 2. Profit and benefit:cost analyses (BCA) for potential Adx enterprises in South Australia (Aust$) [Modified from Williams et al. 2008a]**

| Enterprise | Pre tax $ profit/ha at crop maturity | BCA for volunteer Adx control[1] |
|---|---|---|
| Barmera winery wastewater (40 dry t/ha yield) | | |
| (a) savings from processing wastewater | 3,490 | 175:1 |
| (b) a + carbon credit @ $30/t dry matter | 4,090 | 205:1 |
| (c) b + ethanol price @ $50/t dry matter | 6,090 | 305:1 |
| South-East SA dryland[2] (20 dry t/ha yield) | | |
| (a) ethanol or paper pulp @ $50/t dry matter | 400 | 40:1 |
| (b) a + carbon credit @ $30/t dry matter | 700 | 70:1 |

[1] 50 c/t dry matter levy; BCA based on profit/ha vs. levy cost/ha.

[2] Dryland SE of SA is a warm, temperate climate of c. 500mm rainfall/year.

## Conclusion

Arundo donax (Adx), a second generation biofuel crop, produced high biomass yields (45.2 t/ha of dry tops) /year on saline land using low quality, saline wastewater without pesticides.

There are no shortages of poor quality water and lands in Australia. Ongoing drought in many regions has renewed the interest in alternate uses of saline wastewater rather than disposal to evaporation basins. The Serial Biological Concentration (SBC) system is an integrated sustainable biosystem, which offers communities the potential to generate returns from wastewater flows by growing salt tolerant crops such as Adx rather than discharging to evaporation basins or the ocean.

*Arundo donax* qualified as a valuable, potential carbon credit crop (within a year, it produced 20.6 t of organic carbon /ha of dry tops). If each tonne of C sequestered is valued at Au$30, this would generate Au$618 /ha/year in carbon credits. Our work has shown the potential for Adx to treat saline wastewaters, P, and N rich wastewaters (e.g. sewage or winery wastewaters) and to produce high biomass yields-even on saline soils, by using new irrigation technologies. This biomass can be used as feedstock for combined heat and power factories (to run pumps and other equipment) or to produce ethanol. Since there is no pollen formation and no fertile seed produced by Adx, this excludes the danger of seed-based spread of Adx.

Preliminary economic analysis of Adx enterprises indicated there were viable potential options, namely; for cost saving measures of wastewater mitigation and/or for ethanol or pulp/paper production at Barmera and for ethanol production from dryland systems in the South East of South Australia.

## ACKNOWLEDGEMENT

Financial assistance from the Rural Industries Research and Development Corporation (RIRDC), the Berri Barmera Council and Hardy Wine Company (Australia) is gratefully acknowledged. The authors thank the Griffith City Council and CSIRO for SBC work and Dr Ian Black, Principal Economist, SARDI for conducting the economic analyses. We thank Dr John Virtue, Senior Weed Ecologist, Department of Water Land and Biodiversity Conservation ( South Australia) for weed risk assessment and management strategies.

## REFERENCES

Ayers, R.S., Westcott, D.W., 1985. *Water Quality for Agriculture.* Paper 29 (Rev 1), FAO, Rome, Italy.

Ball, J., Donnelley, L., Erlanger, P., Evans, R., Kollmorgen, A., Neal, B., Shirley, M., 2001. Inland waters, Australia state of the environment report 2001 (Theme report). CSIRO publishing on behalf of the Department of Environment and Heritage, Canberra, pp. 157.

Bell, G.P., 1997. Ecology and management of Arundo donax, and approaches to riparian habitat in Southern California. in: J.H. Brock, M. wade, P. Pysek, D. Green (Eds.), Plant invasions: *Studies from North America and Europe*. Blackhuys Publishers, Leiden, The Netherlands, pp. 103-113.

Biswas, T., 2006. Simple and inexpensive tools for root zone watch. *J. Australian Nutgrower,* 20 14-16.

Biswas, T.K., Jayawardane, N.S., Blackwell, J. and Tull, D. 2002. A land filter system-turning Griffith's sewage into an asset. In: Proc. ANCID Griffith 2002 Conference. *ANCID:* PO Box 58, Berrigan, NSW.

Blackwell, J., Biswas, T.K., Jayawardane, N.S., Townsend, J.T., 2000. Irrigation - getting the edge GRDC irrigation update. *Mathoura Bowling Club, Mathoura, NSW,* pp. 8.

Blackwell, J., Jayawardane, N., Biswas, T.K., Christen, E., 2005. Evaluation of a sequential biological concentration system in natural resource management of saline irrigated area. *Australian Journal of Water Resources.* 9, 169-175.

Boland, A., Hamilton, A., Stevens, D., Ziehri, A., 2006. Opportunities for reclaimed water use in Australian agriculture. in: D. Stevens (Ed.) Growing crops with reclaimed wastewater. *CSIRO Publishing,* Collingwood, Victoria, Australia, pp. 81-90.

Bourne, J.K., 2007. Biofuels: Boon or boondoggle? *National Geographic*. 212, 38-59.

DiTomaso, J.M., Healy, E.A., 2007. Weeds of California and other western states, vol. 2 *Geraniaceae-Zygophyllaceae.* University of California, Oakland, California.

DWLBC, 2008. Department of Water, Land and Biodiversity Conservation. http://www.dwlbc.sa.gov.au.

Flowers, T.J., Hajibagheri, M.A., Clipson, N.J.W., 1986. Halophytes. *The Quarterly Review of Biology*. 61, 313-337.

Jayawardane, N.S., Biswas, T.K., Blackwell, J., Cook, F.J., 2001. Management of salinity and sodicity in a land FILTER system for treating saline wastewater on a saline-sodic soil. *Australian Journal of Soil Research*. 39, 1247-1258.

Jessop, J., Dashorst, G.R.M., James, F.M., 2006. Grasses of South Australia. *Wakefield Press,* Adelaide.

Lewandowski, I., Scurlock, J.M.O., Lindvall, E., Christou, M., 2003. The development and current status of perennial rhizomatous grasses as energy crops in the US and Europe. *Biomass and Bioenergy*. 25, 335-361.

NLWRA, 2002. National Land & Water Resources Audit. Theme six report. NLWRA, Canberra.

NLWRA, 2008. National Land & Water Resources Audit. www.nlwra.gov.au/atlas.

Paul, D., Williams, C., 2006. Arundo donax -plantation establishment and pulp quality in Australia Proceedings of the 5th International Conference of Non-Wood Fibre Pulping and Paper Making Conference, Guangzhou, China, 8-10 November, 2006, pp. 36-39.

Randall, R.P., 2002. A global compendium of weeds. R. G. & F. J. Richardson, Meredith, Victoria, Australia.

Shell, 2007. About Shell Biofuels. (29th November, 2007). http:www.shell.com/home/content/aboutshell-en/what_we_do/refining_selling/fuels/biofuels.html.

Spafford, W.J., 1941. The bamboo reed (Arundo donax) in South Australian Agriculture. The *Journal of the Department of Agriculture of South Australia,* XLV, 77-83.

Williams, C.M.J., Harris, P., Biswas, T.K. and Heading, S. (2006). Use of giant reed (Arundo donax L.) to treat wastewaters for resource recycling in South Australia. *Poster presented at the 5th International Symposium for Irrigation of Horticulture Crops*. Mildura, Australia. http://www.sardi.sa.gov.au/pdfserve/water/ products_and_services/use_of_giant_reed_a4_100dpi.pdf).

Williams, C.M.J., Biswas, T.K., Glatz, P., Kumar, M., 2007. Use of recycled water from intensive primary industries to grow crops within integrated biosystems. *Agricultural Science,* 21, 34-36.

Williams, C.M.J., Biswas, T.K., Black, I., Heading, S., 2008a. Pathways to prosperity second generation biomass crops for biofuels using saline lands and wastewater. *Agricultural Science,* 21, 28-34.

Williams, C.M.J., Harris, P., Biswas, T.K., Heading, S., 2008b. Use of giant reed (Arundo donax) to treat wastewaters for resource recycling in South Australia. *Acta Horticulturae.* 792, 701-707.

Wright, M., Brown, R., 2007. Comparative economics of biorefineries based on the biochemical and thermo-chemical platforms. *Biofuels, Bioproducts, Biorefineries.* 1, 49-56.

In: Technologies and Management for Sustainable Biosystems ISBN: 978-1-60876-104-3
Editors: J. Nair, C. Furedy, C. Hoysala et al. 

*Chapter 2*

# REMOVAL OF SELECTED TRACE POLLUTANTS FROM HOSPITAL WASTEWATERS: THE CHALLENGE OF ON-SITE TREATMENT OPTIMIZATION

***N. Weissenbacher[1], K. Lenz[1], S. Mahnik[1,2] and M. Fuerhacker[1*]***

[1] Institute of Sanitary Engineering and Water Pollution Control,
University of Natural Resources and Applied Life Sciences,
Vienna, Muthgasse 18, A-1190 Austria
[2] Department of Medicine I, Clinical Division of Oncology,
Medical University of Vienna, Austria

## ABSTRACT

The concentrations of pharmaceuticals and their degradation products are relatively high in hospital wastewater compared to municipal wastewater. Many cytostatic agents (CAs) have cytotoxic, mutagenic and/or teratogenic effects. On-site treatment of separated hospital wastewater streams is a possibility to remove specific target pharmaceuticals and their metabolites present at relatively high concentrations. To investigate on-site removal of platinum containing cytostatic agents (CPCs) as well as of 5-flourouracil (5-FU) and anthracyclines from an oncological ward, a bioreactor system was installed on-site. Biodegradability, adsorption and metabolization were of interest for wastewater treatment. The UF-MBR pilot system was operated during a period of one year. $CO_2$ and $NO_x$ were monitored in the off-gas for bio-reactor state estimation and analyses of $NO_x$ from simultaneous nitrification-denitrification (SND). The compact MBR provided good organic carbon elimination (avg. 90% TOC) and nitrogen elimination (max. 90% $N_{tot}$). Whereas 5-FU and the anthracyclines were sufficiently removed, the platinum containing CAs were only partly adsorbed to the activated sludge. Increased CPC removal was achieved by decreased SRT.

---

* Corresponding author: maria.fuerhacker@boku.ac.at; +43 1 360065806

**Keywords:** cytostatic agents, hospital wastewater, MBR, simultaneous nitrification-denitrification, off-gas monitoring, on-site treatment.

## INTRODUCTION

The concentrations of pharmaceuticals and their degradation products are relatively high in hospital wastewater compared to municipal wastewater. Over the last decade, the number of cytostatic agents (CAs) used in cancer therapy has considerably increased. Many antineoplastic agents have cytotoxic, mutagenic and/or teratogenic effects (CMR substances). The European directive on the framework of community action in the field of water policy- the Water Framework Directive (WFD, 2000/60/EC) – defines strategies to restrict the emission of CMR substances to the aquatic environment. The primary source of cytostatic compounds in the environment is the excretion from patients under medical treatment, whereas metabolites of the parent substances may also contribute to the cytotoxic properties of hospital effluents. The occurrence of cytostatic agents in wastewater and in the environment has become an important issue and raised public concern (Mahnik et al., 2002). The total consumption of cytostatic drugs in Austria from June 2000 to June 2001 was 436 kg (IMS, 2001), whereas 5-fluorouracil (5-FU) turned out to be the most relevant compound. Concentrations of the substances may vary over orders of magnitude dependent on the location in the sewer system. Mahnik et al. (2002) assessed the concentration ranges for the most relevant compounds administered at a Viennese hospital based on the yearly cytostatics consumption, dosage and the excretion rates in respect to the amount of wastewater produced. Only 20 % of the dosage is applied by in-patient treatment and is subsequently emitted to the hospital sewer system. The consumption of cytostatic agents in a Viennese hospital and in Austria in total is listed in table 1.

**Table 1. Consumption of selected cytostatic agents in the Viennese hospital and in Austria (Mahnik et al., 2002)**

| Substance | Hospital consumption in kg/a | Consumption in Austria in kg/a |
|---|---|---|
| 5-Fluorouracil | 4.74 | 119 |
| Gemcitabin* | 4.25 | 29 |
| Cyclophosphamide* | 3.1 | 39 |
| Ifosfamide* | 3 | 18 |
| Carboplatin | 0.32 | 3 |
| Epirubicin | 0.13 | - |
| Cisplatin | 0.12 | 1 |
| Oxaliplatin | 0.12 | - |
| Doxorubicin | 0.1 | - |
| Daunorubicin | 0.02 | - |

* Not covered by the project.

On-site treatment of separated hospital wastewater streams would be a possibility to treat specific target pharmaceuticals and their metabolites present at relatively high concentrations. Limited space, user habits and security requirements are basic challenges that have to be faced when it comes to on-site treatment in hospitals. As biological wastewater treatment results only in partial removal of a wide range of emerging trace contaminants in wastewater (Petrovic et al., 2003), the combination with advanced treatment technologies are required. Membrane bioreactor (MBR) systems combine the approved activated sludge treatment with membrane filtration technology. The membrane filtration efficiently separates the treated liquid from the biomass of the activated sludge tank. The combination of biological treatment and membrane processes leads to the following advantages over conventional concepts:

- High biomass concentration and long sludge retention favour the elimination of hardly biodegradable compounds and lead to enhanced nutrient removal.
- High volumetric loading and low sludge loading reduce the required activated sludge tank volume and hence the foot print (indoor applications).
- Effective barrier for suspended particles - reduction of the microbiological contamination dependent on the membrane cut-off (retention of potentially pathogenic bacteria in general).
- Low content of suspended particles in the effluent enhance the efficiency of subsequent treatment (e.g. adsorption, UV radiation).

To achieve effective operation, advanced treatment technologies require monitoring and control strategies based on online monitoring and real time data processing. The information on the current operation state is important for optimization and robust operation.

## OBJECTIVES

To analyse the applicability of biological wastewater treatment for the removal of platinum containing cytostatic agents (CPCs: cisplatin, carboplatin, and oxaliplatin) as well as of 5-FU and anthracyclines (doxorubicin, daunorubicin, epirubicin) from the oncological ward, a pilot plant system was installed on-site. The investigation comprised the following main objectives:

- CA removal: Analyses and optimisation of the elimination of the selected cytostatic compounds from the separated wastewater stream of the oncological ward.
- Nutrient removal: Analyses of the NOx off-gas emissions under the ORP controlled simultaneous nitrification-denitrification (SND) process to investigate the applicability of NOx off-gas monitoring as additional control parameter to the ORP.
- Real time control: Analyses of $CO_2$, NOx off-gas and ORP online monitoring data for real-time bioreactor state estimation under practical conditions - detection of activity changes for operation safety and stable performance.

# MATERIALS AND METHODS

## Pilot Plant System

A UF-MBR pilot system (150 L) with simultaneous nitrification-denitrification (SND) was chosen for application and operated over one year. The system was equipped with a measurement and control system including off-gas monitoring of $CO_2$ and $NO_x$. The storage system (figure 1- Tanks 1 and 2) was capable to collect 24 h mixed samples. 75% of the wastewater of the oncological ward was separated from the central wastewater collection system (18 patients).

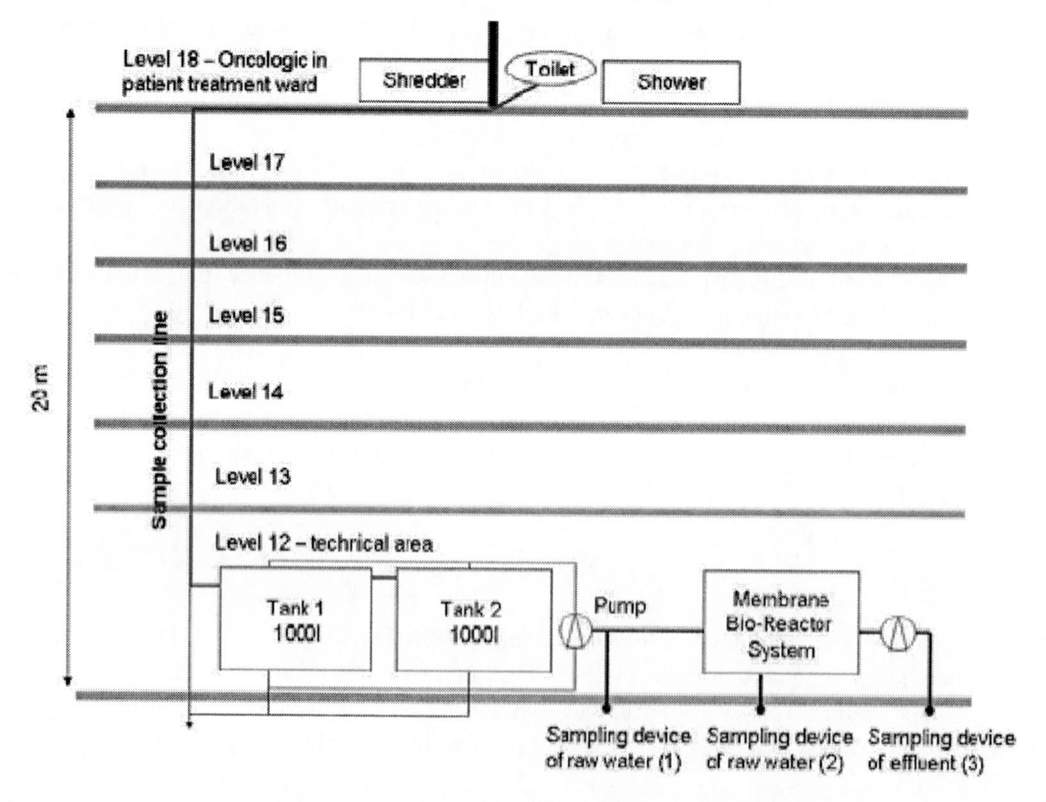


Figure 1. Wastewater collection and storage at the hospital. Wastewater from 18 patients of the oncological ward (toilets and showers) was collected for analyses and treatment (Lenz et al., 2004).

The monitoring devices installed at the membrane bioreactor system are summarized in table 2. Sampling for chemical parameters (organic load, nutrients) was carried out weekly (daily during CA monitoring) for the influent, bio-reactor and effluent, respectively.

**Table 2. Online measurements at the MBR pilot**

| Parameter | Range | Sampling point |
| --- | --- | --- |
| $CO_2$-Conc. in the off-gas | 0-1 [vol%] | Bio-Reactor |
| $NO_x$- Conc. in the off-gas | 0-1000 [ppbv] | Bio-Reactor |
| pH | 2-12 [-] | Influent, Bio-Reactor |
| Oxidation reduction potential | 0-500 [mV] | Influent, Bio-Reactor |
| Conductivity | 0-1999 [μS/cm] | Influent, Bio-Reactor, Effluent |
| Dissolved oxygen | 0-10 [mg/l] | Bio-Reactor |
| Temperature | 0-50 [°C] | Influent, Bio-Reactor |
| Permeate flow rate | 0-45 [l /h] | Effluent |
| Feed flow rate | 0-20 [l /min] | Membrane inlet |
| Air flow rate | 0-6 [m³ /h] | Bio-Reactor |

## Off-Gas Analyses

$CO_2$ off-gas data was used for carbon mass balancing and the analyses of $CO_2$ in relation to the bio-reactor state. To evaluate the influence of the carbonate system in the activated sludge system a mathematical modelling approach was elaborated to analyse the influence of the bicarbonate/$CO_2$ equilibrium on $CO_2$ monitoring for short term (real-time control) and on long term observations (mass balancing, Weissenbacher et al., 2007a). Mass balancing was carried out for a period of 70 days (weekly liquid phase sampling) and a period of 14 days (daily liquid phase sampling).For the analyses of gaseous nitrogen emissions (NO, $NO_2$), the influences of pH changes, different nutrient availability and ORP levels were of interest. Based on investigations on the formation of the nitrogen oxides as denitrification intermediates it had to be cleared, whether gaseous NO and $NO_2$ emissions can be directly related to SND in activated sludge systems or not. Therefore, the ORP threshold level between anoxic and aerobic conditions was changed stepwise (Weissenbacher et al., 2007b).

## Chemical Analyses

For the CPCs, the total amount platinum in all investigated samples was measured using a quadrupole-based inductively coupled plasma mass spectrometer (ICPMS) with a limit of detection (LOD) of 0.01 μg/l. Speciation of the CPCs was performed by online coupling of reversed-phase high performance liquid chromatography with ICP-MS with limits of detection of 0.09, 0.10 and 0.15 μg/l for cisplatin, carboplatin and oxaliplatin, respectively (Lenz et al., 2004). The determination of 5-FU was carried out with solid phase extraction and analysis by capillary electrophoresis with diode array detection. Limit of detection amounted to 1.7 μg/l, the limit of quantification was calculated to be 8.6 μg/l (Mahnik et al., 2004 and 2007). The three selected anthracyclines doxorubicin, epirubicin and daunorubicin were analysed by solid phase extraction and HPLC – FLD at concentrations of 0.1 - 5 µg/l. The limit of detection was calculated to be 0.05 µg/l doxorubicin, 0.05 µg/l epirubicin, and 0.06 µg/l daunorubicin (Mahnik et al., 2006). Due to the substantial efforts for the analyses, CA monitoring was carried out in three monitoring periods (daily sampling) for 74 days in total.

# RESULTS AND DISCUSSION

The compact MBR turned out to be an appropriate system with a compact footprint for applications with limited space like the supplementary installation for on-site treatment of separated wastewater streams in hospitals. The external ultra filtration module was easy to replace for cleaning, an advantage over submerged modules.

## TOC and Nitrogen Removal

The system showed the typical loading conditions of MBR systems - high volumetric loading and low sludge loading. The average loading conditions are compared to levels reported from full-scale MBRs and conventional municipal wastewater treatment plants in table 3.

**Table 3. Loading conditions in comparison to full scale MBRs and conventional activated sludge plants (Weissenbacher, 2006)**

| Plant Type | Volumetric Loading kgCOD/m³/d | Sludge Loading kgCOD/kgTSS/d |
|---|---|---|
| Pilot MBR (Mean) | 1.38 | 0.05-0.30 (avg. 0.10) |
| Full-scale MBRs* | >0.5 | 0.04 |
| Conventional WWTPs** | 0.4-0.5 | <0.10 |

* Kraume et al.(2003), adapted.
** Lenz (2004), adapted.

Stable TOC removal performance was reached after a start up period of four weeks. Despite low dissolved oxygen concentrations below 0.5 mg/l (SND operation), an average TOC removal rate of 90% was achieved over the whole observation time. Maximal nitrogen elimination rates were at about 90 % with an overall average of 60%. Disregarding the nitrogen incorporation into the biomass keeping the biomass constant the elimination of TKN can be seen as nitrification efficiency, which showed a steady increase with increasing ORP over the entire observation period. Denitrification decreased at potentials above 240 mV resulting in a SND optimum of 220-240 mV (Weissenbacher et al., 2007b). The nitrogen removal rates for six ORP ranges are summarized in figure 2.

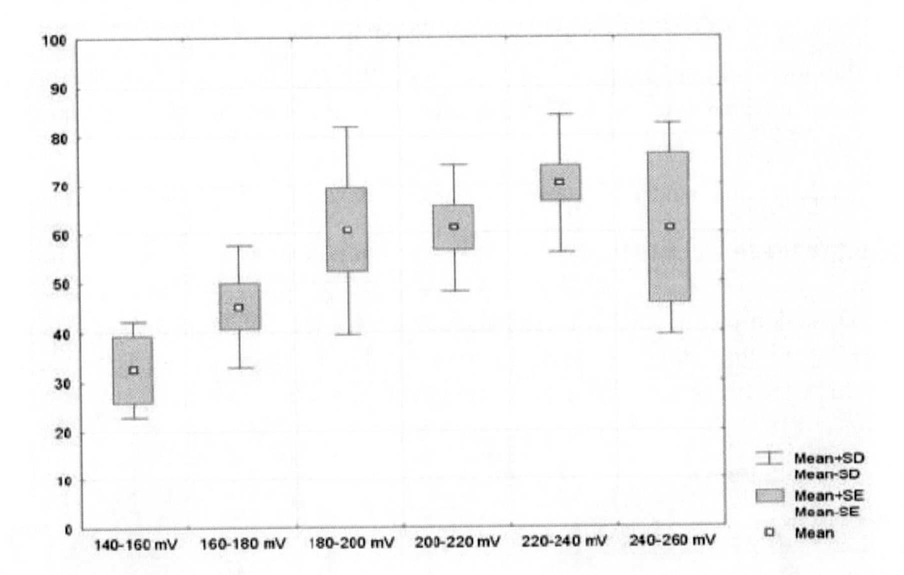


Figure 2. Total nitrogen removal (%) over the whole period of operation for different ORP ranges (Weissenbacher, 2006).

In contrast to ammonium or nitrate batches, relatively small changes of ORP led to a shift of the SND equilibrium causing the gaseous release of denitrification intermediates and the accumulation of nitrite. There were no rapid changes of pH despite the frequent batches of wastewater during the experiments. Chemical conversion was not regarded as reason for the increased gaseous emissions. Pulsing activated sludge with nitrite was reported to lead to an immediate inhibition of the nitric oxide reductase (Schulthess et al., 1995). This was confirmed by nitrite addition during the continuous operation of the pilot plant resulting in a fast response sequence of gaseous NO and $NO_2$ (figure 3).

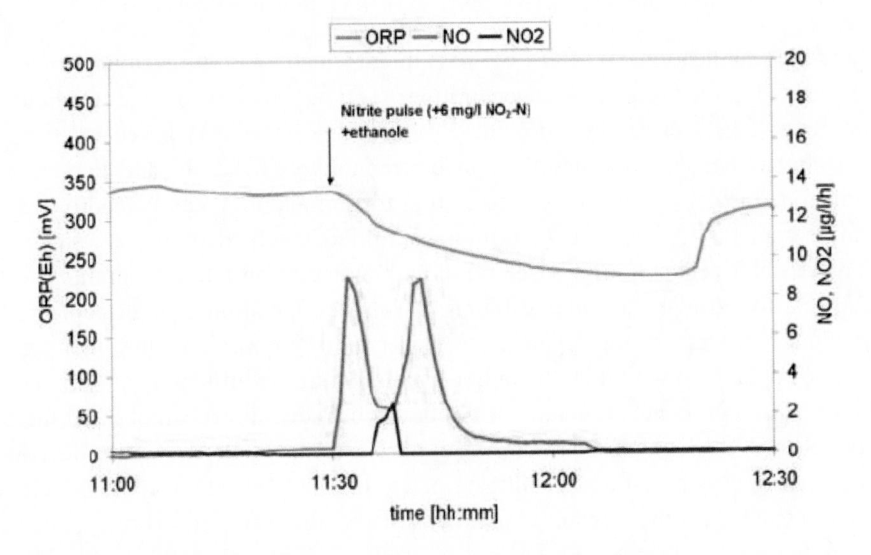


Figure 3. Example for the response of the gas phase parameters NO and $NO_2$ to a nitrite pulse at a high ORP level (MBR at endogenous respiration).

As NO was always present in the off-gas, $NO_2$ was only occurring during nitrite accumulation the liquid phase (Weissenbacher et al., 2007b). This parameter may be used to detect nitrite accumulation with the advantages of off-gas monitoring like reduced maintenance efforts.

## Bio-Reactor State Estimation

The calculation of the carbon mass balance was carried out for a 14 days period with daily sampling for the liquid phase parameters and for 70 days period. The gaseous $CO_2$ turned out to be the major output fraction. The output fractions are shown in figure 4.

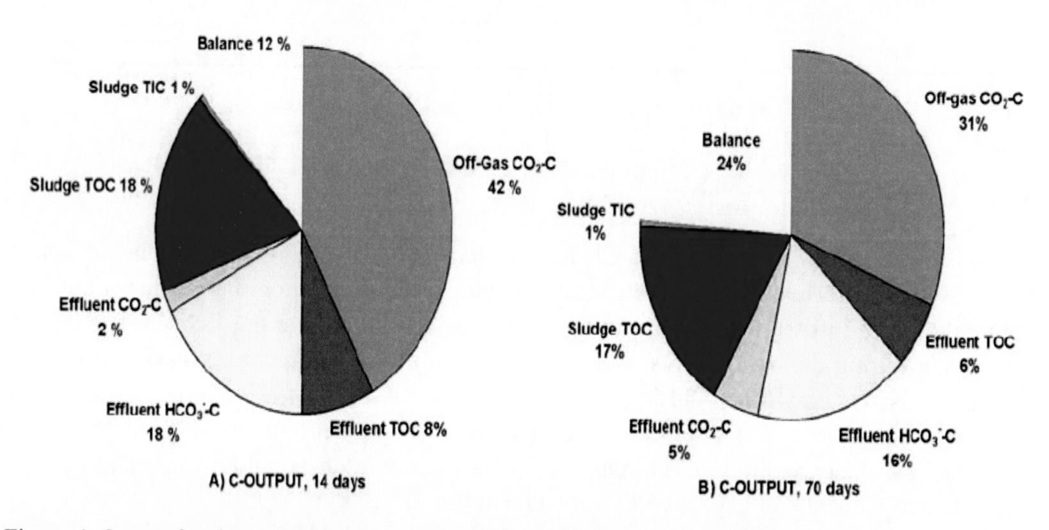


Figure 4. Output fractions of the 14 days (A) and 70 days (B) carbon mass balances.

A possible application of $CO_2$ off-gas monitoring for bio-reactors would be the estimation the organic substrate concentration. The estimation of the Carbon dioxide Evolution Rate (CER) or Carbon dioxide Production Rate (CPR) based on $CO_2$ off-gas measurements has become commonplace in biotechnology (Komives and Parker, 2003). It was demonstrated that the substrate concentration for fermentation can be controlled by such off-gas measurements. Nilsson et al. (2002) controlled the feed of yeast fermentation by estimating the sugar concentration based on of carbon evolution rate measurements. It has to be noticed that the growth rate of the biomass from fermentation is high compared to the activated sludge process resulting in relatively high $CO_2$ emission rates. For wastewater treatment, the application would be the online estimation of the influent quality. In contrast to fermentation with high concentrated, well defined substrates, the substrates and the biomass cannot be clearly distinguished in the case of wastewater. Performance description of wastewater treatment is based on overall parameters like TOC, COD or BOD which describe the organic content of the complex matrix. Hence, the setup of direct stoichiometric relationship to the $CO_2$ production is not possible. Additional problems are caused by substances with adverse effects on the biological community (inhibition) or different biodegradability. The relation between organic load and the emitted mass via the off-gas is of interest. The whole observation period of a year showed that the gaseous emitted $CO_2$-C contributed on average 52% to the eliminated TOC. $CO_2$ off-gas monitoring may be regarded as valuable parameter for the estimation of organic carbon load conditions. It is certainly applicable to detect system failures such as inhibition due to the use of cleaning detergents.

## CA Elimination

Concerning the CA elimination, the main problems faced during the optimization procedure where the flow and load variations from the relatively small ward and fluctuations of the influent concentrations of the targets due to varying application in therapy. Whereas 5-FU was sufficiently removed by biodegradation, the CPCs were only partly adsorbed to the activated sludge (Lenz et al., 2007; Mahnik et al., 2007). The CPC influent concentrations detected were ranging from 3.2 µg/l Pt to 266 µg/l Pt (mean 35.4 µg/l Pt, median 22.0 µg/l Pt). The sludge loading increased during continuous operation up to a maximum of 175µg/g Pt despite the excess sludge removal was increased. The results of the pilot plant indicated that the maximum adsorption capacity had not been reached after the third monitoring. Peak concentrations in the influent were also detected in the effluent, independent of the SRT or CPC sludge loading conditions. A dependency between the liquid phase concentration and the sludge load, like it was observed in lab experiments using pure substance solutions, was not observed (Lenz et al., 2007). This is a further hint that on-site experiments are essential. A comparison of the total platinum content to the sum of individual CPCs identified in the influent and the effluent of the plant showed that a considerable fraction of the CPCs must have been converted to other metabolites. The missing Pt could not be found as Pt(II)chloride or as Pt(IV)chloride. Unidentified metabolites may show a similar (strong) adsorption to activated sludge and to the solid constituents of wastewater as cisplatin. Detailed information on the fate of CPCs in the system is given by Lenz et al. (2007). Platinum mass balances for the three different monitorings show the input and output fractions of CPCs in the activated sludge plant (figure 7). Increased removal was achieved by the reduction of the sludge retention time (table 4).

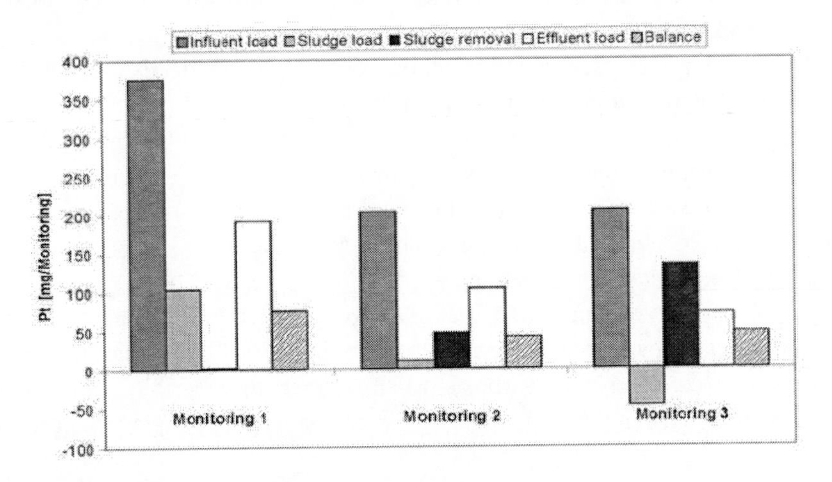


Figure 5. CPC mass balances for the three monitoring periods.

**Table 4. System settings for the three monitoring periods**

| Monitoring | Sampling days [d] | Sludge retention time [d]** | Total suspended solids [g/l] | Flow rate [l/d] | Detention time in the bioreactor [h] |
|---|---|---|---|---|---|
| 1 | 30* | >300 | 11.8 | 304 | 11.8 |
| 2 | 30 | ~100 | 14.4 | 266 | 13.5 |
| 3 | 14 | ~ 50 | 15.2 | 323 | 11.1 |

* No sampling during week-ends.
** Theoretical retention time calculated from the daily excess sludge removal.

Doxorubicin was the only compound of the anthracyclines which could be detected during the cytostatics monitorings in the storage tanks (0.03-0.5 μg/l). The concentrations of the other anthracyclines were below the limit of detection. 5-FU was detected in a range of 1.2-123.5 μg/l (sampling point: storage tanks, figure 1). As the production of biomass in the storage tanks could not be inhibited and suspended solids were present, metabolization or biological degradation might have triggered the elimination processes (Mahnik et al., 2007). None of the compounds could be detected in the bio-reactor effluent. The system was capable for sufficient removal of the anthracyclines and 5-FU.

## CONCLUSIONS

On-site MBR treatment of contaminated hospital wastewater from an oncologic ward showed that:

- The small footprint of the MBR and the external filtration unit simplified the maintenance of the system for the indoor application (e.g. changing modules without contact to wastewater – a safety issue).
- $CO_2$ indicated changes of the biological activity like inhibition (which may be expected from cleaning at the wards) but direct information on COD load could not be derived.
- $NO_2$ reflected increased nitrite formation during periods of disturbed SND and may be used as alarm parameter to avoid nitrite accumulation.
- Optimal nitrogen removal via SND was reached at an ORP of 220-240 mV (Eh), TOC removal was not effected by the low DO conditions.
- The comparison of lab experiments to the performance of the on-site plant with regard to CPC elimination showed that on-site experiments are essential because of matrix effects and metabolization.
- The biological system could be adapted to a certain extend to the removal mechanisms relevant for CA species (especially CPCs) but further treatment may be necessary (e.g. activated carbon). MBRs foster the flexibility of the biological treatment due the higher possible biomass concentration (adaptation, adsorption).

Strong variations in loading rates due to therapy schedules and changing number of patients at the wards led to highly variable operating conditions. Emphasising on robust operation with stable elimination performance for trace contaminants, organic carbon and nutrients, advanced technology such as membrane filtration or real-time monitoring of additional parameters ($CO_2$, $NO_x$) is justified. Adequate personnel for system operation are already available at complex facilities like modern hospitals.

## ACKNOWLEDGEMENTS

This work has been supported by grants from the Austrian Federal Ministry of Agriculture, Forestry, Environment and Water Management (BMLFUW), the Austrian Science Fund (FWF-Project P16089-N03) and the Austrian Industrial Research Promotion Fund (FFF).

## REFERENCES

Ims, 2001. Institut für medizinische Statistik, Med. Univ. Wien [Institute of Medical Statistics, Medical University Vienna]

Komives, K., Parker, R., S., 2003. Bioreactor state estimation and Control. *Curr. Op. in Biotechnology.* 14, 468-474

Kraume M., Bracklow U., 2003. Das Membranbelebungsverfahren in der kommunalen Abwasserbehandlung – Betriebserfahrungen und Bemessungsansätze in Deutschland. [Membrane Bioreactors for Municipal Wastewater Treatment – Operation Experience and Design Criteria in Germany, in German] Membrantechnik in der Wasseraufbereitung und Abwasserbehandlung, 5. Aachener Tagung, Ü2

Lenz K., Hann S., Koellensperger G., Stefanka Z., Stingeder G., Weissenbacher N., Mahnik S.N., Fuerhacker M., 2004. Presence of cancerostatic platinum compounds in hospital wastewater and possible elimination by adsorption to activated sludge. *Science of the Total Environment.* 345, pp141– 152

Lenz K., Koellensperger G., Hann S., Mahnik S.N., Weissenbacher N., Fuerhacker M., 2007. Fate of cancerostatic platinum compounds in biological wastewater treatment. *Chemosphere.* 69 (11), pp 1765-1774

Mahnik S.N., Mader R., Fürhacker M., 2002. Zytostatika im Abwasser. [Cytostatic Agents in Wastewater; in German] *Wiener Mitteilungen.* 178, pp 91-111

Mahnik S.N., Rizovski B., Fuerhacker M., Mader R. M., 2004. Determination of 5-fluorouracil in hospital effluents. *Anal. Bioanal. Chem.* 380: 31–35

Mahnik S.N, Rizovski B., Fuerhacker M., Mader R.M., 2006. Development of an analytical method for the determination of anthracyclines in hospital effluents. *Chemosphere.* 65 (8), 1419-1425

Mahnik S.N., Lenz K., Weissenbacher N., Mader R.M., Fuerhacker M., 2007. Fate of 5-Fluorouracil, doxorubicin, epirubicin and daunorubicin in hospital wastewater and their elimination by activated sludge. *Chemosphere.* 66 (1), pp 30-37

Nilsson A., Taherzadeh M.J., Liden G., 2002. Online estimation of sugar concentrations for control of fed-batch fermentation of lignocellulosic hydrozylates by Saccharomyces cerevisiae. *Bioprocess Biosyst. Eng.* 25, 183-191

Petrovic M., Gonzalez S., Barcelo D., 2003. Analysis and removal of emerging contaminants in wastewater and drinking water. *Trends in Analytical Chemistry.* Vol. 22, No. 10, pp 685-696

Schulthess R. V., Kühni, M. And Gujer W., 1995. Release of nitric and nitrous oxides from denitrifying activated sludge. *Wat. Res.* 29 (1), 215-226

Weissenbacher N., Lenz K., Mahnik S.N., Wett B., Fuerhacker M., 2007a. Determination of activated sludge biological activity using model corrected CO2 off-gas data. *Wat. Res.* 41 (7), 1587-1595

Weissenbacher N., Loderer C., Lenz K., Mahnik S.N., Wett B., Fuerhacker M., 2007b. NOx monitoring of a simultaneous nitrifying – denitrifying (SND) activated sludge plant at different oxidation reduction potentials. *Wat. Res.* 41 (2), 397-405

In: Technologies and Management for Sustainable Biosystems ISBN: 978-1-60876-104-3
Editors: J. Nair, C. Furedy, C. Hoysala et al. 

*Chapter 3*

# 'ENHANCEMENT OF BIOLOGICAL TREATMENT OF INDUSTRIAL WASTE-WATER USING BIOAUGMENTATION TECHNOLOGY'

***M.T. Pandya****

Department of Microbiology
Jai Hind College, Mumbai-400 020

## ABSTRACT

This chapter highlights the applications of specific bacteria developed for the treatment of wastewater containing aromatic and aliphatic organic compounds from petrochemical and bulk drug industries and their successful implementation at plant scale level. Both sources contained organics difficult to degrade. Strains of Micrococcus, Pseudomonas, bacillus and Nocardia degrading phenol, amines, caprolactum and styrene were isolated from soil and sludge. The COD and BOD reductions under the optimum nutritional and physiological conditions were in 70-90% range. The nutrient balance maintained in wastewater were in ratio of COD: N: P as 100:10:1 and pH 7-8. Mixed biomass of these strains showed >95% reduction in COD of effluents containing plastic polymer, phenolics, Nylon-6 and aliphatic amines. The developed biomass of mixed strains added to the bioreactor increased COD and BOD reduction gradually and stabilised at 90-95%. The treated effluent after sand and carbon filtration was suitable for reuse in cooling towers.

**Keywords:** Toxic waste, industrial wastewater, bio-augmentation, bacteria, biodegradation, enrichment.

* Email: mukesh_pandya51@hotmail.com

# 1. INTRODUCTION

Microorganisms are characterized by their incredible metabolic and physiological versatility that permits them to inhabit ecological niche and exploit compounds unpalatable for higher organisms as carbon and energy sources. Species of Pseudomonas, Micrococcus, Nocardia, Bacillus Corynebacterium, Rhodococcus, Arthrobacter, Corynebacterium, Flavobacterium, Cladosporium, Fusarium, widely distributed in soil, sludge, sewage and water, were reported to degrade organic compounds by series of steps including mass transfer, adsorption and biochemical enzymatic reactions leading to growth. Zobell (1964), Davis(1967), Gutnick and Rosenberg(1977), Colwell and Walker(1977), Atlas (1977,1981), Fewson (1981), Singer(1984), Bartha(1984), Timmis et al (1994), Spain(1995) Ryoo et al., (2000), Neumann et al.,(2004), and Toro et al.,(2006)

Bio-augmentation processes using specific bacteria, offer advantages over conventional biological treatments as these microorganisms can completely degrade and oxidise organic compounds to innocuous carbon compounds. Bio-augmented cultures can sustain shock loadings unlike acclimatized bacteria and significantly improve efficiency of both aerobic and anaerobic processes.

This chapter highlights the applications of specific bacteria developed for the treatment of wastewater containing aromatic and aliphatic organic compounds from petrochemical and bulk drug industries and their successful implementation at plant scale level.

# 2. MATERIALS AND METHODS

Wastewater samples after primary treatments were collected from Bulk Drugs (phenolics), Petrochemical units manufacturing Aliphatic amines, ABS plastic polymer and Nylon-6 and polyester yarn and analysed for pH, TDS, BOD and COD. For BOD tests the enriched cultures were used as seed using APHA methods as reported in Standard Methods for Water and Wastewater Analysis (1989).

## 2. A. Isolation and Enrichment of Bacteria

Bacterial strains capable of degrading Phenols, caprolactum, styrene, and Aliphatic amines were enriched and isolated from soil and sludge collected from area nearer to each plant, using Bushnell & Haas medium (1941) liquid medium and enrichment technique reported by Pandya and Doctor (1980). Medium supplemented with nutrients Urea, $KH_2PO_4$, $K_2HPO_4$, $MgSO_4$, CaC12, FeC13, pH 7-7.5 and Carbon substrate (1000ppm) was used for the enrichment. The enriched samples in flasks were incubated at ambient temperature on rotary shaker (250rpm speed) until visible growth of bacteria was observed in 3-7 days period.

The isolated organisms from enriched biomass after 3-4 serial subcultures were screened to select potent degraders and identified up to genus level using Bergy's Mannual of Determinative Bacteriology ($8^{th}$ edition). The selected bacteria strains were studied for their optimal nutritional and physiological parameters for maximum degradation of Phenols,

Caprolactum, Styrene, and Aliphatic amines present in wastewater samples and reduce COD values under the influence of N, P, $Mg2^{+}$ ,$Ca^{2+}$, $Fe^{3+}$ pH and retention time.

## 2. B. Degradation of Phenol

The degradability of the each strain and mixed biomass of (*Micrococcus, Nocardia and Pseudomonas*) were studied for phenol. The rates of degradation of phenol (1000ppm) in modified liquid medium were determined in 2hrs. Residual phenol was analysed using 4-Amino-Antipyrine Method as given in APHA (1989).

## 2. C. Studies in Lab Scale Reactors

The efficiency of Mixed biomass in degradation of organics from wastewater samples from each selected industry were studied in 10, 25 and 100 litres capacity lab scale reactor (made up of plastic carboys) with aeration arrangement. The efficiency of Mixed biomass were then evaluated.

## 2. D. Plant Scale Bio-Augmentation Studies

The augmentation were done in the existing bioreactors of bulk drug ($300m^3$ capacity), petrochemical plant (mixture of amines 100m3 capacity), ABS plastic polymer plant (400m3 capacity) and Nylon-6 plant ($300m^3$ capacity), using selected bacteria strains at the site of each industries. The mixed biomass of cultures developed at the plant using wastewater medium supplemented with nutrients based on BOD value in Pilot plants with capacity of $1m^3$, 5 $m^3$ and $10m^3$ were transferred to the Bioreactor on a continuous basis. Developed biomass was added at a regular interval of one week and the entire program was completed in 3 months.

# 3. Results and Discussion

Bio-augmentation can overcome the limitations of biological treatments like shock loading and the need to operate within a narrow range of physiological parameters. The cultures are more beneficial as compared to acclamatised bacteria (Huban and Plowman, 1997; Boon et al., 2000, 2003) as once the organic loading increases, the activity of acclimatised bacteria reduces, indicating poor sensitivity due to shock loading. Bio-augmentation, fortifies biomass with the microorganisms that have been isolated and selectively adapted to degrade specific compounds.

Wastewater samples from selected industries were collected from the equalization tank and analysed for various parameters. Table 1 shows the results of the analysis of wastewater samples. The samples from bulk drug contained derivatives of phenols with pH 5-9 , TDS in 8000 to 10000ppm range, COD 2000-5000ppm and BOD 1000-3000ppm. The samples of petrochemical unit contained mixtures of mono, di and tri methyl, ethyl and propyl amines with pH 6-8, COD in 3000-5000ppm and BOD in 1000-3000ppm. ABS plastic polymer wastewater contained styrene, ACN (Acrylonitrile), Butadiene and coagulants with pH 6-8, COD 1000-2500ppm and BOD 500-1200ppm. Nylon-6 and polyester yarn effluents mainly contained caprolactum and spin finish oil had pH 6-9, COD 3000-8000ppm and BOD 2000-4000ppm.

**Table 1. Analysis of Wastewater Samples from Industries**

| Parameter | Bulk Drug Salicylic Acids | Petrochemical ABS polymer | Petrochemical Methyl, Ethyl and propyl amines | Petrochemical Nylon-6 (Tyre cord). |
|---|---|---|---|---|
| pH | 5-9 | 8-10 | 6-8 | 6-9 |
| TDS ppm | 8000-10,000 | 1000-5000 | 2000-5000 | 3000-5000 |
| SS ppm | 200-500 | 100-300 | 300-600 | 500-1000 |
| COD ppm | 2000-5000 | 3000-5000 | 1000-2500 | 3000-8000 |
| BOD ppm | 1000-3000 | 1000-3000 | 500-1200 | 2000-4000 |
| Nitrogen | Nil | Nil | 100-300ppm | 50-100ppm |
| Phosphorus | Nil | Nil | Nil | 10-20ppm |
| Products | Phenol derivatives | Mixture of Amines | ACN,Styrene and coagulants | Caprolactum and spin finish |
| Flow Rate | 300m$^3$/day | 100m$^3$ /day | 400m$^3$/day | 300m$^3$/day |

Several different types of microorganisms were isolated from soil and sludge samples, using enrichment culture cultivation technique. Majority of these were bacteria. Table 2 shows the number of potent bacterial isolates with very high activity and growth in effluents. The degradation of organic compound was indicated by heavy growth and reduction in COD. After 3 serial subculture in enrichment medium the reduction in COD values were in 50-70% range.

**Table 2. Number of Bacterial Isolates**

| Samples | Bulk Drug Salicylic Acids | Petrochemical ABS polymer | Petrochemical Methyl, Ethyl and propyl amines | Petrochemical Nylon-6(Tyre cord). |
|---|---|---|---|---|
| Soil | 8 | 16 | 10 | 24 |
| Sludge | 6 | 10 | 6 | 16 |
| Artificially enriched soil | 3 | 12 | 8 | 20 |
| Predominant genera | Pseudomonas Nocardia Bacillus | Micrococcus Pseudomonas Bacillus | Pseudomonas Micrococcus Nocardia | Nocardia Micrococcus Bacillus |

During enrichment a mixed bacterial flora containing both primary and secondary degraders developed and on repeated serial transfers in same conditions further selection of fast growing types of both primary and secondary degraders takes place. Most of these isolates were identified as species of *Pseudomonas, Nocardia, Micrococcus, and Bacillus.* The number of *Pseudomonas* species were *6, Nocardia 4, Micrococcus 4 and Bacillus 4.* These isolates were maintained on glucose yeast extract agar slants, which helped in preserving their essential characteristics.

Among the various compounds used for optimisation of nutritional parameters were urea, ammonium sulfate, ammonium nitrate, di-ammonium hydrogen phosphate, Phosphoric acid, Magnesium sulfate, Calcium chloride and Ferric chlorides (table 3). It was observed that the activity of isolated strains of *Pseudomonas, Micrococcus, Nocardia, and Bacillus* was significantly influenced by N & P source and pH in neutral range. The mix cultures of these strains were superior compared to individual isolates. Uniform & consistent reductions in COD were observed in late exponential and early stationary growth phase of these cultures. The optimal nutrient balance was maintained based on actual BOD value (BOD-100: N-10: P-1.) and pH in 7-8 range.

**Table 3. Optimised growth conditions for Mixed Bacteria**

| Parameter | Bulk Drug Salicylic Acids | Petrochemical ABS polymer | Petrochemical Methyl, Ethyl and propyl amines | Petrochemical Nylon-6 (Tyre cord). |
|---|---|---|---|---|
| pH | 6-8 | 6-8 | 6-8 | 6-8 |
| N* as Urea | 200-300ppm | 200-300ppm | Nil | 50-100ppm |
| P* as Phos-phoric acid | 30-40ppm | 30-40ppm | 30-40ppm | 10—20ppm |
| Magnesium** | 20ppm | 20ppm | 20ppm | 20ppm |
| Calcium** | 10ppm | 10ppm | 10ppm | 10ppm |
| Ferric** | 10ppm | 10ppm | 10ppm | 10ppm |
| Sludge Volume | 20-30% | 20-30% | 20-30% | 20-30% |
| MLSS | 2000-4000ppm | 2000-4000ppm | 2000-4000ppm | 2000-4000ppm |
| DO | 2-4ppm | 2-4ppm | 2-4ppm | 2-4ppm |

Key- * for 1000ppm BOD, ** for 1m3 of wastewater.

Boon et al. (2000) reported the use of induction for degradation of aromatic compounds. The bacteria grown in the presence of compounds such as phenols show excellent degradability of related compounds. Mixed biomass was very effective in the degradation of phenol (1,000ppm), it completely removed phenol (1,000ppm) after 24 hr of incubation in optimal medium. *Micrococcus and Nocardia* degraded phenol (1,000ppm) in 60 hr, while *Pseudomonas* required 72 hr for phenol (1,000ppm), figure 1.

The efficiency of mixed biomass was excellent when tested in the lab scale bioreactors with 10, 30 and 100 litres capacity using wastewater from selected industries. The reduction in COD under the optimised conditions in bioreactors with continuous aeration in wastewater from selected industries were 70-90% for bulk drug in 3-4 days, 90-95% for Petrochemical manufacturing aliphatic amine in 1 to 2 days. 95-98% in 1-2 days for ABS plastic polymer, and 95-98% for Nylon-6 in 3-4days (figures 3 – 5).

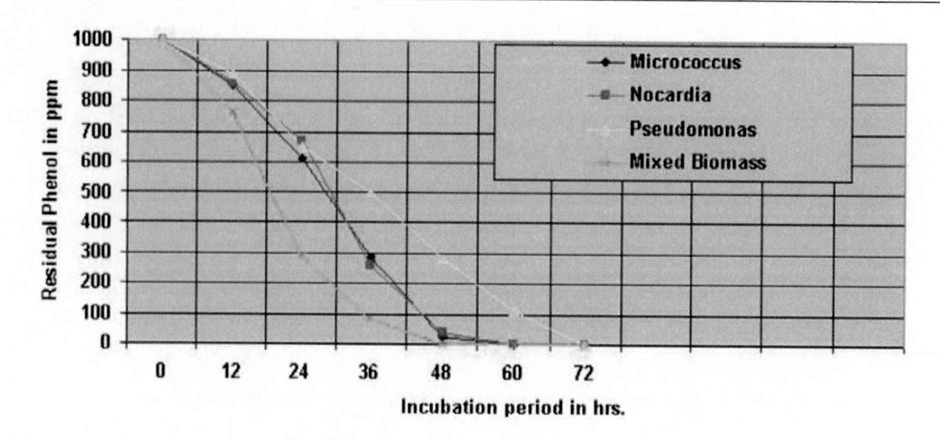


Figure 1. Degradation of Phenol by selected bacterial isolates.

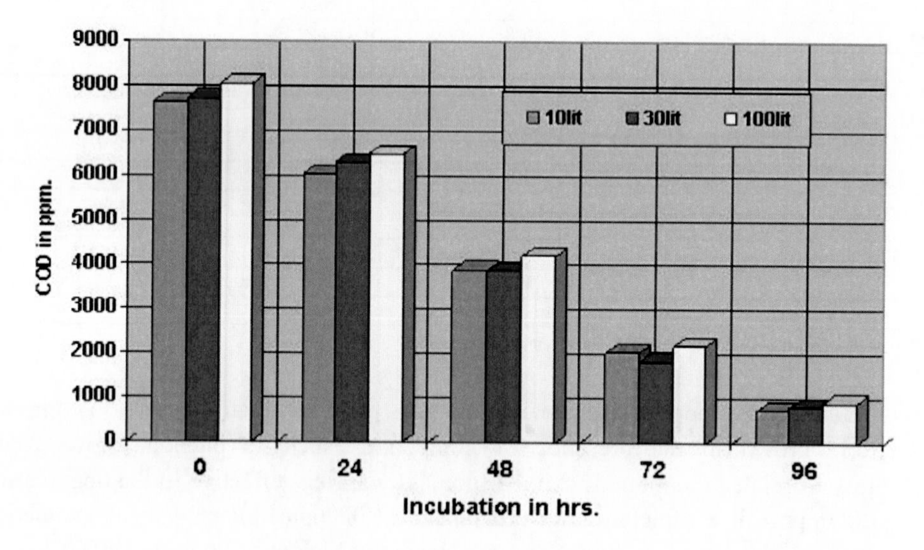


Figure 2. Reduction in COD in Lab Scale Bioreactor using Bulk drug plant wastewater.

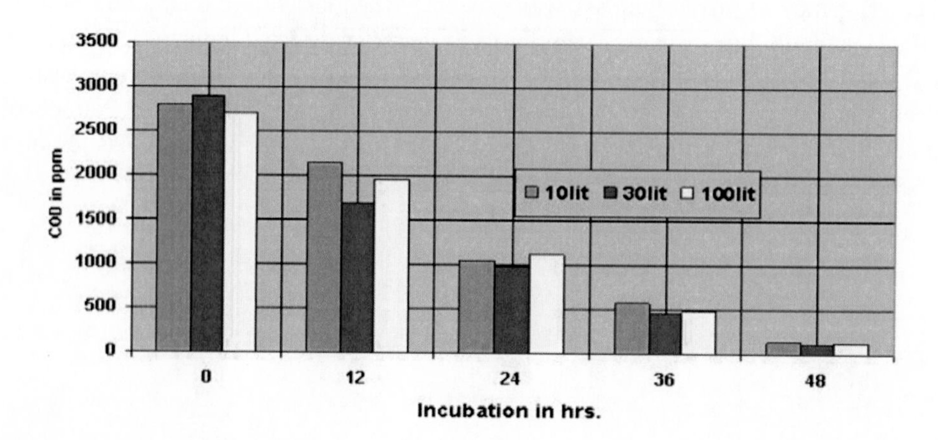


Figure 3. Reduction in COD in Lab Scale Bioreactor using Petrochemical (Aliphatic Amines) plant wastewater.

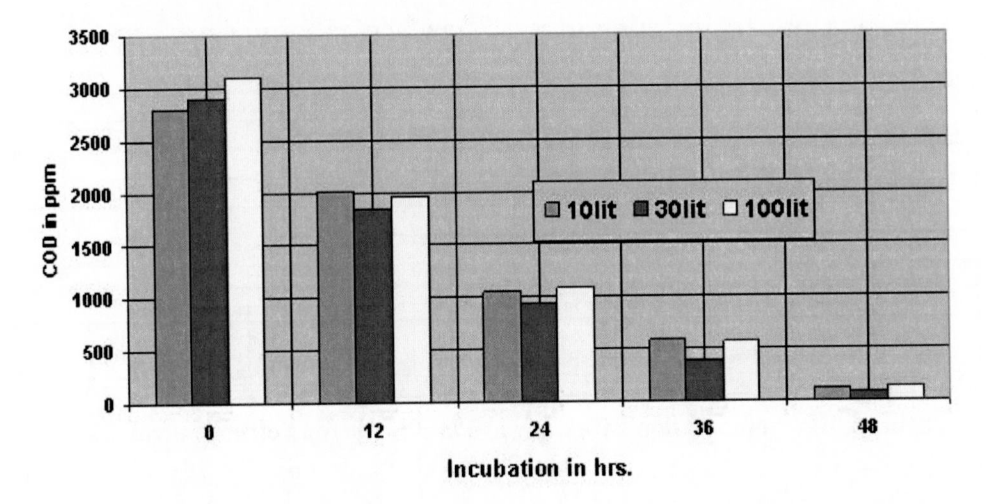


Figure 4. Reduction in COD in Lab Scale Bioreactor using Petrochemical (ABS Polymer) plant wastewater.

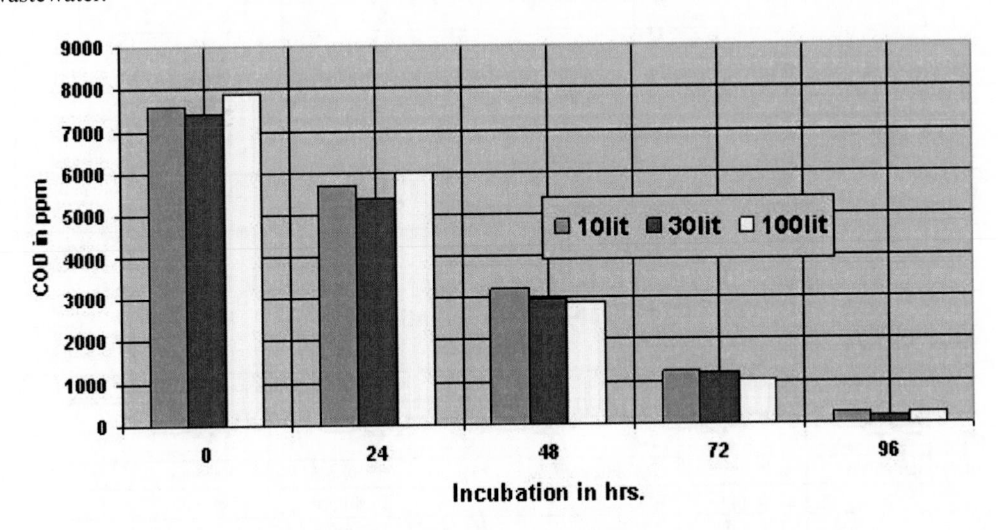


Figure 5. Reduction in COD in Lab Scale Bioreactor using Petrochemical (Nylon-6) paint wastewater.

The efficiency of Mixed Biomass was excellent when tested at plant scale level for treatment of effluents from bulk drug plant as can be seen from the tables 4- 7. The reduction in COD were from 5000ppm to 300-400ppm in 4 days for phenol containing waste water(bulk drug), from 1500ppm to < 100ppm in wastewater from amines plant in 1 day from 2000ppm to <100ppm in wastewater from ABS plastic polymers in 2days, and from 3000ppm to 200-300ppm in wastewater from Nylon-6 plant in 3 days. The stepwise cultivation in the pilot plant and subsequent addition of developed biomass gradually increased the efficiency of the aerobic process and then stabilized. MLSS and sludge volume levels increased consistently. The plant operations were optimized at MLSS 2000-4000ppm and sludge volumes of 20-30%. Settling characteristics improved significantly with increased clarity of treated wastewater from secondary clarifiers.

**Table 4. Bio-Aumentation Efficiency at Plant Scale for Bulk Drug Plant**

| Parameter (Average values) | Before Bioaugmentataion | | After Bioaugmentation | |
|---|---|---|---|---|
| | Inlet | Outlet | Inlet | Outlet |
| pH | 7-8 | 7-8 | 7-8 | 7-8 |
| Suspended solids | <300ppm | 200-300ppm | < 300ppm | < 200ppm |
| COD | 8000ppm | >1000ppm | 8000ppm | 300-400ppm |
| BOD | 3500ppm | >300ppm | 3500ppm | 25-50ppm |
| Sludge Volume | Nil | 20-25% In Bioreactor | Nil | 25-30% In Bioreactor |
| MLSS | Nil | < 2000ppm In Bioreactor | Nil | >3000ppm In Bioreactor |

**Table 5. Bio-Aumentation Efficiency at Plant Scale for Petrochemical Plant (Abs Ploymer)**

| Parameter (Average values) | Before Bioaugmentation | | After Bioaugmentation | |
|---|---|---|---|---|
| | Inlet | Outlet | Inlet | Outlet |
| pH | 7-8 | 7-8 | 7-8 | 7-8 |
| Suspended solids | <300ppm | 200-300ppm | <300ppm | <200ppm |
| COD | 2500ppm | >300ppm | 2500ppm | <100ppm |
| BOD | 1200ppm | >200ppm | 1200ppm | 10- 30ppm |
| Sludge Volume | Nil | 20-25% In Bioreactor | Nil | 25-30% In Bioreactor |
| MLSS | Nil | <2000ppm In Bioreactor | Nil | >3000ppm In Bioreactor |

**Table 6. Bio-Aumentation Efficiency at Plant Scale for Petrochemical Plant (Aliphatic Amines)**

| Parameter (Average values) | Before Bioaugmentation | | After Bioaugmentation | |
|---|---|---|---|---|
| | Inlet | Outlet | Inlet | Outlet |
| pH | 7-8 | 7-8 | 7-8 | 7-8 |
| Suspended solids | <300ppm | 200-300ppm | <300ppm | <200ppm |
| COD | 2000ppm | >300ppm | 2000ppm | <100ppm |
| BOD | 1200ppm | >200ppm | 2500ppm | 10- 20ppm |
| Sludge Volume | Nil | 20-25% In Bioreactor | Nil | 25-30% In Bioreactor |
| MLSS | Nil | <2000ppm In Bioreactor | Nil | >3000ppm In Bioreactor |

**Table 7. Bio-Aumentation Efficiency at Plant Scale for Petrochemical Plant (Nylon-6)**

| Parameter (Average values) | Before Bioaugmentation | | After Bioaugmentation | |
|---|---|---|---|---|
| | Inlet | Outlet | Inlet | Outlet |
| pH | 7-8 | 7-8 | 7-8 | 7-8 |
| Suspended solids | <300ppm | 200-300ppm | <300ppm | <200ppm |
| COD | 8000ppm | >300ppm | 8000ppm | 100-200ppm |
| BOD | 4000ppm | >200ppm | 4000ppm | 20- 40ppm |
| Sludge Volume | Nil | 20-25% In Bioreactor | Nil | 25-30% In Bioreactor |
| MLSS | Nil | <2000ppm In Bioreactor | Nil | >3000ppm In Bioreactor |

## 4. Conclusion

Bio-augmented bacterial cultures isolated from soil/sludge by enrichment culture technique were very useful in studies on treatment of effluents containing variety of organic compounds. Nutrient balance with respect to nitrogen & phosphorous was very important for degradation of organic compounds. Under optimal conditions, the rate of COD and BOD reductions were significant for bulk drug, petrochemical industries. The Bio-augmented bacterial cultures were successfully used in plants.

## References


APHA (1989). Standard Methods for the Examination of Water and Wastewater, 17th Edn, American Public Health Association, Washington, D.C.

Atlas, R.M., (1977) "Stimulated Petroleum Biodegradtion". *Crit. Rev. Microbiol.* 5: 371-386.

Atlas, R.M., (1981) "Microbial Degradation of Petroleum Hydrocarbons: an Environmental Perspective". *Microbiological Reviews.* 45(1): 180-209.

Bartha R., & Bossert I., (1984) "Treatment & Disposal of Petroleum Refinery Wastes". Petroleum Microbiology edited by Atlas R.M. 553-578. Macmilan Publishing Co., New York.

Boon, N., Goris, J., De Vos, P., Verstraete, W. and Top, E.M. (2000). Bio-augmentation of activated sludge by an indigenous 3-chloro aniline degrading Comamonas strain. *Environ. Microbiol.* 66, 2906–2913.

Boon, N., Verstraete, S.D. and Top, E.M. (2003). Bio-augmentation as a tool to protect the structure and function of activated sludge microbial community against a 3-chloro aniline load. *Appl. Environ. Microbiol.* 69(3), 1511–1520.

Bushnell L.D. & Haas H.F., (1941) "The Utilisation of certain Hydrocarbons by Microorganism". *J. Bact.* 41: 653-673.

Colwell, R. R. & Walker J. D., (1977) "Ecological Aspect of Microbial degradation of petroleum in the marine environment" *Crit. Rev. Microbial.* 5:423-445

Davis J. B., (1967) "Microbial Utilisation of Oily Wastes". *Petroleum Microbiology.* edited by Davis J. B. 350-400. Elsevier Publishing Co., New York.

Fewson, C.A., (1981) "Microbial Degradation of Xenobiotics and Recalcitrant compound" edited by Leiseinger T, Cook A.M., Hutter R and Wuesch J. Academic Press. 141-180.

Gutnick D.L. & Rosenberg E., (1977) "Oil Tankers and pollution: a microbiological approach". *Annual Reviews of Microbiology*. 31: 379-396.

Huban C.M. & Plowman R.D., (1997) "Bio-augmentation: Put Microbes to work". *Chemical Engineering*. March. 74-84.

Neumann, G., Teras, R., Monson, L., Kivisaar, M., Schauer, F. and Heipieper, H.J. (2004). Simultaneous degradation of atrazine and phenol by Pseudomonas sp. Strain ADP: Effects of toxicity and adaptation. *Appl. Environ. Microbiol.* 70(4), 1907–1912.

Pandya, M.T. & Doctor, T.R., (1980) "Microbiological Utilisation of Hydrocarbons & Phenolics in Refinery Waste". Proceeding of the seminar on management of environment edited by B. Patel. 287-298.

Ryoo, D., Shim, H., Canada, K., Barbieri, P. and Wood, T.K. (2000). Aerobic degradation of tetra-chloroethylene by toluene-o-xylene mono-oxygenase of Pseudomonas stutzeri OX1. *Nature Biotechnol.* 18, 775–778.

Singer M.E., (1984) "Microbial Metabolism of straight chain and branched chain alkenes". *Petroleum Microbiology.* edited by Atlas R.M. 1-60. Macmillan Publishing Co., New York.

Spain, J.C., (1995) "Biodegradation of Nitroaromatic Compounds". Annual Reviews of Vol. 49. 523-556.

Timmis, K.N., Steffan, R.J. & Unterman, R., (1994) "Designing Microorganisms for the treatment of toxic wastes" *Annual Reviews of Microbiology.* Vol. 48. 525-557.

Toro,S.D.,Zanaroli,G.,and Fava,F., (2006) "Intensification of the aerobic bioremediation of an actual site soil historically contaminated by polychlorinated biphenyls (PCBs) through bioaugmentation with a non acclimated, complex source of microorganisms." *Microbial Cell Factories.* 5:11,Published Online March 20,2006,doi.10.1186/1475-2859-5-11.

Zobell C.E., (1964) "The occurrence, effects & fate of oil polluting the sea, *Adv. Pollution. Res.* 3: 85-118.

In: Technologies and Management for Sustainable Biosystems ISBN: 978-1-60876-104-3
Editors: J. Nair, C. Furedy, C. Hoysala et al. 

*Chapter 4*

# SOLAR PHOTOCATALYTIC TREATMENT OF NON-BIODEGRADABLE WASTEWATER FROM TEXTILE INDUSTRY

*A. Verma and V. Singh* [*]
Thapar University, Punjab, India

## ABSTRACT

The decolorization and mineralization studies were performed under direct solar light using suspension photo catalyst (titanium dioxide). Plenty of research has been done on synthetic compounds used in many industries since last few years, using both UV and visible light. However the use of Advanced Oxidation Processes (AOP's) for live industry effluents and direct solar energy was not studied intensely. Live industry effluent was taken from textile industry and characterized. The effluent was of a non-biodegradable nature besides having intense color and Chemical Oxygen Demand (COD). In this chapter, the optimization of the process has been studied by varying catalyst concentration, variation of pH, addition of oxidant, and treatment under solar irradiation as pre and post treatment to the existing biological treatment for maximum degradation. The analysis of the products by measuring COD, BOD, TSS, TS, TDS, pH, Color and absorbance has been done. The results showed 96% color removal and 90-92% COD reduction along with enhancing the biodegradability of the effluent. This increase in biodegradability favor for further treatment of this effluent into cost effective secondary processes (biological), which were otherwise unsuccessful in the treatment of this effluent. The catalyst used in slurry mode was effectively recycled for 4-5 times to get the reliable results. So the chapter suggests the pretreatment of difficult wastewater with the AOP's to facilitate its treatment in cost effective biological processes. Average solar intensity was 26-30 $W/m^2$during the experiment.

**Keywords:** AOP; Photocatalysis; Biodegradability; COD; BOD; Mineralization; Effluent; $TiO_2$.

[*] *Postal address for co-authors: Assitt. Prof., Panjab Engineering College, Chandigarh, [INDIA]*

## 1. Introduction

One of the characteristics that best defines today's society in what is understood by the developed countries is the production of waste products. There is practically no human activity that does not produce waste products. Approximately 23% of the world's population lives in developed countries consume 78% of the resources and produce 82% of the waste products (Bauer et al., 2001). At present, there are some five million known substances registered, of which approximately 70,000 are widely used worldwide, and it is estimated that 1,000 new chemical substances are added to the list each year. The permitted levels have been vastly exceeded; causing such environmental contamination that our natural resources cannot be used for certain purposes and their characteristics have been altered.

Dyes, phenols, pesticides, fertilizers, detergents, and other chemical products are disposed of directly into the environment, without being treated, controlled or uncontrolled and without an effective treatment strategy. Industries involved in the disposal of these types of compounds are mainly Pharmaceutical industries, Paper and pulp industries, Leather-tanning industries, and Textile industries. The wastewater from these industries is having problems of color, solids, BOD, COD, pH, high nitrogen content, and is mostly non-biodegradable. Out of all these industries, Textile industries are of major concern. The situation worsens by the lack, or insufficiency, of adequate water treatment systems capable of diminishing the concentration of toxic substances that represent a chronic chemical risk.

Textile mills are major consumers of water and consequently one of the largest groups of industries causing intense water pollution. The extensive use of chemicals and water results in generation of large quantities of highly polluted wastewater. According to the U.S. EPA, about 1 to 2 million gallons of wastewater per day are generated by average dyeing facility in the US, reactive and direct dyeing generating most of the wastewater. It is estimated that about 10 % of the chemicals are lost in industrial wastewater (Young and Yu, 1997). The fate of these chemicals varies, ranging from 100% retention on the fabric to 100% discharge with the effluent. As a result, textile industry is confronted with the challenge of both color removal (for aesthetic reasons) and effluent salt content reduction. In addition, reactive dyes are highly water-soluble and non-degradable under the typical aerobic conditions found in conventional biological treatment system (Neppolian *et al.* 2001; Rodriguez *et al.* 2002). $BOD_5$/COD ratio is a count of biodegradability as reported by Marmagne et al. (1996). Dyes generally have high COD Value, moreover as they are having very low biodegradability factor (BOD/COD) they are difficult to be treated with the help of biological treatment processes. Toxicity is the main problem encountered by bacteria engaged for the degradation of Dyes, which makes it very difficult to treat it in biological processes.

To overcome the shortcomings in the existing treatment techniques research and development in innovative technologies during the last decade have shown that advanced oxidation processes (AOP's) that are combination of powerful oxidizing agents like UV light, UV/$TiO_2$, $O_3$/UV, $H_2O_2$ to mention a few are highly promising for the remediation of complex organic compounds which are present in contaminated water /effluent systems without generating any sludge or solid material of hazardous character. Destruction or mineralization of organic compounds by these processes is based on oxidative degradation by free radical attack, particularly by the hydroxyl radical, which is far more powerful than many other oxidants. Heterogeneous and homogeneous photocatalytic detoxification methods

($TiO_2/H_2O_2$, $Fe^{3+}/H_2O_2$) have shown recently great promise in the treatment of industrial wastewater, groundwater. A general description of heterogeneous photocatalysis under artificial or solar irradiation was presented in several excellent review articles by Taicheng An et al. (2003), Herrmann J.M. (1999), Hoffman et al. (1995). Solar advanced oxidation processes (AOPs) have the advantage over other advanced oxidation technologies of using natural sunlight and having as its main characteristic, that they are environmentally friendly technologies. In principle, the process involves a mild catalyst working with mild oxidants under miid conditions. AOPs coupled with biological treatment are seen as viable solution for economic treatment of biorecalcitrant organic compounds in wastewater. The objective of this chapter was to study the increase in the biodegradability of actual industrial (textile) wastewater using solar photo catalysis.

## 2. Materials and Methods

Raw effluent sample was collected from equalization tank of S.R. Industries (Textile Unit), Dera Bassi (Punjab [INDIA]). The sample was checked for some initial parameters/characters and then it was optimized for process parameters (concentration of photo catalyst, operating pH and conc. of oxidant) to carry out efficient treatment.

### Reagents and Chemicals Used

The photocatalyst was $TiO_2$ P-25 (a mixture of Anatase and Rutile form of titanium dioxide in the ratio of 70:30, procured form Degussa company, Indian Branch, Bombay with a BET surface area of $50\pm15\ m^2g^{-1}$ and average particle size of 30 nm.). Hydrogen peroxide (Ranbaxy laboratories) was used as an oxidant. BOD, COD, TKN, Color, TDS, and TSS were analyzed according to the procedures in standard methods (APHA Standards for examination of water and wastewater). All chemicals were used as received.

### Apparatus Used

**Radiometer-** Solar UV Intensity was measured hourly during experimental days with Eppley (model no.-33013) radiometer.

**pH meter-**The pH of the solution was adjusted with the help of HCl (0.1N) and NaOH (0.1N) and measured with the help of pH meter (ELICO).

**Filtration-** Samples after the photocatalytic treatments were filtered through Injection filters (millipore size 0.45μm), Whattman's filter paper (No.42).

**COD Digester-**COD digester was used for the digestion of samples in the process of COD determination. Semi automatic autoclave (EQUITRON) was used for the same.

## Shallow Pond Slurry Type Reactor

For the photocatalytic process reactors used were either cylindrical in shape, made of borosil glass, which has a diameter of 5 cm and is 6cm in height with a capacity of approximately 1000ml or rectangular type having dimensions 2'x1'x1.5'm. The experiments were carried out in a shallow pond batch slurry reactor.

## Preparation of the Sample

Wastewater collected from the textile industry was highly concentrated. So to get the values within range, the sample (100ml) was diluted in the ratio of 1:1. Distilled water was used for all the dilutions. Initial pH was checked and varied to get the optimized value of the pH with the help of HCl (0.1N) and NaOH (0.1N). Catalyst was added in a wide range from 0.1% to 1% to optimize the process for maximum pollutant degradation. $H_2O_2$ (30%)was added in the range of 1ml-8.0 ml/200 ml to check optimum volume for the process. Finally the reaction vessel was kept over the magnetic stirrer under sunlight.

## Procedure

Sample was treated in natural sunlight/artificial UV source, as shown in figure 1. Intensity of UV radiations was measured continuously with the help of radiometer. Samples were withdrawn after every one hour and filtered. COD of the samples were then measured as per the standard methods.

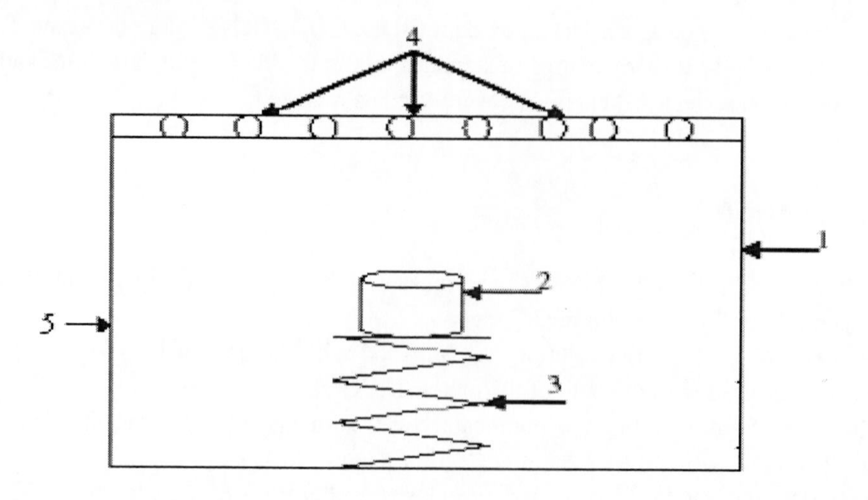


Figure 1. Schematic diagram of lab scale set up: (1) UV chamber, (2) Reactor, (3) Lab jack, (4) UV source (natural/artificial), (5) Holes.

## 3. RESULTS AND DISCUSSION

### Wastewater Characteristics

Wastewater taken from flow equalization tank and analyzed for its various parameters (the results of the various parameters), are shown in table 1.

**Table 1. Characteristics of raw textile effluent from S.R.Industries, Dera Bassi, Punjab [INDIA]**

| S.No. | Parameter | Prevailing Range (mg/ L) |
|---|---|---|
| 1. | COD | 700-800 |
| 2. | TDS | 8500-9000 |
| 3. | TS | 9000-9800 |
| 4. | TSS | 600-800 |
| 5. | BOD/COD | 0.2-0.25 |
| 6. | COLOR | > 5000 Pt CO |
| 7. | pH | 10-11 |
| 8. | TKN | 30 |
| 9. | Temperature | $40^0$C |

These parameters show that the wastewater is highly polluted. So some pretreatment is required so as to safely discharge the water.

### Solar Photocatalytic Pretreatment

After characterization of the waste sample, its photocatalytic treatment was done. Photocatalytic treatment depends upon the various factors like catalyst concentration, operating pH, oxidant addition etc. So depending upon these factors, optimized reaction conditions were calculated and used throughout the process.

### Radiation Conditions in Punjab during summers

The amount of UV radiation, which can be used for $TiO_2$ photocatalysis on a summer day (March, April, May) with a maximum of about 30-37 $W/m^2$. Figure 2 shows the variation of solar intensities with time during experimental days.

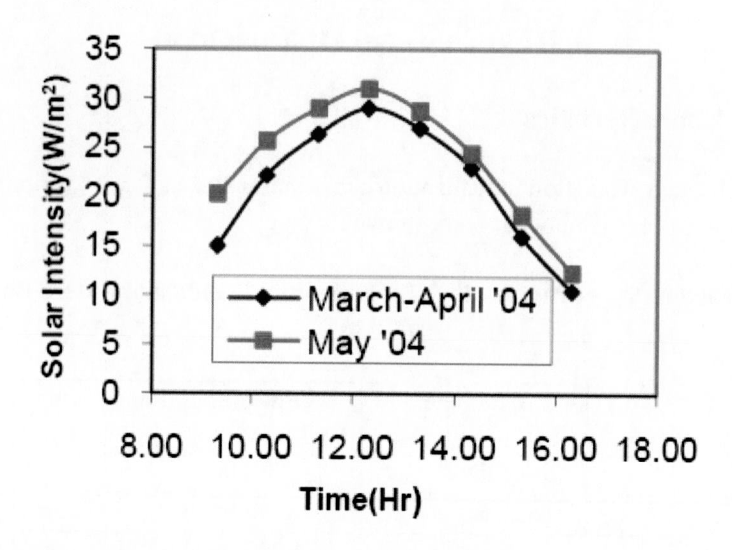


Figure 2. Intensities of solar radiations during experimental days.

## Catalyst Concentration

Catalyst ($TiO_2$) concentration was varied from 1.0 g/l to 10 g/l (i.e. 0.1% to 1.0%) during reactions in sunlight. It was observed that rate increases with increase in catalyst concentration and becomes constant above a certain level as shown in figure 3. The reasons for this decrease in degradation rate are:

Aggregation of $TiO_2$ particles at high concentrations causing a decrease in the number of surface active sites and

Increase in opacity and light scattering of $TiO_2$ particles at high concentration leading to decrease in the passage of irradiation through the sample.

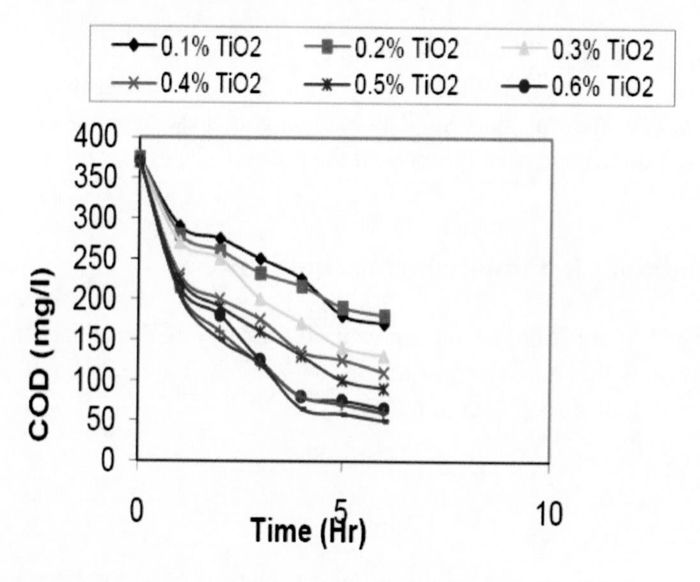


Figure 3. Effect of catalyst concentration on the COD reduction.

In this context, it was observed that the COD continuously decreases from 360 to 50 mg/L on increasing the catalyst concentration from 0.1% to 1.0 %. An optimum of catalyst concentration has to be taken when the decrease in COD level are to be within acceptable limits and at the same time use of higher concentration of catalyst will increase the cost of the process and decrease the permeability of sunlight. Thus accordingly 0.1% catalyst concentration was selected for process optimization.

## Effect of Operating pH

The wastewater from textile industries usually has a high pH values (nearly 11). Further, the generation of hydroxyl radicals (AOP's) is also a function of pH. Thus pH plays an important role both in the characteristics of textile wastes and generation of hydroxyl radicals. Hence, attempts have been made to study the influence of pH in the degradation of dye in the range 3 – 11 as shown in figure 4.

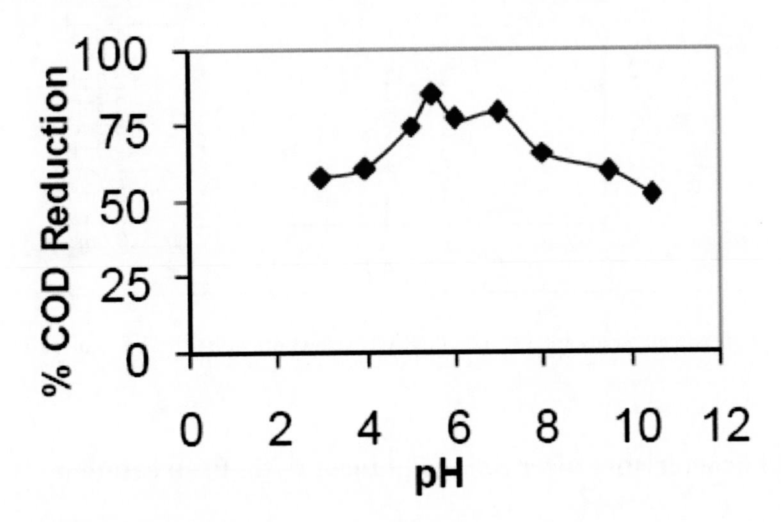


Figure 4. Effect of pH on the % degradation of the effluent at 0.1% $TiO_2$.

In our experiments, maximum degradation was observed at pH near 5.5. Other pH values also responded to good degradation rates but the final pH after photocatalytic treatment which was the deciding factor for determining the optimum pH, which was found to be in the range of 6.8-7.0 as shown in table 2. This is important because after photocatalytic treatment, the water is to be subjected to biological treatment where the pH should be in the range of 6-8.

Table 2. Change in pH values after the photocatalytic treatment

| pH before Treatment | 3<br>.0 | 4<br>.0 | .0 | 5<br>.5 | 6<br>.0 | 7<br>.0 | 8<br>.0 | 9<br>.0 | 1<br>0.0 |
|---|---|---|---|---|---|---|---|---|---|
| pH after Treatment | 3<br>.2 | 4<br>.23 | .3 | 7<br>.0 | 8<br>.1 | 8<br>.45 | 8<br>.5 | 8<br>.95 | 9<br>.5 |

## Effect of Oxidant Addition

One possible way to increase the reaction rate would be to increase the concentration of OH radicals because these species are widely considered to be promoters of photocatalytic degradation. The addition of hydrogen peroxide to the heterogeneous system increases the concentration of OH radical, since it inhibits the electron-hole recombination, according to equation (i):

$$TiO_2(e^-) + H_2O_2 \rightarrow TiO_2 + OH^- + OH^\bullet \quad \text{(i)}$$

From the experiments conducted by varying the hydrogen peroxide concentration from 1.0 to 8.0 ml per 200ml of the effluent, the best results were obtained when oxidant addition came out to be 1 ml/200.ml of effluent and have been taken as the optimum amount required for maximum effective degradation of pollutants as is clear from figure 5.

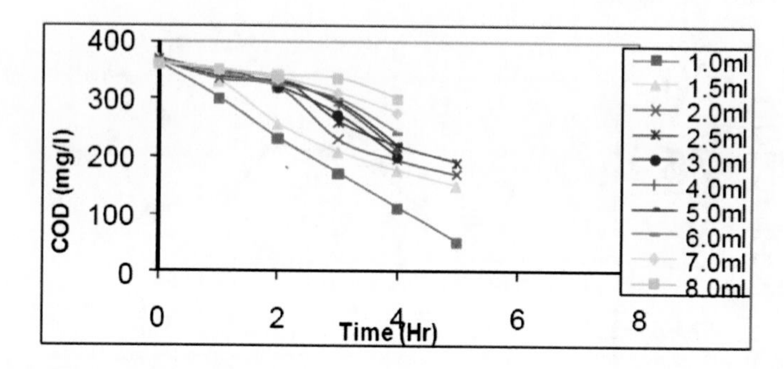


Figure 5. Effect of oxidant addition ($H_2O_2$) on the COD reduction with 0.1% of catalyst loading at 5.5 pH.

## Effluent Characteristics after Solar Photocatalytic Pretreatment

Effluent characteristics were determined after photocatalytic pretreatment under optimized conditions i.e. at $TiO_2$ concentration of 0.1%, 1 ml of oxidant addition and at 5.5 pH and shown in table 3.

**Table 3. Characteristics of the wastewater after solar Photocatalytic treatment (pH 5.5, $H_2O_2$= 1ml, $TiO_2$ = 0.2 gm/200ml (0.1%), Average sun Intensity = 30W/m$^2$)**

| S.No | Parameter | After Photocatalytic Treatment (mg/ L) (Optimized Conditions) |
|---|---|---|
| 1. | COD | 50-60 (after4.5 hrs) |
| 2. | BOD/COD | 0.7-0.8 |
| 3. | TDS | 4100 |
| 4. | TS | 4110-4120 |
| 5. | TSS | 10 |
| 6. | COLOR | 50 Pt- Co |
| 7. | pH | 7.0 |
| 8. | TKN | 18.8 |

## Performance of the Solar Reactor vs. the Artificial UV Reactor

Degradation is related to the photons, which are available for the total volume of the reactor, the efficiency of the Solar Reactor (SR) is higher than that of the Artificial UV reactor (AUV). Of course, a higher concentration of radiation can result in a faster degradation but the dependence is not linear. The electric energy, which is needed to produce photons in the lamp, is not used with the highest efficiency. The SR reactor is very efficient in using the available photons and that opens up possibilities for the implementation of the technology in industrial processes as clear from figure 6.

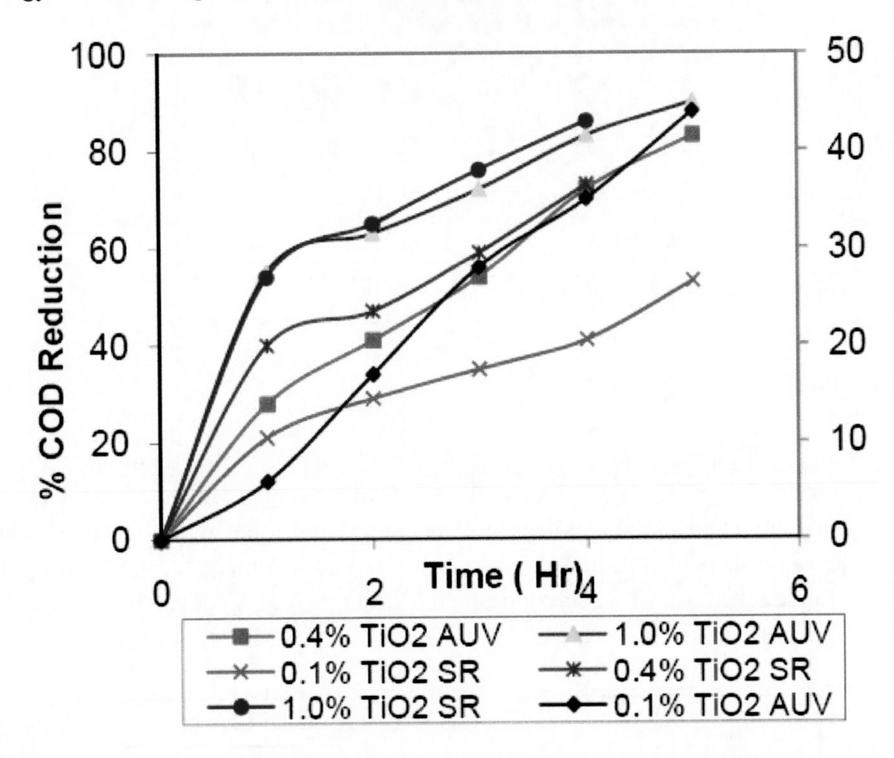


Figure 6. Comparison between Solar reactor (SR) and Artificial UV reactor (AUV).

## Color Removal

Color is usually the first contaminant to be recognized in wastewater. Many azo dyes, constituting the largest dye group, may be decomposed into potential carcinogenic amines under anaerobic conditions in the environment. Color removal from wastewater is often more important than the removal of soluble colorless organic substances, a major fraction of which contribute to the COD and BOD besides disturbing the ecological system of the receiving waters.

The traditional techniques used for color removal are activated carbon (charcoal), filtration and coagulation. Each method has few advantages and disadvantages. For example, the use of charcoal is technically easy but has high waste disposal cost. Although filtration potentially provides pure water as the final product, it is possible for low molar mass dyes to pass through the filter system. Coagulation using alums, ferric salts or limes is a low cost process, but all these methods have a major disadvantage of simply transferring the pollutants to another phase rather than destroying them. Biological treatment is a proven technology and cost effective, however it has been reported that the majority of dyes are only adsorbed on the sludge and not degraded. Figure 8 shows the results for color removal after photocatalytic treatment, thus clearly indicating the efficacy of AOP's in color removal applications.

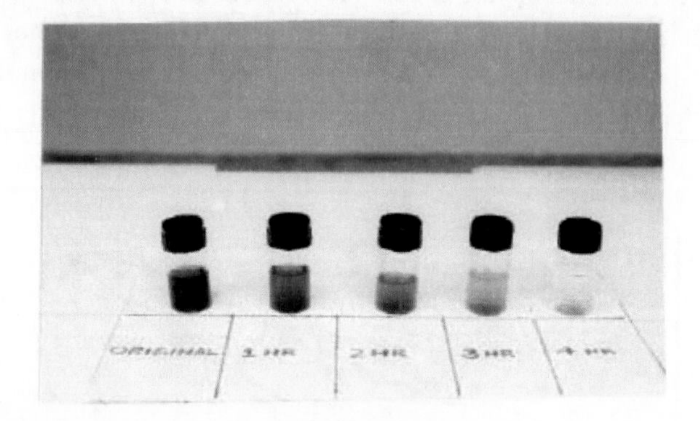

Figure 7. Color removal after Photocatalytic treatment (PKT).

## Reuse of $TiO_2$

The catalyst's lifetime is an important parameter of the photocatalytic process, due to the fact that its use for longer period of time leads to a significant cost reduction of the treatment. For this reason, the photocatalytic experiment was repeated 4 times with the same amount of $TiO_2$ P-25 as catalyst. After the first day of photocatalytic treatment for 4 Hrs at optimized conditions, the catalyst was recovered through filtration. The Catalyst recovered by this method was activated at $100^0$C before using.

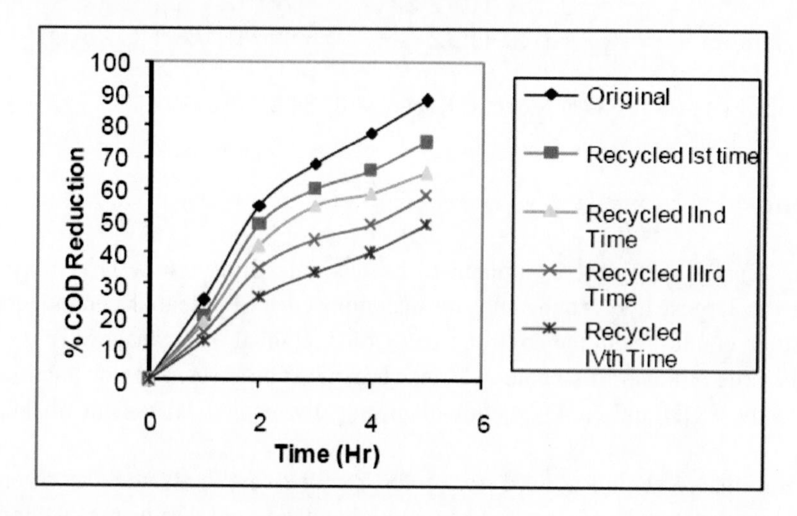


Figure 8. Recyclability of $TiO_2$ P-25 with optimized conditions (pH 5.5, $H_2O_2$= 1ml, $TiO_2$ = 0.2 gm/200ml (0.1%), Average sun Intensity = 30W/m$^2$).

Then process was repeated until reasonable COD reduction was achieved (figure 8).

## Photocatalytic Coupled Biological Treatment

The effluent after photocatalytic treatment subjected to the biological process i.e. coupled photocatalytic and biological treatment. Positive results indicate that effluent being fed to biological reactor is having suitably high value of biodegradability and in this case the effluent which is fed to the biological process is having biodegradability value of 0.70-0.80, thus, clearly highly biodegradable. This effluent is subjected to the aerobic biological treatment. Small intermediates left during photocataytic treatment are completely degraded in biological process as confirmed by evaluating various parameters as shown in table 4.

**Table 4. Characteristics of effluent after Coupled treatment (Photocatalytic followed by Biological treatment)**

| S.No. | Parameter | Range ($mg.\ L^{-1}$) |
|---|---|---|
| 1. | COD | 30-40 |
| 2. | TDS | 3800 |
| 3. | TS | 3800 |
| 4. | TSS | Nil |
| 5. | BOD/COD | > 0.85 |
| 6. | COLOR | Nil |
| 7. | pH | Neutral |

M.P.Reddy et al. (2003) used coupled photocatalytic and biological processes to treat some common industry effluents. GC analysis shows reduced no. of peaks after photocatalytic and coupled treatment.

## Perspectives for the Coupling of Photochemical and Biological Processes at Field Pilot Scale

Direct solar light is promising and economically attractive source of UV irradiation, which has been applied in combination with oxidants and catalyst for the complete mineralisation of a wide range of organic pollutant in water .We have performed a series of experiments in order to find out whether photo-AOP using solar radiation's could be coupled with a biological system.

Interesting results were obtained using suspended $TiO_2$-photoassisted process since this catalyst is not sacrificed and can be separated from the treated water as confirmed by increase in the biodegradability shown in figure 9.

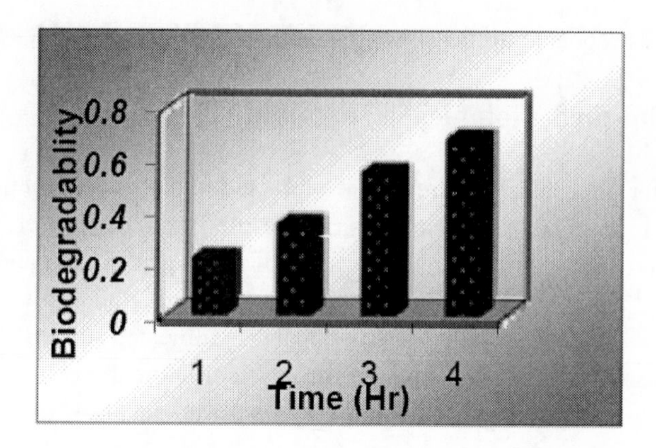


Figure 9. Biodegradability of the effluent during the photocatalytic treatment.

This is very important when an industrial application is contemplated since $TiO_2$ is recycled and clean treated water could be drained away. Different procedures of filtration or sedimentation have been published for this purpose. These encouraging results open new possibilities for the coupling of $TiO_2$-photoassisted and biological processes, at pilot scale, employing $TiO_2$ and bacteria.

## Decolorization and Biodegradability of Textile Wastewater

To verify the relationship between decolorization and biodegradability of Textile wastewater, the wastewater from Textile industry was taken for study purposes. The properties of the wastewater are shown in table 1. Clearly, it cannot be biodegraded according to the ratio BOD/COD = 0.26. However, the color and COD of the wastewater were removed by $TiO_2$ photocatalytic treatment.

The complete decolorization of wastewater with $TiO_2$ slurry is achieved for about 130-150 min. irradiation. At the same time, the COD conversion of wastewater was 70-80% shown in figure 10.

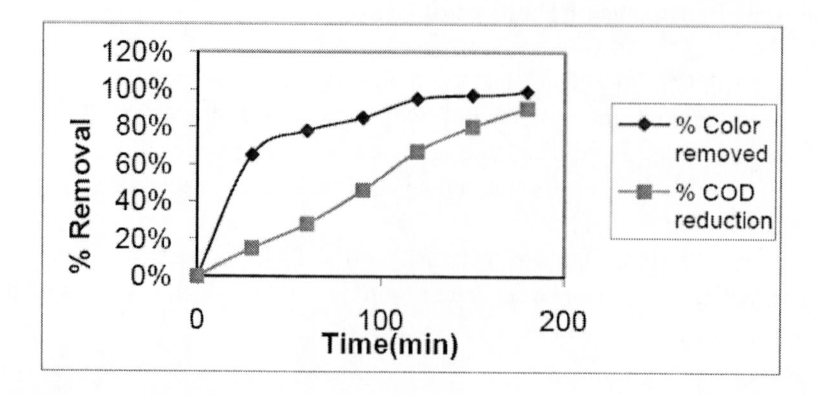


Figure 10. Graph showing % Color removed and % COD reduction with time of Actual Industry effluent (pH 5.5, $H_2O_2$= 1ml, $TiO_2$ = 0.2 gm/200ml (0.1%), Average sun Intensity = 30W/m$^2$).

Furthermore, the COD complete removal can be obtained after 240-min.irradiation. Initial dyes chemical had inhibition for the biodegradation; however, the ratio of BOD/COD of most dyes solution is more than 0.3 when the color disappeared completely. The byproducts arising from decolorization reaction were biodegraded by biological processes.

## 4.0. CONCLUSION

Coupling of Photocatalytic and biological processes is a good alternative to minimize the treatment cost of wastewater containing biorecalcitrant and/or toxic pollutants. The chemical process could be used as pretreatment in order to increase the biodegradability of the wastewater or as a post-treatment to remove the non-biodegradable compounds. First of all, due to the high cost of chemical treatments, biodegradability test should be carried out since for biodegradable compounds classical biological treatments are, at present, the cheapest and most environmentally compatible processes. Solar photo catalysis can be used for increasing biodegradability of the textile wastewater, which is supposed to be non biodegradable. This pretreated wastewater can be easily mineralized in cost effective biological units, due to its increased biodegradability. Thus coupling of Photocatalytic and biological processes is a good alternative to minimize the treatment cost of wastewater containing biorecalcitrant and/or toxic pollutants.

## ACKNOWLEDGMENT

The authors of this paper are very thankful to-

Dr. Anita Rajor (Lecturer, Thapar University, Patiala [INDIA]) M/s. S.R. Industries, Dera Bassi, Punjab, [INDIA] for providing the effluent from time to time.

Dr. Susheel Mittal (Head, School of Chemistry and Biochemistry, Thapar Uni, Patiala).

## REFERENCES

Bauer C, Jacques P, Kalt A. Photooxidation of an azo dye induced by visible light Incident on the surface of $TiO_2$. *J. Photochem. Photobiol. A: Chem.* 2001; 140: 87-92.

Herrmann J.M. (1999), Heterogeneous photocatalysis: fundamentals and applications to the removal of various types of aqueous pollutants, *Catal Today.* 53(1), 115–129.

Hoffman M, (1995), Environmental applications of semiconductor photocatalysis, *Chem. Rev.* 95(1), 69–96.

Marmagne, O., Coste C., April 1996 "Color Removal from Textile Plant Effluents" *American Dyestuff Reporter.*

Neppolian, B., Sakthivel, S., Arabindoo, B., Palanichamy, M. and Murugesan, V. (2001) "Kinetics of photocatalytic degradation of reactive yellow 17 dye in aqueous solution using UV irradiation", *J. Environ. Sci. Health.* Part A. Tox Hazard Subst. *Environ. Eng.*, 36: 203-213.

Rodriguez, C., Dominguez, A. and Sanroman, A. (2002) "Photocatalytic degradation of dyes in aqueous solution operating in a fluidised bed reactor", *Chemosphere.* 46: 83-86.

Reddy M.P. et al., (2003), *Indian Journal of Environ. Protection.* 23 (4), 438-445.

Taicheng An , Haofei Gu , Ya Xiong , Weiguo Chen , Xihai Zhu , Guoying Sheng , Jiamo Fu, (2003), Decolourization and COD Removal from Reactive Dye-containing Wastewater using Sonophotocatalytic Technology, *J. Chem. Tech. Biotech.* 78, 1142 – 1148.

Young, L. and Yu, J. (1997) "Ligninase-catalysed decolorization of synthetic dyes", *Water Res.* 31: 1187-1193.

In: Technologies and Management for Sustainable Biosystems ISBN: 978-1-60876-104-3
Editors: J. Nair, C. Furedy, C. Hoysala et al. 

*Chapter 5*

# MICROALGAE AN ALTERNATIVE TO COAL FOR POWER GENERATION

***M. T. Pandya****
Department of Microbiology
Jai Hind College, Mumbai-400 020.INDIA

## ABSTRACT

Constant increases in the petroleum prices over the last few years have brought Micro-algae as a novel renewable energy resource, from which both oils and biomass can be obtained. Micro-algae are remarkable and efficient biological factories capable of converting energy in the form of $CO_2$ to a high density liquids and heavy biomass. The long term demand for coal brings with it a demand for technologies that can mitigate the environmental problems associated with coal. Energy security, Climate change, synergy of coal and micro-algae and aquatic biomass are some of the benefits, which interest newer research and development activities in the field of algae technology.

Species of green and blue algae from cooling water deposits, lake water and soil were isolated by enrichment culture cultivation methods. Enrichment of samples were carried out in mineral salt media for 7-10 days and after repeated cultivation algal biomass developed was isolated on solid media. Strains were cultivated using glucose as carbon source to develop biomass and compare the growth rate under autotrophic and hetertrophic conditions. These were identified as species of *Cholrella, Nostoc, Chlamydomonas Oscillitoria, Anabena* and few were Diatoms. Growth yields were 5-10gm per lit. Selection of culture media improved the yields. Carbon content of these algae was in 40-55% range. Selected algal strains were cultivated in 5lit and 10lit capacity open and flat reactors with artificial light source. Biomass was separated by centrifugation at 5000 rpm for 10mins and then dried in oven at 105°C Dried biomass with about 10% moisture was tested for calorific value by standard methods and the same were in a range of 3000-3500 kcal per kg of biomass.

---

* mukesh_pandya51@hotmail.com

# 1. Introduction

Micro-algal photosynthesis has increasingly received attention as a means of reducing the emission of $CO_2$ from atmosphere (Kodama et al 1993, Yang and Chang 1997, 2003, Yang et al 2000). Micro-algal biomass generated can be used as a fuel to supplement coal in the generation of electricity (Oliver Danielo, 2005). Shortage due to depletion of fossil fuels and constant increase in the price of fossil fuels over the last few years has brought into focus micro-algae as a renewable energy resource, which can be used to obtain both oil and biomass (Thomas, 2006).

The energy crisis of the 1970s was followed by increased concern for the environment, with particular emphasis on protection of the global ecosystem. The Research Association for Petroleum Alternatives Development (RAPAD) was established in Japan, one of its main tasks was to investigate the development of technologies for biomass conversion and utilization, in particular, the production of ethanol from cellulosic biomass.

Micro-algae are a diverse group of prokaryotic and eukaryotic photosynthetic microorganisms that grow rapidly due to their simple structure. They can potentially be employed for the production of bio-fuels in an economically effective and environmentally sustainable manner (Yanqun et.al, 2008). Microalgae have been investigated for the production of a number of different biofuels including bio-diesel, bio-oil, bio-syngas, and bio-hydrogen. The production of these bio-fuels can be coupled with flue gas $CO_2$ mitigation, wastewater treatment and the production of high-value chemicals. Micro-algal farming can also be carried out with seawater using marine micro-algal species as the producers. Developments in micro-algal cultivation and downstream processing (e.g., harvesting, drying, and thermo-chemical processing) are expected to further enhance the cost-effectiveness of the bio-fuel from micro-algae strategy.

This chapter highlights developing mix biomass of micro-algae which can be cultivated at industrial site using wastewater and flue gases generated during the operations and then use it as fuel for generating power.

# 2. Materials and Methods

## 2.1. Enrichment, Isolation and Identification

Water samples from freshwater bodies, cooling water deposits, soil and sludge were enriched in CHU no-3 medium. Flasks were incubated at ambient temp for 10 days in chamber with artificial light (white) source. Four serial subcultures were made in the fresh medium and algal biomass developed was cultivated under the same conditions. Biomass after serial subculture was isolated on CHU-no-3 solid media plates and incubated for 7 days in chamber with artificial light sources. The isolates were identified using microscopic observation.

Micro-algal species were cultivated in various culture media like Beneck's liquid medium, CHU no 10, Kuhl and Lorenzen medium, Erddekokt and Salze, Pringsheim medium and Czurda medium (Aaronson 1970; Vonshak 1986)

Micro-algal strains were grown in CHU-No-3 medium supplemented with glucose at 1% level. The flasks were incubated at ambient temperature for 10 days in chamber with artificial light.

## 2.2. Cultivation in Bioreactor

Microalgae species were grown in flasks containing CHU No-3 enrichment culture medium for 7 days and the biomass separated was inoculated in 1 and 5 litre rectangular open bioreactor and in 5 litre photobioreactor and incubated at ambient temperature in chamber with light sources.

Biomass was allowed to settle and the sediment collected was centrifuged and dried at 105°C. Dried biomass was analysed for TOC (Dichromate reflux method) protein (colorimetric method), lipid contents (Soxhlet extraction) and calorific value ( using Bomb calorimeter) as given in Standard Methods for Water and Waste Water Analysis (1989).

# 3. Results and Discussion

The derivation of energy from algal biomass is an attractive concept in that unlike fossil fuels, algal biomass is rather uniformly distributed over much of the earth's surface, and it's utilization would make no net contribution to increasing atmospheric $CO_2$ levels. Although algal biomass is regarded as a low-grade energy source owing to its high moisture content, through biological processes, it may be converted to modem gaseous and liquid fuels such as hydrogen, methane, ethanol, and oils.

Various strains of Blue green algae, Green Algae and Diatoms were isolated from water, soil and cooling water deposit samples using enrichment culture cultivation techniques. These strains showed excellent growth in mineral salt media supplemented with carbonates as source of carbon. The growth was induced significantly in artificial light source as compared to the natural sunlight. Cooling water deposits showed wide variety of algal species. All the selected media supported moderate to heavy growth of both green and blue green algae (table-1). The growth in CHU no-3 media was very good so for further studies this medium was used (table-2) the biomass was in a range of 4-8gm/lit. The organisms isolated were identified by microscopic methods as species of Diatoms, Chlorella, Chlamydomonas, Nostoc Oscillitoria and Anabena. Growth of diatoms and chlorella was more compared to other strains. Both Nostoc and Anabena showed ability to fix nitrogen and grew in media without nitrogen source. Nitrogen and Phosphorus concentrations influenced the composition of cellular biomass. At the limiting levels the lipid contents increased. Medium supplemented with 1% glucose showed very heavy growth of Chlorella. The biomass was in a range of 10gm per lit in 10 days of incubation period.

**Table 1. Growth of Microalgae in culture media**

| Culture Medium | Selected Microalgae strain | | | | | |
|---|---|---|---|---|---|---|
| | Chlamydomonas | Chlorella | Oscilitoria | Anabena | Nostoc | Diatoms |
| Beneck's | ++ | +++ | +++ | ++ | ++ | ++ |
| Chu-No-10 | ++++ | ++++ | +++ | +++ | ++ | ++ |
| Kuhl & Lorenzen | +++ | +++ | | | | |
| Erddekokt & Salze | +++ | ++ | +++ | ++ | +++ | ++ |
| Pringsheim | +++ | +++ | +++ | +++ | +++ | ++ |
| Czurda | +++ | ++ | ++ | ++ | ++ | ++ |

**Table 2. Growth yield in Chu-10 Medium**

| Strain | Biomass gm/lit |
|---|---|
| Chlamydomonas | 5 |
| Chlorella | 8 |
| Oscillitoria | 4 |
| Anabena | 4 |
| Nostoc | 6 |
| Diatoms | 4 |

The growth response in 5 litre photo-biorecator was better as compared to the rectangular bioreactor. Artificial light too induced better growth. Bioreactor had mixing arrangement preventing growth of algae on the surfaces of the reactor. Growth in rectangular reactor was more on the sides and surfaces. In a period of 7days the growth yields were in a range of 5-10gm/lit in optimized CHU-no 10 medium. Algae contained fat, carbohydrate, and protein, some contained up to 60 % fat (Sawayama et al 1985; Kurano et al 1995; Chang and Yang, 2003; Yue and Chen, 2005; Michele and Jorge 2007). The protein content of these strains was in a range of 40-60%. The Total Organic Carbon content were in a range of 40-55%. The lipid contents were 12-18% (table-3).

Micro-algal biomass generated can be used as a fuel to supplement coal in generation of electricity (Kadam and Sheehen, 1996). Preliminary results have shown that the co-firing of 7 % biomass, on a heat input basis, with crushed, pulvarised coal can lower $NO_x$ and $CO_2$. The biomass was analysed for its calorific value. The dried biomass with about 10% moisture had calorific values in a range of 3000-3500Kcal/kg vide table-4.

**Table 3. Chemical Composition of Selected Strains**

| Micro-Algae Strain | Protein | Total Organic carbon | Lipid |
|---|---|---|---|
| Chlamydomonas | 48 | 50 | 12 |
| Chlorella | 60 | 55 | 17 |
| Oscillitoria | 47 | 45 | 16 |
| Anabena | 46 | 42 | 16 |
| Nostoc | 52 | 49 | 15 |
| Diatoms | 42 | 40 | 18 |

**Table 4. Calorific Value of Selected Alagal biomass**

| Micro-Algae Strain | Calorific Value |
|---|---|
| Chlamydomonas | 3100 |
| Chlorella | 3500 |
| Oscillitoria | 3300 |
| Anabena | 3300 |
| Nostoc | 3000 |
| Diatoms | 3100 |

## 4. Conclusion

Different green algae, blue green bacteria and diatoms are isolated from various samples Concentration of $CO_2$ which is a major greenhouse gas can be reduced by generation of biomass which can be then used for production of electricity. The algal species Diatoms, Chlorella, Chlamydomonas, Nostoc and Anabena had good calorific values and lipid contents. The algal species had calorific values although lower than coal but can still find its utility as biomass for fuel. Large scale cultivation in suitably designed photobiorecator can not only provide biomass, but also reduce emission of flue gases which can be used for cultivation.The developed biomass can be dried and then used as a supplement for coal. Microalgae development can give tailor made solutions for industries with uninterrupted power supply at reasonable cost.

## References

APHA (1989). Standard Methods for the Examination of Water and Wastewater, 17th Edn, American Public Health Association, Washington, D.C.

Aaronson Sheldon (1970), *Experimental Microbial ecology.* 1970.Academic press.

Kodama, M.,H.Ikemoto, and S.Miyachi, 1993, *J. Mar. Biotech.* 1:21-25.

Kadam,K.L. and Sheehen,J.J.1996.Microalgae technology for remediation power plant flue gas. *World Resource Review.* [8:4],pp493-504.

Kurano,N., H.Ikemoto, H.Miyashita, T. Hasegawa, and S.Miyachi, 1995a, *J. Mar. Biotech.* 3:108-110.

Kurano,N., H.Ikemoto, H.Miyashita, T.Hasegawa, H.Hata,and S.Miyachi, 1995a, *Energy Consers. Mgmt.* 36: 689-692.

Lihong Yue and Weigong Chen, 2005, *Energy Conservation & Management',* 46(11-12) 1868-1876.

Michele Greque de Moris and Jorge Alberto Vieira Costa, 2007 Jour. Bact, 129(3) 439-445.

Oliver Daniel, May 2005, ' *An Algae based fuel, Biofutur.* No 255.

Sawayama, S.S. Inoue,Y.D.Dote, and S.Y. Yokoyama, 1995, *Energy Convers. Mgmt.* 36: 729-731.

Thomas Schultz, 2006 'The Economics of micro-algae production and processing into biodiesel', Farming Systems, Department of Agriculture and Food, Western Australia Ted Atwood Report from Global Greenlife institute

Yang, S.S., E.H. Chang, J.Y.Lee,Y.Y. Horng and C.R.Lan, 1997, *Biol. Fertil. Soils.*25:245-251.

Yang,S.S., and Chang,E.H., 2003, Microalgae for biofixation for Carbon Dioxide., *Bot. Bull. Acad. Sin.* 44:43-52.

Yang, S.S., E.H. Chang, J.Y.Lee,Y.Y. Horng and C.R.Lan, 2000, Month, *J. Taipower's Eng.* 624:65-82.

Yanqun Li, Mark Horsman, Nan Wu, Christopher Q. Lan, and Nathalie Dubois-Calero, 2008, *Biotechnol. Prog.,* Article 10.1021/ 371-2

Vonshak,A .1986.Laroratory techniques for the cultivation of microalgae. In A.Richmond [ed], *'Handbook of Micro-algal mass culture'.* CRC Press, FL, 117-199.

In: Technologies and Management for Sustainable Biosystems ISBN: 978-1-60876-104-3
Editors: J. Nair, C. Furedy, C. Hoysala et al. 

*Chapter 6*

# COMPARISON OF PATHOGEN DIE-OFF PATTERNS OF TOMATOES GROWN IN TWO HYDROPONICS SYSTEMS

***Noraisha Oyama*[*1], Jaya Nair[*1] and Goen Ho[*1]***

[1] Environmental Technology Centre, Division of Science and Engineering, Murdoch University, 90 South Street, Murdoch, 6150 WA, Australia

## ABSTRACT

Due to water shortages in most parts of the world, alternative water sources are required for daily activities, such as agriculture and domestic uses. Treated domestic wastewater reuse is gaining acceptance around the world, mainly for non-human contact use. Research is being conducted in using treated domestic effluent to grow edible food crops. However, one of the major concerns with wastewater reuse for food production is the risk of pathogen contamination to the edible parts of the food and to the people exposed to irrigation.

Wastewater application in horticulture using hydroponics technology should minimise the exposure and contamination risk to the workers. Since the edible parts of the plant, with the exception of root crops, may not be in direct contact with the wastewater, contamination to the edible parts may also be reduced. This chapter examined two hydroponics systems, nutrient film technique and water culture (without aeration), for their efficiency in causing pathogen die-off. Three treatments, secondary treated domestic wastewater, control medium (commercial hydroponics medium) and pathogen spiked control medium were tested in triplicate. *S.typhimurium* (ATCC14028) and *E.coli* (WACC4) were used to spike one of the treatments (spiked control medium). The experiment was conducted over four months with the medium changed every fortnight. The results showed that there was a general decrease of pathogens over seven days (>40%) in the medium and complete die-off was observed after 14 days (99%), in

* n.oyama@curtin.edu.au
* j.nair@murdoch.edu.au
* g.ho@murdoch.edu.au)

both types of hydroponics systems. In both systems, there were no pathogens detected in the fruits. The hydroponics techniques for domestic effluent reuse, is a viable option for edible crop production as it reduces the risks of bacterial pathogen contamination.

**Keywords:** Wastewater reuse; hydroponics; pathogen; contamination; domestic wastewater; tomatoes.

## INTRODUCTION

The risk of pathogen contamination of edible food crops has limited the use of treated wastewater for crop production. The risk to humans through contact with wastewater reuse, involves transmission of pathogens including infectious enteroviruses (Mignotte *et al.*, 1999), especially in edible food crops. When growing crops in effluent, it is necessary to consider the risks involved, especially if the crops are for sale or consumption (Ottoson *et al.*, 2005). Amahmid *et al.* (1999) found *Giardia* cysts and *Ascaris* eggs on crops irrigated with raw wastewater, however, not on crops irrigated with treated wastewater.

Different types of irrigation systems can be used in reducing the risk of transferring contaminants to plants and workers. Irrigation systems such as sprinkler and open irrigation, where humans are in contact with the effluent are not widely accepted due to the associated health risks. There may be risks involved with using effluent for soil irrigation as it may contaminate edible food crops through direct contact with the plants (Rosas *et al.*, 1984). Soil and groundwater contamination with pathogens and parasites is also possible through soil irrigation.

The nutrient film technique (NFT) and water culture (WC) are forms of hydroponics that may reduce these risks compared to other irrigation systems. If the hydroponics technique is used for growing leafy and fruit crops the edible parts of the plant are not in contact with the wastewater because a physical barrier is placed between the plant parts and medium. The other advantage of this system is that it is a type of intensive agriculture where farmers/communities are able to grow substantial amount of crops in limited space.

This chapter looked at the possibility of bacterial contamination of plants if wastewater was used to grow edible crops in hydroponics systems using the NFT and WC system. The suitability of tomatoes grown in secondary treated effluent for human consumption was determined. It also examined the bacterial pathogen die off rate in the solution at different nutrient solution retention times. Although parasites are the main concern in wastewater irrigation in most developing countries, their level in secondary treated effluent from Australia are negligible therefore only bacterial pathogens were tested

## MATERIALS AND METHODS

The nutrient film technique (NFT) experiment was set-up as shown in figure 1 and the water culture (WC) experiment was set-up as shown in figure 2.

## Wastewater and Control Medium

The secondary treated domestic wastewater was collected from a domestic wastewater treatment plant (Perth, Western Australia) in 200L drums for the experiments. The control medium used was a commercially available hydroponics nutrient solution (Ag-grow) for fruits and vegetables. This nutrient solution was chosen for this study as it was readily available and it was the most popular in the store. The control treatment was prepared as per the specified ratio of 5mL of hydroponics medium to 1L of water, as recommended by the manufacturer.

## Experiment Design

The experiment was conducted in a greenhouse to provide uniform conditions throughout the growth phase. In the NFT experiment, secondary treated wastewater was pumped from a 42L reservoir to the channel where plants were grown. The 295cm x 12cm x 12cm channels had an inlet and outlet leading to the reservoir. The volume of solution in channels at any one time was approximately 7L. The effluent was drained by gravity flow back into the reservoir. Each tray was set up as shown in figure 1. Uniform-sized plant seedlings were purchased from a commercial nursery and planted in a pot containing expanded clay balls, which were used as the bedding material and inserted into the eight planting slots of each channel. Pumping from the reservoir into the channels and recirculation of effluent was considered to provide adequate aeration.

In the WC system, the plants were grown in tubs containing the nutrient medium. It was a closed-system without recirculation of the nutrient medium. The WC nutrient solution retention time was 14 days. The tomato seedlings were planted in 10cm x 10cm nursery pots, filled with expanded clay balls and then suspended into the tubs to allow the roots to grow into the nutrient solution.

Both experiments were conducted in triplicate.

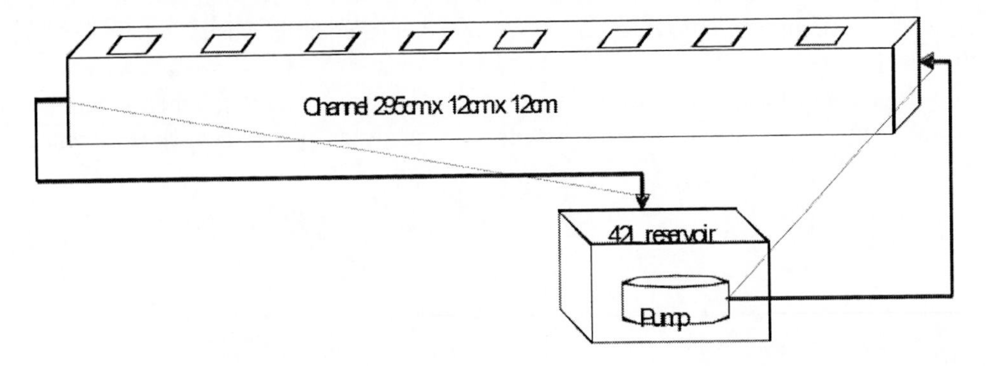


Figure 1. NFT experiment design.

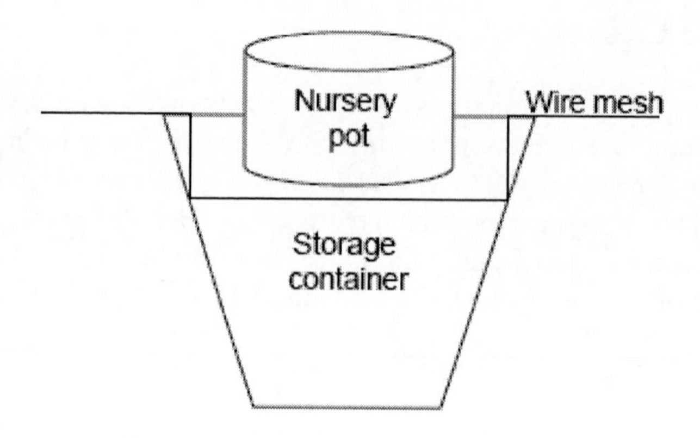


Figure 2. WC experiment design.

The commercial nutrient solutions were inoculated with *S.typhimurium* and *E.coli*. Pure cultures of *Salmonella typhimurium* (ATCC14028) and *Escherichia coli* (WACC4) used for this experiment were grown in buffered peptone water and lauryl tryptose broth respectively. Serial dilutions were prepared according to Standards Australia (1991), method DR88082. The pathogens from the pure culture were pipetted (1mL) into Mcartney bottles containing 9mL sterile water, and was then spiked to the treatment (storage containers). Sampling of wastewater (WW), control medium (CM) and spiked control medium (CMS) was conducted every 7 days and after every 14 days when the medium was changed and pathogens were inoculated. The experiment was conducted for 35 days and was spiked three times.

For analysis, the water samples were collected in sterile 250mL schott bottles and tested following the methods by Standards Australia for both *E.coli* (Standards Australia., 1995a), method DR93215 and *S.typhimurium* (Standards Australia., 1995b), method DR93222.

When the plants were ready for harvest, the edible parts of the plants were separated and were tested. Organically grown tomatoes were purchased from a supermarket for comparison of the quality. The edible parts of the plants were washed with sterile water and the wash water was analysed for *E.coli* (Standards Australia., 1995a) and *S.typhimurium* (Standards Australia., 1995b). Then the edible parts of the plants were analysed for *E.coli* (Standards Australia., 1992), method DR91155 and *S.typhimurium* (Standards Australia., 2004), method DR07430CP.

To determine whether there was a significant difference between the media samples and the plant samples, results of the experiments were analysed using Independent-Samples T Test and One-Way ANOVA.

## RESULTS

### Solution

Overall, NFT had a better pathogen reduction rate than the WC within the first seven days (table 1). There were no detectable pathogens after 14 days in both systems (table 2).

**Table 1. Percentage reduction of bacterial pathogens in NFT and WC systems in 7 days**

| Spike | | Percentage reduction in pathogens | | | | | |
|---|---|---|---|---|---|---|---|
| | | NFT (7 days) | | | WC (7 days) | | |
| | | WW | CM | CMS | WW | CM | CMS |
| 1st spike | *E.coli* | 100 | 100 | 100 | 69 | 100 | 73 |
| | *S.typhimurium* | 100 | 100 | 100 | 87 | 100 | 78 |
| 2nd spike | *E.coli* | 100 | 100 | 100 | 99 | 100 | 97 |
| | *S.typhimurium* | 100 | 100 | 100 | 95 | 100 | 97 |
| 3rd spike | *E.coli* | 88 | 100 | 99.9 | n/a | n/a | n/a |
| | *S.typhimurium* | 67 | 100 | 100 | n/a | n/a | n/a |

n/a – not applicable.

**Table 2. Percentage reduction of bacterial pathogens in NFT and WC systems between 7 and 14 days**

| Spike | Pathogen | Percentage reduction in pathogens | | | | | |
|---|---|---|---|---|---|---|---|
| | | NFT (14 days) | | | WC (14 days) | | |
| | | WW | CM | CMS | WW | CM | CMS |
| 1st spike | *E.coli* | 100 | 100 | 100 | 100 | 100 | 100 |
| | *S.typhimurium* | 100 | 100 | 100 | 100 | 100 | 100 |
| 2nd spike | *E.coli* | 100 | 100 | 100 | 100 | 100 | 100 |
| | *S.typhimurium* | 100 | 100 | 100 | 100 | 100 | 100 |

## Harvest

### *Pathogen Concentration in Wash Water*

The wash water from WW, CM, CMS and organically grown tomatoes had no pathogens detected (table 3).

**Table 3. Number of bacterial pathogen contamination in silver beet and tomato wash water (100mL) in NFT and WC (WW, CM, CMS), organically grown (O)**

| Pathogen | NFT | | | WC | | | Supermarket |
|---|---|---|---|---|---|---|---|
| | WW | CM | CMS | WW | CM | CMS | O |
| *E.coli* | n.d. | n.d. | n.d. | n.d. | n.d. | n.d. | n.d. |
| *S.typhimurium* | n.d. | n.d. | n.d. | n.d. | n.d. | n.d. | n.d. |

n.d. – not detected.

***Pathogen Concentration on the Edible Parts of the Plants***

There were no detectable concentrations of *E.coli* or *S.typhimurium* in any of the edible parts of the plants grown in WW, CM, CMS and O tomatoes.

## DISCUSSION

The die-off rate was quite significant in most cases (tables 1 and 2). Pathogens weresignificantly reduced in most samples by the $7^{th}$ day, which is less time (between 14 – 21 days) than was noted by Oyama *et al.* (2008). The die off rate was higher in the NFT than WC, this may be due to the aeration in the NFT system. The microbial effluent quality after 14 days was within the World Health Organisation guidelines (WHO, 1989). The revised guidelines state that drip irrigation with treated wastewater should contain less than 1000cfu/100mL *E.coli* for low growing crops and less than 100000cfu/100mL *E.coli* for higher growing crops (WHO, 2006). The Australian (ARMC (Australia) *et al.*, 2000) guidelines for raw edible food crops in contact with treated effluent should have thermotolerant coliform count of less than 10cfu/100ml and crops not in direct contact with treated effluent should have a thermotolerant coliform count of less than 1000cfu/100ml. The effluent quality of the medium was well below these guidelines. The validity of using faecal coliforms as an indicator of pathogens has been questioned, however, Harwood *et al.* (2005) found that it was an adequate indicator to use in order to protect human health.

A study conducted by Teltsch and Katzenelson (1978) showed a strong possibility that enteric bacteria and viruses can be spread in the air via spray irrigation. Studies have shown the possibility of contaminated water used for spray irrigation may play a big role in contaminating vegetables (Islam *et al.*, 2004; Rosas *et al.*, 1984) which, could be due to contaminated water being in contact with the vegetables.

It was observed that crops like lettuce and parsley irrigated with raw wastewater were more contaminated compared to crops like tomatoes and pimento (Melloul *et al.*, 2001). The most likely reason given was that leafy vegetables that develop at the soil surface have more foliage, which offers more area for contamination from water spray (Melloul *et al.*, 2001; Rosas *et al.*, 1984). Samples of lettuce and radish grown in soil fertilised with manure and fertiliser are observed to be contaminated with faecal coliforms which could have been due to contact of the vegetables with soil (Machado *et al.*, 2006).

Using alternative methods for wastewater irrigation can reduce the exposure of edible parts to pathogens (NRMMC *et al.*, 2006). This study revealed that growing vegetables using the nutrient film technique in hydroponics system could reduce the exposure risk. However, a study showed that hydroponic tomatoes grown in an inoculated nutrient solution was found to take up *S.typhimurium* to the stems and leaves of young plants (before fruit maturation) (Guo *et al.*, 2002). In this study (table 3), tomatoes grown in WW and CMS were not contaminated with the pathogen. One reason may be because there is no contact between the edible parts of the plant and the contaminated medium.

Tomato fruits have the ability to promote salmonella growth, depending on the handling methods (Zhuang *et al.*, 1995). Abdul-Raouf *et al.* (1993) found that it was possible to contaminate salad vegetables with *E.coli* O157:H7 during production, harvest, processing, marketing and preparation. As a result care should be taken when handling the vegetables.

Another possible source of contamination of food crops is through the water sprinkled on vegetables in order for them to look fresh (Hamilton *et al.*, 2006) for marketing purposes.

As bacterial survival depends on climatic conditions, the results may vary from region to region (Vaz da Costa-Vargas *et al.*, 1991). In this study, pathogen contamination of vegetables using secondary treated domestic wastewater is significantly low and as a result can be recommended for safe consumption. The effluent after passing through the hydroponics system, for a period of 7 – 14 days showed complete elimination of *E.coli* and *S. typhimurium,* which made the final effluent safe for open irrigation.

## CONCLUSION

This chapter has shown that edible parts of tomatoes are safe from bacterial pathogens when wastewater or a solution containing pathogens is used as the nutrient medium. The effluent after passing through the hydroponics system was completely deprived of *E. coli* and *Salmonella* sp. between 7-14 days. The effluent after going through these systems (nutrient film technique and water culture) can be used safely for further irrigation. However, it has to be noted that the safety depends on the system hygiene.

## REFERENCES


Abdul-Raouf, U. M., Beuchat, L. R., & Ammar, M. S. (1993). Survival and Growth of *Escherichia coli* O157:H7 on Salad Vegetables. *Applied and Environmental Microbiology, 59*(7), 1999-2006.

Amahmid, O., Asmama, S., & Bouhoum, K. (1999). The effect of waste water reuse in irrigation on the contamination level of food crops by *Giardia* cysts and *Ascaris* eggs, *International Journal of Food Microbiology. 49*, 19-26.

ARMC (Australia), ANZECC, & NHMRC (2000). *Guidelines for Sewerage Systems: Use of Reclaimed Water*. Canberra, ACT, Agriculture and Resource Management Council of Australia and New Zealand: Australian and New Zealand Environment and Conservation Council: National Health and Medical Research Council.

Guo, X., van Iersel, M. W., Chen, J., Brackett, R. E., & Beuchat, L. R., (2002). Evidence of Association of Salmonellae with Tomato Plants Grown Hydroponically in Inoculated Nutrient Solution. *Applied and Environmental Microbiology, 68*(7), 3639-3643.

Hamilton, A. J., Stagnitti, F., Premier, R., Boland, A.-M., & Hale, G. (2006). Quantitative Microbial Risk Assessment Models for Consumption of Raw Vegetables Irrigated with Reclaimed Water. *Applied and Environmental Microbiology, 72*(5), 3284-3290.

Harwood, V. J., Levine, A. D., Scott, T. M., Chivukula, V., Lukasik, J., Farrah, S. R., & Rose, J. B. (2005). Validity of the Indicator Organisms Paradigm for Pathogen Reduction in Reclaimed Water and Public Health Protection. *Applied and Environmental Microbiology, 71*(6), 3163-3170.

Islam, M., Morgan, J., Doyle, P., Phatak, S. C., Millner, P., & Jiang, X. (2004). Fate of *Salmonella enterica* Serovar Typhimurium on Carrots and Radishes Grown in Fields

Treated with Contaminated Manure Composts or Irrigation Water. *Applied and Environmental Microbiology, 70*(4), 2497-2502.

Machado, D. C., Maia, C. M. M., Carvalho, I. D., Fontoura da Silva, N., Cláudia, M., André, D. P. B., & Serafini, Á. B. (2006). Microbiological quality of organic vegetables produced in soil treated with different types of manure and mineral fertilizer. *Brazilian Journal of Microbiology, 37,* 538-544.

Melloul, A. A., Hassani, L., and Rafouk, L. (2001). *Salmonella* contamination of vegetables irrigated with untreated wastewater. *World Journal of Microbiology & Biotechnology, 17,* 207-209.

Mignotte, B., Maul, A., & Schwartzbrod, L. (1999). Comparative study of techniques used to recover viruses from residual urban sludge. *Journal of Virological Methods, 78,* 71-80.

NRMMC, EPHC & AHMC (2006). *Australian Guidelines for Water Recycling: Managing Health and Environmental Risks. (Phase 1).* 21.

Ottoson, J., Norstrom, A., & Dalhammar, G. (2005). Removal of micro-organisms in a small-scale hydroponics wastewater treatment system. *Letters in Applied Microbiology, 40,* 443-447.

Oyama, N., Nair, J., & Ho, G. E. (2008). Utilising an integrated wastewater hydroponics system for small scale use. *Water Management Series,* 13-23.

Rosas, I., Báez, A., & Coutiño, M. (1984). Bacteriological Quality of Crops Irrigated with Wastewater in the Xochimilco Plots, Mexico City, Mexico. *Applied and Environmental Microbiology, 47*(5), 1074-1079.

Standards Australia. (1991). *Australian Standard. Food Microbiology: Method 1.2: General procedures and techniques - Preparation of dilutions.*

Standards Australia. (1992). *Australian Standard. Food Microbiology: Method 2.3: Examination for specific organisms - Coliforms and Escherichia coli.*

Standards Australia. (1995a). *Australian Standard. Water Microbiology: Method 7: Thermotolerant coliforms and Escherichia coli - Membrane filtration method.*

Standards Australia. (1995b). *Australian Standard. Water Microbiology: Method 14: Salmonellae.*

Standards Australia. (2004). *Australian Standard. Food Microbiology: Method 10: Microbiology of food and animal feeding stuffs-Horizontal method for the detection of Salmonella spp.*

Teltsch, B., & Katzenelson, E. (1978). Airborne Enteric Bacteria and Viruses from Spray Irrigation with Wastewater. *Applied and Environmental Microbiology, 35*(2), 290-296.

Vaz da Costa-Vargas, S. M., Mara, D. D., & Vargas-Lopez, C. E. (1991). Residual faecal contamination on effluent-irrigated lettuces. *Water Science and Technology, 24*(9), 89-94.

WHO (1989). *Health Guidelines for the use of wastewater in Agriculture and Aquaculture.* Technical Report Series 778. Geneva: World Health Organisation.

WHO (2006). *Guidelines for the safe use of wastewater, excreta and greywater: Wastewater use in agriculture.* Geneva, World Health Organisation.

Zhuang, R.-Y., Beuchat, R., & Angulo, F. J. (1995). Fate of *Salmonella montevieo* on and in Raw Tomatoes as Affected by Temperature and Treatment with Chlorine. *Applied and Environmental Microbiology, 61*(6), 2127-2131.

In: Technologies and Management for Sustainable Biosystems ISBN: 978-1-60876-104-3
Editors: J. Nair, C. Furedy, C. Hoysala et al. 

*Chapter 7*

# The Uptake of Zinc and Copper by Tomato Plants Grown in Secondary Treated Wastewater

***Jason Levitan and Jaya Nair****
Environmental Technology Centre, Murdoch University,
Perth, WA 6150, Australia

## Abstract

The effects of copper (Cu) and zinc (Zn) on the growth of tomato plants when grown in hydroponics were tested; whether the Cu and Zn is taken up into the tomato fruits to toxic levels; and to determine the suitability of both secondary treated wastewater and Cu and Zn contaminated water as a nutrient source for tomato plants. Good yields of tomato fruit were observed on the plants grown in the control media (CM), the Cu spiked media (0.5mg/L and 1.5mg/L Cu spiked) and the Zn spiked media (2.2mg/L and 2.6mg/L Zn spiked), whereas the secondary treated wastewater (WW) did not produce adequate yields and the plant growth was stunted. The stunted growth in the WW plants was found to be more likely due to lack of macronutrients rather than the presence of Cu and Zn. In all media, accumulation of Cu and Zn was observed to be more in the roots than in the fruits. The use of Cu and Zn contaminated nutrient sources could be feasible, but there is still cause for concern as the ability of the tomato plants to accumulate potentially toxic levels of Cu and Zn is possible.

Keywords: **Secondary treated effluent; heavy metals; hydroponics.**

## Introduction

The increase in water demands from population growth, industrial development and urbanisation are being noticed worldwide, especially within Africa and the Middle East

* Corresponding author. Tel.: +61 8 9360 7322; fax: +61 8 9310 4997. *E-mail address:* j.nair@murdoch.edu.au

(Kivaisi, 2001 and Blumenthal *et al*, 2005). It has been reported that by 2050 around 50 countries will be under water stress as indicated by the water stress index (US EPA, 2004). This index is measured as the annual renewable water resources per capita that are available to meet domestic, industrial and agricultural needs. The increase in demand has come from a number of reasons that include water scarcity, drought, environmental protection, changing lifestyle and socio-economic factors. The US EPA (2004) has identified that in many of the world's arid regions using wastewater to irrigate the land is common. As a consequence the water quality is compromised and as such the quality of the crops is compromised (Avnimelech, 1993). The issue of public health protection is a priority and some countries limit the use of secondary treated wastewater to irrigate only non-edible crops; such as in Israel 56% of irrigation is with reused wastewater; nearly all of this is limited to only crops such as cotton due to the possibility of exposing the public to a variety of pathogens (Avnimelech, 1993). The National Water Quality Management Strategy (2000) for Australia identified that there is a major risk of human contact with infectious viruses, bacteria and protozoa when wastewater is re-used. Vazquez-Montiel *et al.* (1996) stated that despite the benefits of being both a water and nutrient source for plants, the chemicals and other compounds contained within wastewater may be potentially harmful depending on the intended use and management practices of the wastewater reuse system. Rayment (2005) suggested that the long-term agricultural sustainability and quality of food are being endangered by chemicals. Some heavy metals are phytotoxic if accumulated in the fruit and can impose a health risk to humans (Omran *et al.*, 1988).

The elevated concentrations of heavy metals found in plants grown in wastewater are significant as they could pose a bio-accumulation hazard which could have detrimental effects on any organism that consumes the plant (Peltier *et al.,* 2002). The threat comes from the persistence of and the survival longevity of heavy metals within the environment at concentrations high enough to affect the health of the surrounding human populations and the consumers of the food products.

Epstein (2003) stated that the importance of both zinc (Zn) and copper (Cu) to human health is significant, as both elements are required for a healthy lifestyle but in excess they can be detrimental. Out of all the various heavy metals, Zn and Cu are the most studied due to their presence in most wastewaters in high quantities. This presence has led to a risk of accumulation of Zn and Cu within edible crops where wastewater is used for irrigation (Chunilall *et al.*, 2004; Gothberg *at el.*, 2002).

This chapter looks at the effect Cu and Zn have on tomato plants in regards to their growth and the accumulation of heavy metals in the tomato fruits. The efficiency of secondary treated wastewater for growing tomatoes was also researched in a hydroponics set-up.

The research objectives of this chapter were to:

- Determine the suitability of utilising Cu and Zn contaminated water for the irrigation of tomato plants;
- Determine the suitability of secondary treated wastewater for the irrigation of tomato plants;
- Identify whether Cu and Zn accumulate in the fruits of tomato plants to levels that cause consumer risk;
- Understand the effect of Cu and Zn accumulation on tomato plant growth.

# METHODS

A closed hydroponics set-up was used to grow the tomato plants. The nutrient medium was replaced every 14 days to prevent nutrient deficiencies in the medium. The set-up used buckets with 20 litre capacities fitted with lids that allowed for a plant pot to be suspended over the nutrient medium. The plant pots were packed with clay beads to provide support to the tomato plants in the absence of soil. The height of the tomato plant seedlings were between 10 – 12.5 centimetres. Six nutrient media were tested using this hydroponics system; 1) Secondary treated wastewater (WW), 2) Commercial Hydroponics Medium (CM), 3) 2.2mg/L Zn spiked commercial medium, 4) 2.6mg/L Zn spiked commercial medium, 5) 0.5mg/L Cu spiked commercial medium, 6) 1.5mg/L Cu spiked commercial medium. These levels of Cu and Zn have been selected to be higher than their recommended levels in WW (ACT Wastewater Reuse for irrigation, 1999; WHO., 1996). A popular brand for hydroponics (Ag-Grow) was used as the commercial hydroponics medium.

The forms of Cu and Zn used for this research were copper sulphate ($CuSO_4$) and zinc sulphate ($ZnSO_4$) respectively.

The study lasted 84 days with sampling of the nutrient medium, measurements of plant heights and flower and fruit counts of each plant occurring every 14 days. The sampling began at 0 weeks and the nutrient medium samples, plant measurements and flower/fruit counts were taken before the nutrient medium was replaced. The duration of the study (84 days) was decided by the time taken for the fruit of the plants to mature.

The nutrient media were tested to establish the levels of dissolved Zn and Cu, pH, DO and EC. The Zn and Cu levels were determined using Inductively Coupled Plasma Spectroscopy, whilst the pH, DO and EC were determined using pH meters, DO meters and conductivity meters respectively.

The data was statistically analysed using an analysis of variance test (ANOVA) using Microsoft Excel. Standard errors were also calculated and recorded where suitable.

# RESULTS AND DISCUSSION

## Water Quality

The original CM nutrient medium has higher levels of total phosphorus, nitrate and total nitrogen than the WW nutrient medium which had higher levels of Cu and Zn (table 1).

Figure 1 shows that the WW had an erratic pH ranging from 8.37 – 5.4, whilst the remaining media held a fairly stable pH of 6.59 – 7.87 throughout the experiment. These values are above the recommended pH values of 5.5 - 6.5 for tomato plant growth given by the Department of Agriculture WA (2006). However Marschner (1995) found that root growth is not much affected by pH values between the wider range of 5 - 7.5.

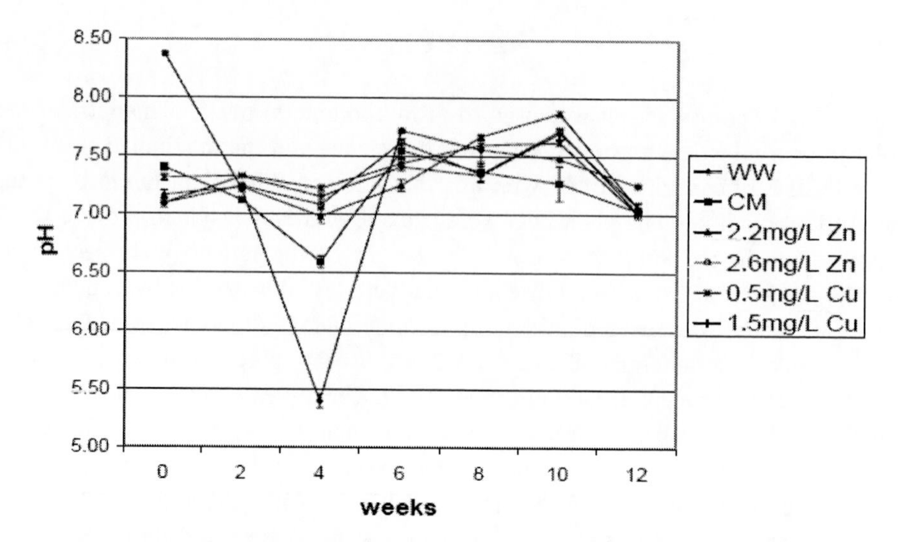


Figure 1. The average pH of the nutrient media (± standard errors) taken at 2 week intervals for a period of 12 weeks.

The DO levels for all the media were between 13.89mg/L and 3.86mg/L during the experiment (figure 2.), indicating that there should be more than enough oxygen present to sustain crop growth (Angelakis *et al.*, 1999) despite there being no constant aeration. Nair *et al.* (2007) observed very similar trends in the DO levels in a study using similar methods with silver beet. There was a general decreasing trend of the DO levels over the course of the study which decreased in a negative correlation with the increase in plant size and could have been due to a larger root mass forming in the nutrient medium.

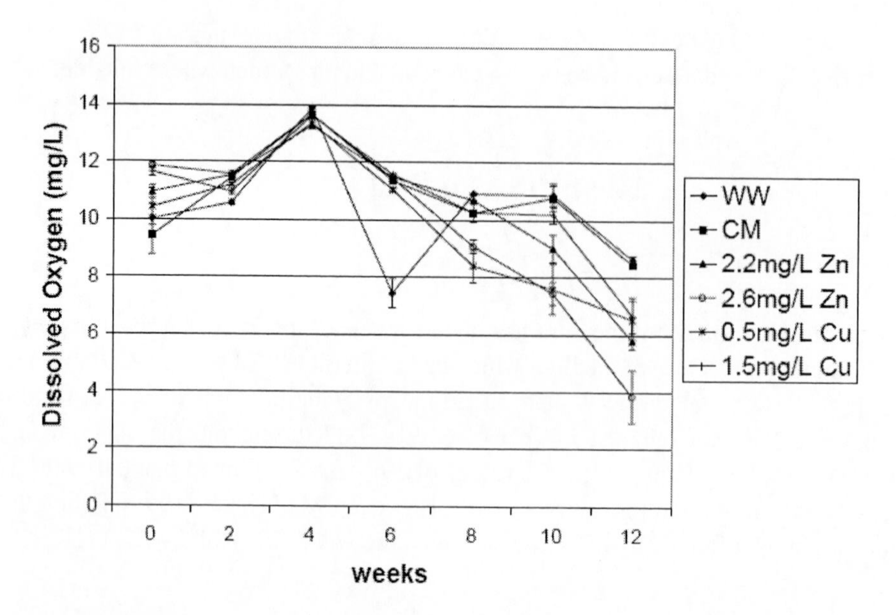


Figure 2. The average dissolved oxygen of the nutrient media (± standard errors) taken at 2 week intervals for a period of 12 weeks.

Figure 3 shows that the EC of the WW nutrient medium (0.49mS – 0.90mS) was consistently below the EC of the other media (0.85mS – 1.89mS). Ideally in a closed hydroponics system, the EC should be between 2.5 and 3.5mS (Department of Agriculture WA, 2006). Due to the nature of intensive growth and high uptake of the nutrient solution by plants in a closed hydroponics systems, it is very common for the levels of some nutrients to accumulate and thus the EC to increase significantly if the nutrient solutions are not re-stocked regularly (NSW Department of Primary Industries, 2005). High EC is a problem that has been associated with continuous soil irrigation with wastewater (Maurer *et al.*, 1995). This phenomenon can have a significant effect on tomato plants as it can cause a reduction in both shoot and root growth (Kutuk *et al.*, 2004). The tomato plant has a threshold EC value of 2.5 deci-siemens per metre at which its fruit yield is affected (Marschner, 1995). In this research the highest EC value obtained was 1.89mS, which is significantly below the value at which high salinity will affect the tomato plants.

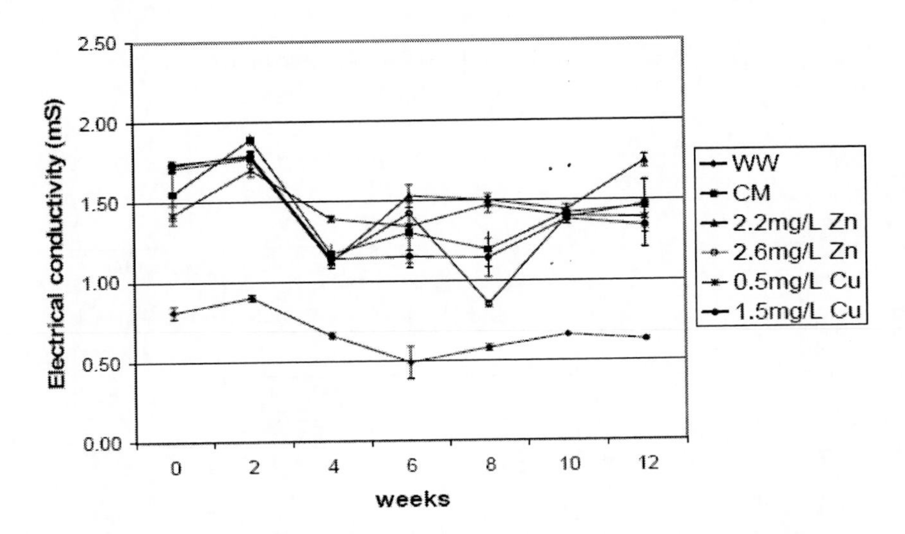


Figure 3. The average electrical conductivity of the nutrient media (± standard errors) taken at 2 week intervals for a period of 12 weeks.

The levels of Zn and Cu in the WW and CM media remained steady throughout the experiment (figure 4 and 5). The Cu spiked media fluctuated over the last few samples, however the general trend throughout the experiment remained stable (figure 4) with the average concentrations being 0.52mg/L and 1.30mg/L for the 0.5mg/L Cu medium and the 1.5mg/L Cu medium respectively. The Zn spiked media also remained stable throughout the experiment (figure 5) with average concentrations of .19mg/L and 2.58mg/L for the 2.2mg/L Zn medium and the 2.6mg/L Zn medium respectively.

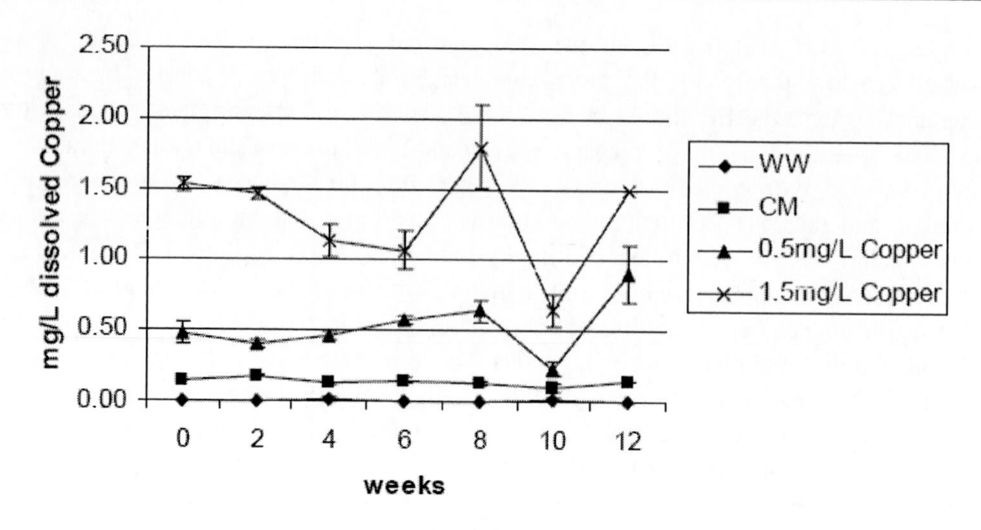


Figure 4. The levels of dissolved copper in the nutrient media (± standard error).

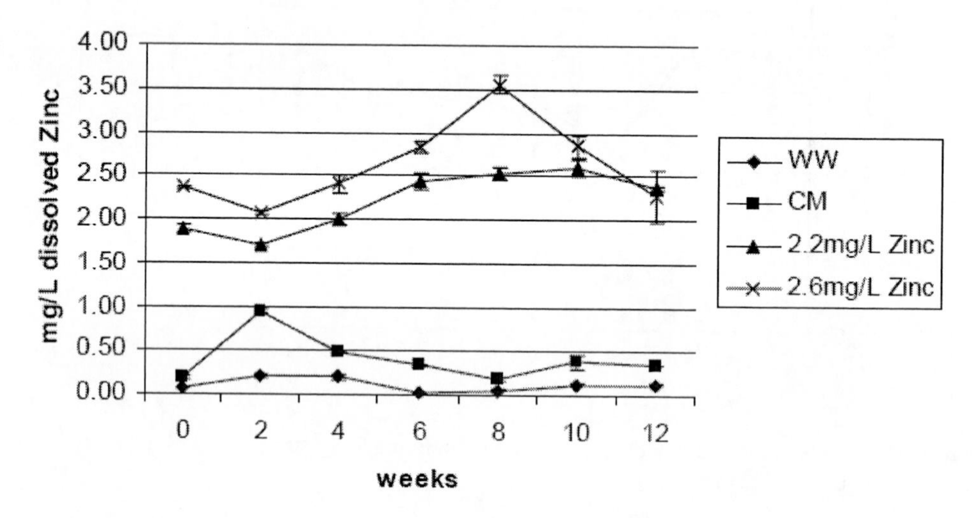


Figure 5. The levels of dissolved zinc in the nutrient media (± standard error).

## Plant Growth

The heights of the WW plants were consistently below all the plants in the other set-ups, reaching between 60 and 70cm (figure 6.). The tomato plants grown in the WW had a final average height of less than 50cm. The difference in the final height of the CM and the WW tomato plants was significant ($p < 0.05$) whilst the difference in heights between the heavy-metal spiked plants and the CM plants was not significant ($p > 0.05$). After the first 2 weeks, it was noticed that the WW tomato plants had yellowing leaves. Nitrogen deficiency can lead to yellowing leaves (Department of Agriculture WA, 2006) and based upon the values of the nitrogen levels compared to the CM (table 1), it is possible that there is nitrogen deficiency associated with the WW as a nutrient source. A deficiency in nitrogen can reduce the fruit production in tomato plants and phosphorus deficiencies will stunt the growth of plants (Weir and Cresswell, 1993) which would suggest that WW is not suitable for tomato production. The Department of Agriculture WA (2006) states that plants grown in low EC levels could become spindly and thus are providing a possible explanation for the stunted growth of the tomato plants (figure 3).

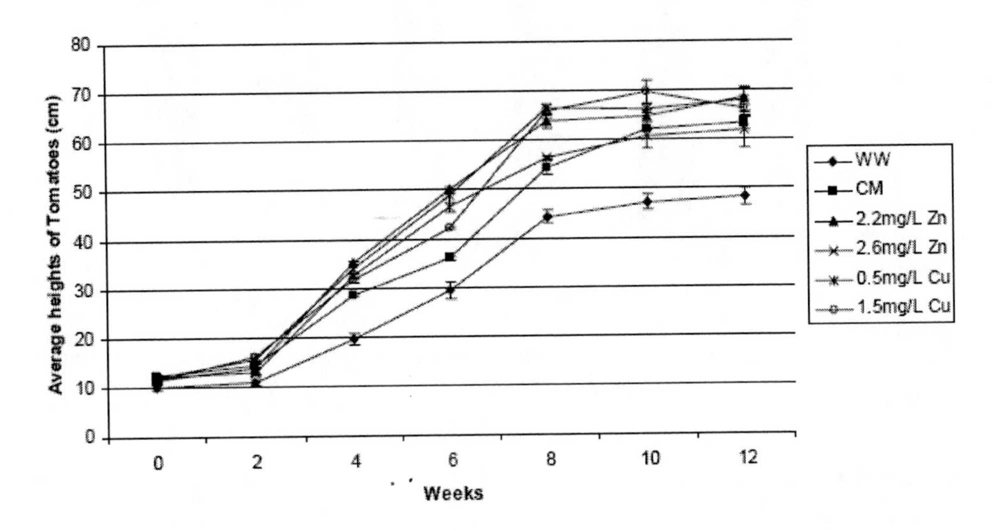


Figure 6. The average heights of the Tomato plants in the different nutrient media (± standard error).

The lengths of the roots of the plants in the CM, WW and the heavy-metal spiked tomato plants did not differ significantly ($p>0.05$) (figure 7.). It was observed by Nair *et al.*, (2007) that the higher the concentration of heavy metals the longer the roots in silver beet. However this study showed that no other medium produced longer roots than the CM nutrient medium.

The WW tomato plants had the most successful conversion of flowers into fruits and the lowest conversion rate was seen in the CM tomato plants (table 2). However, the fruits from the CM, Zn and Cu spiked media were significantly heavier ($p<0.05$) (table 3) and more numerous (table 2) than in the WW medium. The average wet weight of the fruits grown in the Cu and Zn spiked media were greater than the CM (table 3). The WW plants had the smallest biomass of all the media suggesting that WW is less suitable than the CM medium for tomato production. This study has found that Cu and Zn spiked media produced similar quantities of fruit as the CM medium plants. Nair *et al.* (2007) observed that Cu and Zn spiked media gave better a harvest of silver beet than the CM and WW despite there being studies that have recorded results in contrast to this study (Chunilall *et al.*, 2004; Gothberg *et al.*, 2002; Uveges *et al.*, 2002). Oyama *et al.* (2006), Marr (1994) and Nair *et al.* (2007) concluded that the low level of total nitrogen in the WW compared to the CM could be the reasoning behind the stunted growth.

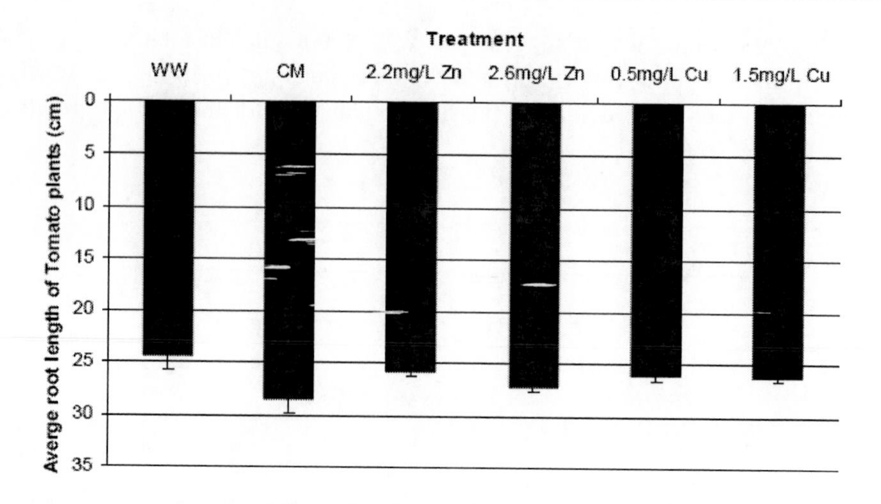


Figure 7. The average root length of the tomato plants at the end of the experiment (showing standard error).

## Heavy Metal Uptake

In all the biomass divisions of the plants, the fruits had the lowest levels of Zn and Cu (table 4). However, it was found that the WW fruits accumulated more Cu (7.03mg/kg dry wt) than both the spiked media (6.13 and 6.79 mg/kg dry wt.) and had the second highest accumulation of Zn of the media (table 4). All of these values exceeded the permissible limit set by WHO (1996) and Raven *et al.* (1999) for Cu in plants at 5.0mg/kg dry wt and 6.0mg/kg dry wt respectively. Only the CM medium produced fruits with levels of Cu below the permissible limit with a value of 4.69mg/kg dry wt. According to the Jordan standards (ISM, 2001), the permissible limit of Zn and Cu in tomato juice is 5mg/kg. The CM plants also accumulated the least amount of Zn (15.93mg/kg dry wt) in the fruits, however all the values still exceeded the permissible value set by WHO (1996) of 10mg/kg but not those set by Raven *et al.* (1999) of around 20mg/kg dry wt.

This research found that both Cu and Zn accumulated more in the roots than in the above-ground biomass, a result also found in other studies (Demirezen and Aksoy, 2004; Deng *et al.*, 2004; Nair *et al.*, 2007; and Pinto, 1996).

Kabata-Pendias and Pendias (1984) found that a concentration of 27-150mg/kg dry wt of Zn and 5-30mg/kg dry wt of Cu was sufficient for plants. If these guidelines are compared to the levels within the fruits then all the fruits are acceptable. Al-Lahham *et al.* (2007) noticed that the accumulation of heavy metals in tomatoes irrigated with treated wastewater is dependant on the cultivar. They concluded that the accumulation of Cu, Zn, Fe, Ni and Mn were within the standard limits and therefore that treated wastewater is suitable for irrigating tomato plants. However no production and growth measurements were taken during the study meaning no comparison with tomatoes grown in a control medium could occur.

The finding that the Cu and Zn accumulated more in the roots, stems and leaves than in the fruits indicates that an increase in Cu and Zn in the nutrient medium may not be a concern for annual plants especially when using a hydroponics system for producing edible crops. However, both the CM and WW media had lower Cu levels than the 0.5mg/L Cu medium yet higher uptakes of Cu were observed. This suggests that the Cu uptake by the tomato plants may not be related to the Cu concentrations of the media. Further research is suggested on this aspect.

## Conclusion

- Both Cu and Zn accumulated more in the roots of the tomato plants, with the fruits accumulating the least;
- According to a range of guidelines, the Cu and Zn levels of the nutrient medium at 0.5mg/L Cu and 2.2mg/L Zn will produce tomato fruit very close to the limits likely to cause detrimental human health in a hydroponics system and therefore lower levels of the heavy metals are recommended;
- The accumulation of Cu and Zn does not seem significant in relation to plant growth; however the lack of macronutrients (especially nitrogen) is a possible explanation for the stunted growth. The low EC of the wastewater is also a potential problem for the growth of tomato plants;
- Secondary treated wastewater may not be an ideal growth medium for tomato production due to the low level of nitrogen and phosphorus. If supplemented with these macronutrients, secondary treated wastewater could be ideal and the level of Cu and Zn may not be an issue in the produce of and quality of tomato fruits.

## References

ACT Wasterwater Reuse for Irrigation. 1999. Environment Protection Policy. *Environment. ACT.* Canberra

Al-Lahham, O., El Assi, NM., and Fayyad, M. 2007. Translocation of heavy metals to tomato (Solanum lycopersicom L.) fruit irrigated with treated wastewater. *Scientia Horticulturae.* 113, 250-254

Angelakis, AN., Marecos do Monte, MHF., Bontoux, L. and Asano, T. 1999. The status of wastewater reuse practice in the Mediterranean basin: Need for guidelines. *Wat. Res.*, 33(10), 2201-2217

Avnimelech, Y. 1993. Irrigation with Sewage Effluents: The Israeli Experience. *Environmental Science and Technologies.* 27(7), 1278-1281

Blumenthal, UJ., Mara, DD., Peasey, A., Ruiz-Palacios, G., and Stott, R. 2005. Guidelines for the microbiological quality of treated wastewater used in agriculture: Recommendations for revising WHO guidelines. *Bulletin No 78(9)*, 1104-1116

Chunilall, V., Kindness, A., and Jonnalagadda, SB. 2004. Heavy Metal Uptake by Spinach Leaves Grown on Contaminated Soils with Lead, Mercury, Cadmium and Nickel. *Journal of Environmental Science and Health: Part B – Pesticides, Food Contaminants, and Agricultural Wastes.* B39(3), 473-481

Demirezen, D., and Aksoy, A. 2004. Accumulation of heavy metals in Typha angustifolia (L.) and Potamogeton pectinatus (L.) living in Sultan Marsh (Kayseri, Turkey). *Chemosphere.* 56(7), 685-696

Deng, H., Ye, Z., and Wong, MH. 2004. Accumulation of lead, zinc, copper and cadmium by 12 wetland plant species thriving in metal-contaminated sites in China. *Environmental Pollution.* 132(1), 29-40

Department of Agriculture WA. 2006. Hydroponic production of tomatoes. Farmnote 136. Department of Agriculture, Western Australia

Epstein, E. 2003. Land Application of Sewage Sludge and Biosolids. Lewis Publishers, USA

Gothberg, A., Greger, M., and Bengtsson, B-E. 2002. Accumulation of Heavy Metals in Water Spinach (*Ipomoea aquatica*) cultivated in the Bangkok region, Thailand. *Environmental Toxicology and Chemistry.* 21(9), 1934-1939

Institution for Standards and Metrology (ISM). 2001. Jordan Standard for Drinks and Juices – Tomato Juice preserved exclusively by physical means, second ed. JS 283:2001, Amman, Jordan

Kabata-Pendias, A., and Pendias, H. 1984. Chapter 4: Soil Constituents. *Trace Elements in Soils and Plants*. CRC Press, USA

Kivaisi, AK. 2001. The Potential for constructed wetlands for wastewater treatment and reuse in developing countries: a review. *Ecological Engineering.* 16, 545 – 560

Kutuk, C., Cayct, G., and Heng, LK. 2004. Effects of increasing salinity and 15N-labelled urea levels on growth, N uptake, and water use efficiency of young tomato plants. *Australian Journal of Soil Research.* 42, 345-351

Marr, CW. 1994. *Hydroponic Systems*. Kansas State University Agricultural Experiment Station and Cooperative Extension Service, Kansas University

Marschner, H. 1995. Mineral Nutrition of Higher Plants, 2nd Edition. Academic Press, London

Maurer, MA., Davies, ES., and Graetz, DA. 1995. Reclaimed wastewater irrigation and fertilisation of mature "Redblush" grapefruit trees on spodosols in Florida. *J. Am. Soc. Hort. Sci.* 120(3), 394-402

Nair, J., Levitan, J., Oyama, N., 2007. Zinc and copper uptake by silver beet grown in secondary treated effluent. *Bioresource Technology.* 99, 2537-2543

National Water Quality Management Strategy. 2000. *Guidelines for Sewerage Systems: Use of Reclaimed Water*. Commonwealth of Australia, Canberra

NSW Department of Primary Industries. 2005. Guidelines for the development of controlled environment horticulture: Planning greenhouse and hydroponic horticulture in NSW. NSW Department of Primary Industries

Omran, MS., Waly, TM., AbdElnaim, EM., and Nashar, BMB. 1988. Effect of sewage irrigation on yield tree components and heavy metal accumulation in Navel orange trees. *Biol. Wastes.* 23, 17-24

Oyama, N., Nair, J., Ho, G.E., 2006. Utilisation of an integrated wastewater hydroponics system for small scale use. International Conference for Decentralised Water and Wastewater Systems. Fremantle, July, 2006, pp.10-11.

Peltier, EF., Webb, SM., and Gaillard, J-F. 2002. Zinc and lead sequestration in an impacted wetland system. *Advances in Environmental Research.* 8(1), 103-112

Pinto, FC. 1996. Slurry as a nutritive substance. In *ISOSC Proceedings. 1996*, 349-364

Qu, RL., Li, D., Du, R., and Qu, R. 2003. Lead uptake by roots of four Turfgrass species in hydroponic cultures. *HortScience.* 38(4), 623-626

Raven, PH., Evert, RF., and Eichhorn, SE. 1999. Biology of Plants, 6th Edition. WH Freeman/Worth Publishers, New York

Rayment, GE. 2005. Cadmium in Sugar Cane and Vegetable Systems of Northeast Australia. *Communications in Soil Science and Plant Analysis*. 36, 597-608

US EPA. 2004. Water Reuse Outside the US. In *Guidelines for Water Reuse*. US EPA.

Uveges, JL., Corbett, AL., and Mal, TK. 2002. Effects of lead contamination on the growth of *Lythrum salicaria* (purple loosestrife). *Environmental Pollution*. 120(2), 319-323

Vazquez-Montiel, O., Horan, NJ., and Mara, DD. 1996. Management of domestic wastewater for reuse in irrigation. *Water Science and Technology*. 33(10-11), 355-362

Weir, RG., and Cresswell, GC. 1993. Plant Nutrient Disorders 3: Vegetable Crops. *New South Wales Agriculture*. Inkata Press

World Health Organisation. 1996. Permissible Limits of Heavy Metals in soils and plants. Geneva

# SECTION TWO: INTEGRATED BIOSYSTEMS

In: Technologies and Management for Sustainable Biosystems ISBN: 978-1-60876-104-3
Editors: J. Nair, C. Furedy, C. Hoysala et al. 

*Chapter 8*

# SAFE REUSE OF HUMAN WASTES FROM PUBLIC TOILETS THROUGH BIOGAS GENERATION: A SUSTAINABLE WAY TO PROVIDE BIO-ENERGY AND IMPROVE SANITATION

***Pawan Kumar Jha****

Sulabh International Academy of Environmental Sanitation,
Mahavir Enclave, Palam Dabri Road, New Delhi-110045, India

## ABSTRACT

For the provision of sanitation in slums and at market / public places, operation and maintenance of public toilets on pay and use basis is a sustainable option. To overcome the challenge of safe disposal of human wastes from public toilets, an efficient design of biogas plant linked with public toilet has been developed. Produced biogas from human wastes is being used for cooking, lighting through mantle lamps by neighborhood community and even for electricity generation. A simple and convenient technology has been developed to make effluent of biogas plant colourless, odourless, pathogen free, and lowering its BOD (Biochemical Oxygen Demand) to around 10 mg/l – suitable for agriculture purposes or discharge into any water body without health or environmental risk. Operation and maintenance costs of the system can be met out of the part of the amount received from users' charge of public toilet.

Keywords: **Human wastes, biogas, waste water treatment, sanitation.**

## INTRODUCTION

Quantum of energy utilized is regarded as the yardstick of socio-economic status of any society. Fossil fuels will not continue longer to meet our energy needs. There is lack of

* Tel. No. +91-11-25038179 Fax. +91-11-25055952; Mob. +91-9811964887; Email. drpkjha@yahoo.com

availability of adequate energy to meet the minimum needs of people in most of the developing countries. Due to lack of fuels, people, mostly from rural areas, spend most of their valuable times to collect fire woods for cooking. Countless days are lost resulting in stagnation of progress and productivity in such communities. Additionally, safe & hygienic way of disposal of human wastes is an increasing problem in most of the developing countries resulting into high mortality, morbidity and decreasing community health, sanitation and consequently productivity. Even after several attempts made by local governments and several international agencies the progress to meet the Millennium Development Goals on sanitation is far below the level of satisfaction in most of the developing countries (JMP Report 2008). It is an unaffordable task for the local governments / bodies to provide adequate sanitation system due to the fact that available sewerage technology has high operational & maintenance costs and that too without economic return out of the system. Further, specialized man power is required for the system for operation. Moreover, sludge management from sewerage system remains a major problem as it contains numerous heavy metals whose negative synergistic effects is incontrollable when used as 'manure' for agriculture purpose.

There are altogether 921 Class I (population > 100,000) and Class II (population 50,000-100,000) towns in India, out of which only 252 towns have sewage treatment plant facility (CPCB 2003). Treatment capacities of such plants are much below the rated or required capacity. Untreated or semi-treated sewage finally leads to adjoining river or water bodies of city/town, causing severe problems to aquatic lives, community health & hygiene and environment. Many important rivers in India are getting practically converted into sewage drains. Decentralized treatment of human wastes through biogas generation can solve the problem of waste management. Biogas can be produced from a variety of sources like animal wastes, vegetable wastes, human wastes etc; However, reuse of human wastes or biogas generation has additional advantages- it helps improve sanitation and provides good quality manure. Generation and utilization of biogas from human wastes remained unnoticed due to the fact that there is psychological taboo associated with it and no design of biogas plant was available to meet the socio-cultural and economic needs of the targeted society. Utilization of human wastes for biogas generation from public toilets at public places, and in slum areas where people generally do not have individual household toilets, is the best option to improve community health & sanitation with resource recovery.

Sulabh is the pioneering organization in India in operation and maintenance of public toilets on pays & use basis in slums, markets and at public places. The system has helped a lot to local municipality in maintaining sanitation and providing employment to local people. In order to reuse human waste from public toilets for biogas generation the organization developed a more efficient design of biogas plant that has been approved by the Ministry of Non-conventional Energy Sources, Government of India for its implementation through state nodal agencies (MNES Final Report 1992). This chapter describes about design of public toilet linked biogas plant, different uses of biogas and on-site treatment of effluent from biogas plant for its safe reuse for agricultural purpose or discharge in any water body.

## PUBLIC TOILET LINKED BIOGAS DIGESTER

Human excreta fed biogas plant system; especially those linked with community toilets have a number of limitations:

1. These are used by people from different socio-economic and cultural backgrounds whose food habits and toilet habits are different.
2. Human excreta are malodorous and associated with psychological and religious taboos.
3. It contains full spectrum of pathogens causing health hazards if not carefully handled.
4. Variation in the number of users leads to variation in loading rate of the digester.
5. Wide variation in the frequency and quantity of water used for cleaning the pans and toilet floor, although the amount of water used for personal cleaning does not vary much.
6. With direct gravity feeding arrangement, the feeding of the digester can at best be termed as intermittent or semi-continuous depending upon the frequency of use.
7. The public conveniences are generally constructed in congested and busy areas where space is often limited.
8. Energy input in the form of heating, mixing, pumping etc. has to be kept to the minimum.

Keeping in view of the above limitations, some basic criteria needs to the considered for the night solid based biogas plant:

a. There should not be any direct handling of excreta.
b. Aesthetically it should be free from odour. Human excreta should not be visible at any stage.
c. Cleaning water should not be more than two liters per use.
d. Use of disinfectants for cleaning latrines should not be permitted.
e. There is no direct control over the concentration of the feed material, loading rate, Hydraulic Retention Time (HRT), temperature etc. The design criteria have to take all these into account, and the design parameters have to be flexible to accommodate the variations.

Sulabh biogas digester is underground structure made up of Reinforced Cement Concrete (R.C.C) with arched bottom, domed top and vertical side walls. The outlet chamber is extended to form displacement chamber. Gas storage space is provided under the dome with 50% of produced biogas per day. The daily gas production capacity of these digesters varies from 30 cum to 60 cum depending upon the number of users of the toilets and consequently size of the plant. Based on this design Sulabh International has constructed 190 biogas plants in different states of India. Human wastes from public toilet flow into digester under gravity without any manual handling. Hydraulic retention time of feed is maintained at 30 days. Due to underground biogas plant, temperature inside digester remains more or less constant resulting almost constant gas production throughout the year – irrespective of variation of atmospheric temperature during winter or summer seasons (figure 1).

Figure 1. A Public toilet linked biogas plant.

The rate of biogas production is one cubic foot per user per day of toilet complex i.e. 1000 cft (30 cum) of biogas is produced from a toilet complex visited by 1000 users daily.

## FUNCTIONAL DESIGN OF THE HUMAN EXCRETA FED DIGESTER

| | |
|---|---|
| 1. Volume of feed material per user per day (Excreta +ablution and flushing water + occasional cleaning water) | 4 litres |
| 2. Volume of digested sludge per user per day | 0.00021 cum |
| 3. Average hydraulic retention time (HRT) | 30 days |
| 4. Cleaning (desludging) interval for half yearly digested sludge | |
| 5. Expected average biogas production per user per day | 30 litres |
| 6. Pressure of biogas inside the digester | 20 cm |

7. Slurry level inside the digester
(a) maximum (highest) — 1.0 meter below crown of the top dome
b) Normal — 0.2 meter below maximum slurry level
8. Diameter: depth ratio — 1.5 : 1.0
9. Rise of top dome (h1) — D/5
10 Rise of bottom dome (h2) — D/2
11 Position of inlet pipe — H/3 below the top ring beam.
12 Position of outlet pipe — middle height of the cylindrical wall.

## BIOGAS PRODUCTION

Before the human wastes are allowed to flow into biogas plant, cow dung mixed with water (1:1) is put once into the plant as inoculums. Volume of such inoculums is kept around 10% of the estimated daily quantity of feed materials from public toilet. Production of biogas starts on 20$^{th}$ day of start of feeding. However, gas production is stabilized after 40 days. Biogas is utilized daily for cooking, lighting and electricity generation.

## CHEMICAL ANALYSES OF EFFLUENT

For the analyses of different parameters like, BOD, COD, Total nitrogen, Phosphate, Potash and MPN from treated and untreated effluent, were carried out as per the Standard Methods, APHA (1989). All the parameters were monitored twice a week under full operating condition of the plant for over 6 months (50 samples).

Methane is the only combustible constituent which is utilized in different forms of energy. A thousand cft (30 cum) of biogas is equivalent to 600 cft of natural gas, 6.4 gallons of butane, and 5.2 gallons of gasoline or 4.6 gallons of diesel oil. The important uses of biogas are cooking, lighting through mantle lamp, electricity generation and body warming in winter.

## USE FOR COOKING

Biogas can efficiently be used for cooking purpose (figure 2). Its calorific value is 26 MJ/cum and burns with blue flame without any soot and odour. In rural areas where cow dung and wood burning are the only source of fuel, biogas is a boon. Depending upon the size a cooking burner consumes 8-25 cft biogas per hour.

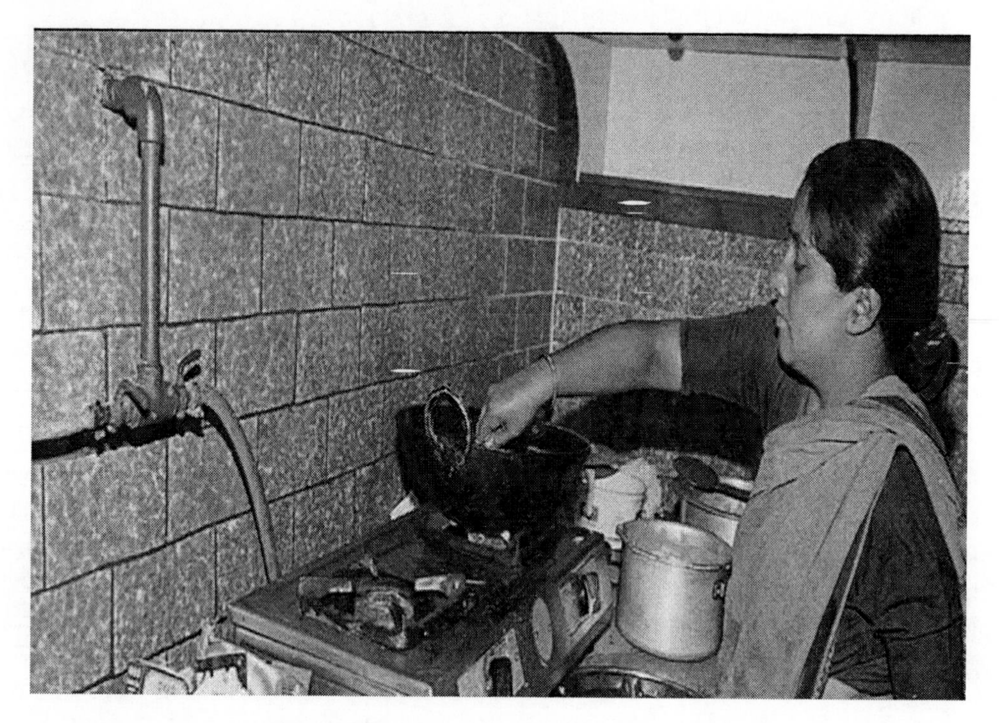

Figure 2. Biogas being used for cooking purpose.

A comparative thermal value of biogas from that of dung shows that 5 kg of fresh dung will generate 16 MJ heat on burning and will give net heat of 1.3 MJ in open earthen oven at 8% efficiency (Advisory Board of Energy, Govt. of India, 1985). On the other hand 5 kg of fresh cattle dung will yield approximately 0.25 cum of biogas having thermal value of 5MJ. Since gas is being utilized in gas burners at 60% efficiency, the net heat value for cooking is 3MJ. Thus, net available heat through conversion into biogas is nearly 2.5 times higher than that of burning dung cake in open earthen oven.

## Use for Lighting through Mantle Lamp

Biogas can be utilized for mantle lamp lighting. A mantle lamp consumes 3 cft of biogas per hour. Its illumination capacity is equivalent to 40 watt mercury bulb at 220 volt.

## Use for Electricity Generation

Biogas can be used for generating power through dual fuel generator coupled to alternator. Consumption of biogas by the engine is @ 15 cft of biogas per BHP of engine per hour. Biogas completely replaces diesel in producing electricity. From a 30 cum of biogas one genset of 10 BHP can be run for 6 hours a day producing 32 units of power per day. Produced electricity can be used for lighting or any other electric appliances.

## Biogas Plant Effluent Treatment System for Safe Reuse of Effluent

During biogas generation there is remarkable reduction (up to 85%) of Biochemical Oxygen Demand (BOD) of effluent of biogas plant in comparison to its affluent value. In absolute term the BOD of effluent is around 135 mg/l .Similarly, pathogen count is still much higher ($6 \times 10^7$ / 100 ml) -not suitable for use or discharge in any water body. Such effluent contains good percentage of nitrogen, potash, phosphate and micronutrients for plants, but its aesthetically bad odour, yellowish colour, high BOD and pathogen contents limit its reuse for agriculture/horticulture or safe discharge in water body.

To make effluent safe reusable, a simple and convenient technology has been developed (figure 3). The technology is based on sedimentation and filtration of effluent through sand and activated charcoal followed by ultraviolet rays. The system consists of an overhead sedimentation tank with bottom conical shape fitted with valve. Effluent from outlet chamber of biogas plant is lifted to this tank and left for half an hour to settle. It is passed through the sand filter column under gravity through liquid flow meter. From sand column effluent passes through an activated carbon column vertically upward where carbon contact time is maintained for 5-6 minutes from where it passes through Ultra-Violet (UV) channel that helps eliminate bacteria and other pathogens. The channel is made up of galvanized iron sheet having flat bottom to minimize the depth of effluent to get maximum effect of UV. Its exposure time is maintained at 4 minutes. The treated effluent is colourless, odourless, pathogen free having BOD around 10mg/l (table 1). Such treated waste water is quite safe for aquaculture, agriculture/horticulture purposes or discharge into any water body without causing pollution.

Figure 3. Effluent treatment technology.

**Table 1. Comparative analyses of treated and untreated effluent of biogas plant**

| Sl. No. | Parameters | Effluent of biogas plant | |
|---|---|---|---|
| | | Untreated | Treated |
| 1 | MPN / Coliform counts | 6 x $10^7$/ 100 ml | Nil |
| 2 | Total Suspended Solid (TSS) | 300 | 35 |
| 3 | Nitrogen as N | 80 | 57 |
| 4 | Potassium as K | 49 | 14 |
| 5 | Phosphate as $PO_4$ | 59 | 11 |
| 6 | BOD | 135 | 10 |
| 7 | COD | 285 | 35 |

Except MPN, all values are in mg/l and average of 50 results.

## Recurring Expenditure

The system requires 1 H.P. of electric motor to lift effluent to the overhead tank for maximum 3 hours a day i.e., 2.5 units of power per day is required. For operation of 3 nos. of UV (15 watts each) about one unit of power will be consumed per day. Such low consumption of electricity can easily be obtained through biogas. Expenditure incurred on the periodical replacement of activated carbon is very low that can easily met out of the user's charges of the toilet complex.

## Advantages of the System

1. No manual handing of human excreta
2. Aesthetically and socially accepted
3. Technically appropriate and financially affordable
4. Operational & Maintenance cost very low
5. Biogas is used for different purposes.
6. Treated effluent is safe to reuse or discharge into any water body.
7. In draught prone areas treated effluent can be used for cleaning of floor of public toilets.
8. Direct economic return by using effluent in agriculture and aquaculture.

## Discussion

There are reports on generation and utilization of biogas from human wastes (Moulik, T K 1981, Lichtman RJ, 1983, Leech, G, 1987 Sinha A 1988). However, such utilization for community purpose from public toilets, particularly for electricity generation has not been reported. There is no report for on- site treatment of effluent of biogas plant for its safe reuse. For the treatment of effluent, it is left for half an hour to sediment in the overhead settling tank, most of the settlable solids settle at the conical bottom of the tank, that is emptied once a month through bottom valve and put into a compost pit to make it free from pathogens before it is used for agriculture purpose. The suspended particles from the effluent, when passed through the sand column, are removed. Depth and diameter of the sand column is 3 feet and 2 feet respectively. To have homogeneous vertical flow of effluent from all over the surface area of the sand column, it is covered with plastic sheet with several small perforations. This helps reduce per unit area hydraulic load, resulting in better effluent. . There is provision for backwash of the sand filter. Depending upon the organic load in the sand filter back wash is done in 7-10 days. Activated carbon is known for removing organic colour and organic matters as well, through adsorbing. Different carbon grades (based on the Iodine Value) were used for the experiments. Taking into account the present requirement and economic aspect, the activated carbon of 800 Iodine Value was optimized for the experiment and field applications. The use of UV to kill pathogens/ bacteria in water and waste water industries is in practice since long. However to make effluent of human excreta based biogas pant free from pathogens was a challenge due to much high count of pathogens. The design of the UV channel consists of flat bottom that minimizes the depth of effluent passing through it. Thus effective exposure of UV on effluent is maximized causing elimination of total pathogens. Such effluent is suitable for reuse in agriculture, horticulture, or discharge in any water body without causing any health or environment risk.

## ACKNOWLEDGEMENT

The author is thankful to Dr. Bindeshwar Pathak, Founder, Sulabh International Social Service Organisation , New Delhi for providing various support and facility to carry out R&D and field implementation of the public toilet linked biogas plant and effluent treatment system.

## REFERENCES

JMP Report, 2008. Joint Monitoring Report of World Health Organization & UNICEF on Progress on Drinking Water and Sanitation: Special Focus on Sanitation

Leach, G. 1987 Household energy in South Asia. *Biomass.* 12 ( 1987) 155-184

Lichtman, RJ. 1983: Biogas systems in India. Vita (Volunteers in Technical Assistance). Virginia, USA

Moulik, TK. 1981: Biogas the Indian Exoerience: Technical Feasibility alone is no guarantee of success. *Unesco Courier.* No. 7,pp33-34

MNES- Sulabh Report 1991: “Techno-economic evaluation of human excreta based biogas plants for community purposes and evaluation of plant designs, process control and pre-treatment feedstock for optimization of and standardization for mixed feed” the report of the Ministry of Non-Conventional Energy Sources, Govt. of India

Sinha A, 1988: Socio-economic problems of community biogas plants in Punjab. *Journal of Rural Development.* Vol.7 (5) pp 591-59.

In: Technologies and Management for Sustainable Biosystems ISBN: 978-1-60876-104-3
Editors: J. Nair, C. Furedy, C. Hoysala et al. 

*Chapter 9*

# RECYCLING OF FISHPOND WASTE FOR RICE CULTIVATION IN THE CUU LONG DELTA, VIETNAM

***Cao van Phung*[*,1], *Nguyen be Phuc*[2], *Tran kim Hoang*[2] and *R.W. Bell*[3]**

[1] Cuu Long Rice Research Institute, O'Mon, Cantho Province, Vietnam
[2] An Giang University, Long Xuyen, An Giang Province, Vietnam
[3] School of Environmental Science, Murdoch University, Murdoch 6150, Australia

## ABSTRACT

Catfish (*Pangasianodon hypophthalmus*) production has expanded to over one million tonnes in 2007 from ponds that cover about 5,000 ha in the Cuu Long delta, Vietnam. From these ponds, large quantities of liquid and solid waste are discharged to waterways without treatment. Consequently, the pollution of canals or rivers by loading of fishpond waste, rich in nutrients (especially nitrogen, phosphorus and potassium) has emerged as a major concern. A survey in the dry season 2007 of 16 paired fields showed that rice yield in 8 paddies receiving waste from fishpond was 1 t/ha higher than in another 8 paddies that did not use wastes. A field experiment was conducted in the wet season 2007 using three doses of solid wastes (1, 2 and 3 tonne/ha) in combination with 1/3 or 2/3 of the recommended inorganic fertiliser rate (60N-17P-24K in kg/ha). Rice yields were more or less the same in all treatments, suggesting that the fishpond waste replaced 1/3 to 2/3 of the fertiliser normally applied. Another experiment was carried out using liquid waste from fishponds for irrigating rice together with inorganic fertilisers at 2/3 of the recommended farmer dosage. Rice yields were also the same in all treatments. These results confirmed that solid and liquid wastes from fishponds can be recycled for rice culture to mitigate pollution of waterway and reduce fertiliser costs.

Keywords: **Catfish, fishpond waste, pollution, nutrients.**

* Cuu Long Rice Research Institute, O'Mon district, Cantho city-Vietnam. Phone No (84) 710861452. Fax: (84) 710861457. Email: caovanphung@hcm.vnn.vn

# I. INTRODUCTION

Catfish culture in the Cuu Long Delta has been practiced for a long time but this industry became important for export only after the year 2000 with an annual growth rate of about 15-20 %. Total catfish production in the Cuu Long delta has increased from 265,000 tonnes in 2004 to 1.5 million tonnes in 2007. In the production of these large quantities of fish, it is estimated that about 450 million cubic metres of solid and liquid waste from fishponds is discharged annually directly to water sources (Phuong, 1998). As a result, pollution due to fishpond waste contains high organic carbon and nutrients (Pillay, 1992). The quantity of waste produced depends upon the quantity and quality of feed (Cowey and Cho, 1991). However, integration of aquaculture into existing agricultural systems has been reported to improve productivity and ecological sustainability through better management and improved soil fertility arising from waste recycling (Bartone and Arlosoroff; 1987). Moreover, properly managed composts can reduce the need for fertilisers (Falahi-Ardakani et al. 1987).

The present study aims at recycling solid and liquid wastes from fishponds for rice cultivation to make use of nutrients and the organic content in wastes in order to reduce inorganic fertiliser application by farmers and to reduce pollution of surface water bodies from discharge of fishpond wastes.

# II. MATERIALS AND METHODS

Solid waste from fishponds in the form of sludge, with up to 60 % water content, was drained and then incubated to allow partial composting. The composition of compost for the wet (Cw) and dry season (Cd) is shown in table 1. Inorganic fertilisers used for field experiments were urea, superphosphate and muriate of potassium.

**Table 1. Nutrient concentration on dry weight basis of compost made in the wet season (Cw) and dry season (Cd) from fishpond solid waste**

| Sample | N % | P % | K % | Mg % | Fe % | Mn (%) | Ca (mg/ kg) | Cu (mg/ kg) | Zn (mg/ kg) |
|---|---|---|---|---|---|---|---|---|---|
| Cw | 0.280 | 0.067 | 0.75 | 0.371 | 5.557 | 0.122 | 42.0 | 61.7 | 120 |
| Cd | 0.551 | 0.108 | 1.99 | 0.532 | 4.121 | 0.097 | 42.7 | 128 | 255 |

Field experiments on recycling of solid waste were carried out during the wet season 2007 and dry season 2008 at the Cuu Long Rice Research Institute farm at O Mon, Can Tho Province (soil type Umbri-EndoOrthiThionic-Gleysols). Soil characterisation is given in table 2. Treatments comprised inorganic fertiliser (T1-control) at the recommended dosage of 60N-17.44P-49.8K/ha for wet season and 80N-17.44P-49.8K/ha for dry season crops, respectively. Fishpond sludge compost was applied at 1, 2 or 3 tonnes/ha on dry weight basis in combination with inorganic fertiliser dosages of 1/3 (referred to as treatments T2, T3, T4) or 2/3 (referred to as treatments T5, T6 and T7, respectively) of those applied in the control.

A survey on the beneficial use of fishpond waste for rice cultivation on farmers' fields was carried out in the dry season 2007 at Chau Phu and Phu Tan districts of An Giang province. Soil characterisation is shown in table 2. In every district, 16 fields were selected comprising 8 which used waste water from fishponds and the other 8 paired sites which had levees to prevent waste water flowing in. Rice samples were harvested in 5 $m^2$ with 3 replications for yield evaluation.

**Table 2. Soil characterization of experiments at CLRRI and on farmers' fields in An Giang province**

| Location | Soil name (FAO/UNESCO) | pH (1:5 $H_20$) | Org. C % | Total (%) | | |
|---|---|---|---|---|---|---|
| | | | | N | P | K |
| CLRRI | Eutric Gleysol | 4.8-5.2 | 2.29 | 0.268 | 0.021 | 0.915 |
| Chau Phu | Umbric Fluvisol | 5.6-6.2 | 0.8-1.1 | 0.161 | 0.047 | 1.556 |
| Phu Tan | Thionic Fluvisol | 4.9-5.5 | 0.9-1.3 | 0.198 | 0.035 | 1.368 |

Experiments on recycling of waste water for rice production were carried out during the wet season 2007 and dry season 2008 at Chau Phu district. Nutrient composition of wastewater is shown in table 3. There were 5 treatments for experiments using chemical fertilisers (N-P-K rates in kg/ha given in parentheses) as follows: T1( 90-26.16-49.8); T2( 60-13.08-24.9); T3( 30-0-24.9); T4( 30-26.16-24.9); and T5( 0-13.08-49.8). These experiments were laid out in a randomized complete block design with 3 replications. Irrigation with wastewater was done 5 times for the wet season and 10 times for dry season rice crop. The volume of wastewater applied at each irrigation event was 1000 $m^3$/ha, supplying the following rates of nutrients: 54 N, 84.6 P, and 44.7 K kg/ha in dry . Likewise, nutrients contributed by wastewater in wet season were only 50% of those in dry season.

**Table 3. Nutrient composition in wastewater at An Giang province**

| Location | pH | EC (µS/cm) | $NH_4$-N (mg/L) | $NO_3$-N (mg/L) | Total N (mg/L) | Total P (mg/L) | Total K (mg/L) |
|---|---|---|---|---|---|---|---|
| Chau Phu | 7.13 | 234 | 3.4 | 0.418 | 5.40 | 8.46 | 4.47 |

Organic carbon is determined by wet digestion; analysis of nutrients (N, P, K, Ca, Mg, Fe, Cu, Zn, Mn) followed standard methods for soil (Page et al. 1982), plant and water analysis (Chapman and Pratt, 1961). Statistical analysis was completed with IRRISTAT software version 5.1 by applying a balanced one-way ANOVA.

# III. Results and Discussion

## III.1. Experiment on Recycling Of Fishpond Sludge

In the wet season 2007, rice yields were not significantly different between different treatments. The reason might be due to low yield ranging from 2.04 to 2.40 t/ha so nutrients were not limiting factors. However; there was a significant difference between treatments in the dry season 2008 (figure 1). Treatment T1 (100 % inorganic fertiliser at recommended rate for dry season) achieved the highest yield. However, it was not significantly higher than treatments T2, T5, T6 and T7. This indicated that using composted fishpond sludge at 1-3 t/ha can save 1/3 to 2/3 of the inorganic fertilisers recommended for rice cultivation.

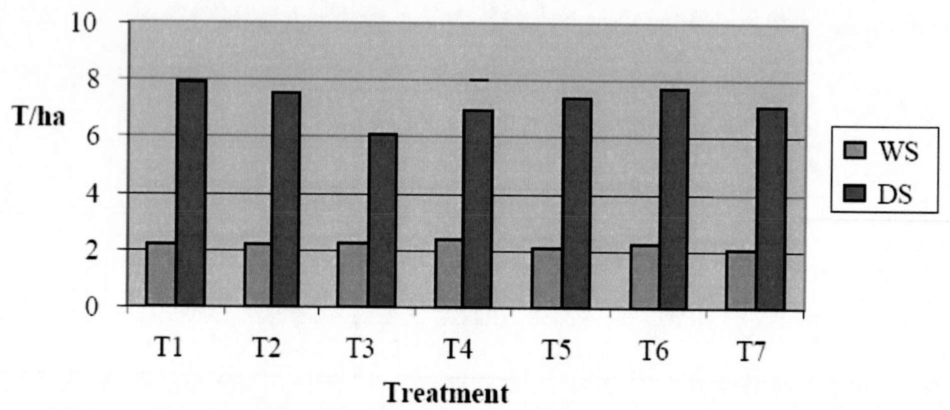


T1: 60N-17.44P-49.8K kg/ha (WS); 80N-17.44P-49.8K/ha (DS);
T2: 1 ton sludge compost/ha + 1/3 T1
T3: 1 ton sludge compost/ha + 2/3 T1
T4: 2 tonnes sludge compost/ha + 1/3 T1
T5: 2 tonnes sludge compost/ha + 2/3 T1
T6: 3 tonnes sludge compost/ha + 1/3 T1
T7: 3 tonnes sludge compost/ha + 2/3 T1.

Figure 1. Analysis of soil, straw and grain for concentration of macro, secondary and micronutrients showed no variation among treatments for both crops. This indicated that the use of fishpond sludge for rice cultivation did not negatively effect rice growth.

## III.2. Use of Liquid Waste in Farmers' Fields in an Giang Province

Results from the survey showed that rice yields in fields using wastewater from fishponds for irrigation were higher than in paddies without recycling of wastewater. Yield difference between the two methods was about 1 t/ha (table 4). This indicates that wastewater can help to further increase in rice yield.

**Table 4. Mean rice yields in paired farmers' fields at Chau Phu and Phu Tan districts supplied with fishpond wastewater for irrigation or river water. Values are means from 8 fields**

| Treatments | Chau Phu | Phu Tan |
|---|---|---|
| Irrigation with wastewater | 7,920 a | 7,436 b |
| Irrigation with river water | 6,898 b | 6,613 c |
| CV% | 6.1 | 6.1 |

Analysis of soil samples at harvest time showed that total nitrogen, phosphorus and potassium in paddies with wastewater application were significantly higher than plots without wastewater application but organic carbon was lower (table 5). Wastewater is rich in nitrogen, phosphorus, potassium (e.g. see table 3) and bacteria which is likely why soils receiving it have higher nutrient contents. By contrast, the high bacterial loading in waste water may accelerate decomposition of organic matter leaving lower organic C levels but higher mineralized nitrogen.

**Table 5. N, P, K and organic carbon in soils after harvesting rice in fields with and without application of wastewater to crops**

| Soil properties | Chau Phu | | Phu Tan | | CV% |
|---|---|---|---|---|---|
| | + waste water | - waste water | + waste water | - waste water | |
| Org C % | 1.59b | 2.60a | 2.24ab | 3.05a | 37 |
| N % | 0.380b | 0.155c | 0.469a | 0.156c | 8.9 |
| P % | 0.369a | 0.224b | 0.354a | 0.211b | 9.2 |
| K % | 2.375b | 0.948c | 2.620a | 0.874c | 10.3 |

Note: Values in same row with the same letter were not statistically different $P < 0.05$.

The survey also recognized that farmers usually added zeolite, lime and dolomite while cleaning fishponds after harvesting. This resulted in high contents of calcium and magnesium in paddies receiving wastewater. Besides that, iron and manganese were also significantly higher in wastewater treated fields (table 6).

**Table 6. Ca, Mg, Fe and Mn in soils after harvesting rice in fields with and without application of wastewater to crops**

| Soil properties | Chau Phu | | Phu Tan | | CV% |
|---|---|---|---|---|---|
| | + waste water | - waste water | + waste water | - waste water | |
| Ca (mg/kg) | 55.0a | 31.0b | 49.8a | 30.6b | 22.8 |
| Mg (%) | 0.11 a | 0.06b | 0.12a | 0.06b | 9.5 |
| Fe (%) | 3.32a | 2.82b | 3.29a | 2.72b | 5.1 |
| Mn (mg/kg) | 332a | 187c | 262b | 157c | 21.8 |

Note: Values in same row with the same letter were not statistically different, $P < 0.05$.

## III.3. Experiments on Recycling of Wastewater for Rice Cultivation at Chau Phu

Results of field experiments at Chau Phu indicated that rice yields of all treatments in the wet season 2007 were not statistically different. However, rice yields of T1 and T2 were highest and were statistically different to the other treatments (T3, T4 and T5) in the dry season 2008 (table 7). The higher yields in T1 and T2 are attributed to the acidity of soils in which phosphorus is a key factor for crop growth (Cong et al. 1995). This explains why yields in T3 were low. Besides that, nitrogen in T3, T4 and T5 was low and not sufficient to achieve potential yields for the dry season. Rice yield in the wet season is usually lower than in dry season in the Cuu Long Delta due to lower solar radiation (Hung et al., 1995)

**Table 7. Rice yields (t/ha) in Chau Phu district for the wet season (WS) 2007 and dry season (DS) 2008. Values are means of three replicates. All plots were watered with fishpond wastewater at 7-10 day intervals in the wet season and 4-5 day intervals in the dry season (see table 3 for composition of waste water applied)**

| Treatments (N-P-K kg/ha) | WS2007 | DS2008 |
|---|---|---|
| T1(90-26.16-49.8) | 3.99 | 5.59 |
| T2(60-13.8-24.9) | 4.38 | 5.58 |
| T3(30-00-24.9) | 3.91 | 4.21 |
| T4(30-26.16-24.9) | 3.96 | 4.32 |
| T5(00-13.8-60) | 3.91 | 4.62 |
| LSD5% | NS | 0.885 |
| CV% | 14.0 | 11.8 |

Analysis of soil, straw and grain samples at harvesting time showed no significant difference among treatments in concentrations of N, P and K (data not shown).

## IV. CONCLUSION

- Wastewater from fishponds can help to increase rice yield because it contains sufficient quantities of nutrients, especially nitrogen, phosphorus, calcium and magnesium, for rice production;
- The use of fishpond waste, either in solid or liquid forms, can save a significant amount of nitrogen, phosphorus and potassium currently applied to crops as inorganic fertiliser;
- Recycling of waste from fishponds for rice cultivation can alleviate water pollution by reducing the quantity discharged directly to water sources;
- No phytotoxicity to rice plants was observed on application of waste from fishponds to paddies.
- Continued monitoring of fields under treatment with fishpond waste is necessary to determine longer term effects on nutrient budgets, soil quality, rice yields and environmental water quality.

## ACKNOWLEDGEMENTS

This research was financially supported by CARD project VIE/023/06. The assistance of staff in the Soil Science Department and a student of An Giang University to carry out this study are greatly appreciated. Thanks also to Cuu Long Rice Research Institute and the Ministry of Agriculture & Rural Development, Vietnam for the facilities and services granted to complete this investigation.

## REFERENCES


Bartone, C.R. and Arlosoroff, S. 1987. Irrigation reuse of pond effluents in developing countries. *Wat. Sci. Tech.,* 19(12), 289-297.

Chapman, H.D., and P. F. Pratt. 1961. Methods of analysis for soil, plant and water. *Division of Agricultural Sciences,* University of California, Riverside.

Cho, C.Y., Hynes. J.D., Wood. K.R. and Yoshida.H.K., 1991. Quantification of fish culture wastes by biological and chemical (limnological) methods. In : C.B. Cowey and C.Y. Cho (Editors), *Nutritional Strategies and Aquaculture Waste.* University of Guelph, Canada.

Cong, P.t, Sat, C.D, Castillo, E.G, and Singh, U. (1995) Effect of phosphorus and growing season on rice growth and nutrient accumulation on acid sulfate soils pages 123-135. In *Vietnam and IRRI: A Partnership in Rice Research.* International Rice Research Institute, P.O. Box 933, Manila 1099, Philippines

Falahi-Ardakani, A., J.C. Bouwkamp., F.R. Gouin, and R.L.Chaney. (1987) Growth response and mineral uptake of vegetable transplants grown in a composted sewage sludge amended medium. *Journal of Environment Horticulture.* 5: 559-602.

Guong, V.T, Lap, T.T, Hoa, N.M, Castillo, E.G. Padilla, J.L, and Singh, U. (1995) Nitrogen use efficiency in direct seeded rice in the Mekong River Delta: varietal and phosphorus response. Pages 150-159. In *Vietnam and IRRI: A Partnership in Rice Research.* International Rice Research Institute, P.O. Box 933, Manila 1099, Philippines

Hung, N.N, Singh, U., Xuan, V.T, Buresh, R.J., Padilla, J.L., Lap, T.T and Nga, T.T. (1995) Improving nitrogen use efficiency of direct seeded rice on alluvial soils of the Mekong River Delta. Pages138-149. In *Vietnam and IRRI: A Partnership in Rice Research.* International Rice Research Institute, P.O. Box 933, Manila 1099, Philippines.

Page, A.L., R.H. Miller and D.R. Keeney (eds) 1982. *Methods of Soil Analysis.* Number 9 (part2). Madison, Wisconsin USA.

Phuong, N.T. 1998. Cage culture of Pangasius catfish in Mekong delta, Vietnam: current situation analysis and studies for feed improvement. Unpublished Ph.D thesis, National Institute Polytechnique of Toulouse, France.

Pillay, T.V.R., 1992. Aquaculture and Environment. Blackwell Scientific Publication Inc., Cambridge, England.

In: Technologies and Management for Sustainable Biosystems ISBN: 978-1-60876-104-3
Editors: J. Nair, C. Furedy, C. Hoysala et al. 

*Chapter 10*

# AN INTEGRATED WATER MANAGEMENT PLAN (IWRMP) FOR OMAN USING AN EXPERT SYSTEM TECHNIQUE

***Nassereldeen Ahmed Kabbashi*[*1], *Al Khabouri Abdulbaqi*[2] and *Suleyman Aremu Muyibi*[1]**

[1] International Islamic University Malaysia, Kulliyyah Engineering, Department of Biotechnology Engineering, Bioenvironmental Engineering Research Unit (BERU), Jalan Gombak P. O. Box 10, 50728 Kuala Lumpur Malaysia
[2] P.O. Box: 1059, Postal Code: 112, Rui – Sultanate of Oman

## ABSTRACT

Due to the rapid development during the last thirty years in Oman, the growing economy has brought an increase in urbanization with a high demand for quality water supplies. The system incorporates the benefit over cost ratio and the social implication of each supply/demand option to give an expert advice on the best available management scenario. The application of the proposed IWRMP for Wadi Ma'awil showed that there is a lot of saving in water. Switching over from conventional flood irrigation systems to a modern irrigation system (MIS) will save 9.13 $Mm^3$/yr and changing crop pattern will save 8.8 $Mm^3$/yr. The control of abstraction from agricultural wells by installing water meters can save 11 $Mm^3$/yr. Reuse from treated wastewater for various purposed can save 4.9 $Mm^3$/yr. The savings in desalination is estimated to be 5.77 $Mm^3$/year. Change in land use through land purchase can help to reduce water demand in the catchment. If all the measures recommended above are applied, the water deficit may ultimately be surmounted around 2015.

**Keywords:** Wadi Ma'awil, Expert system, IWRMP, socio study, MIS.

---

* E-mail: nasreldin@iiu.edu.my. Tel: +60-3-61964524, Fax: +60-3-61964442

## INTRODUCTION

The Sultanate of Oman is situated in the south eastern part of the Arabian peninsula, surrounded by United Arab Emirates (U.A.E) from north west, Saudi Arabia form the west, Gulf of Oman and Arabian sea from the east and south east. Oman, with an area of almost 310,000 km$^2$, is located in the northern tropical arid zone. Daytime temperatures are high, generally above 30°C and seasonally above 40°C. The mean annual rainfall is low and highly variable, exceeding 350 mm in the mountains of Northern Oman and Dhofar, but reducing to 100 mm in the foothills and to less than 50 mm at the coast and in the desert interior. Potential evapo-transpiration varies from 1,660 mm/year on the Salalah plain, with its lower temperatures and seasonal fogs, to 2,200 mm/year in the interior part of Oman which encounter hot and dry climate MWR, (2000). The growing economy has brought an increase in urbanization with a demand for high levels of service and quality for water supplies. The accompanying requirement for foodstuffs has led to a major expansion of well-based agriculture, beyond that of the traditional aflaj areas (singular, falaj: a channel dug into the earth or running along the earth's surface that is used to collect groundwater), as a result water demand in the agricultural sector has tripled. The renewable water resources have primarily been developed for irrigated agriculture and consumption now exceeds the water availability MWR (2000).

Over abstraction has resulted in depletion of groundwater in aquifer storage and saline intrusion in coastal areas. Desalination has been introduced to augment natural resources for township water supplies. The desalination is primarily of sea water for Muscat and some other coastal towns but also of brackish water in the Interior part of Oman. The complementary collection and treatment of wastewater has not developed in an integrated way. Despite high evapo-transpiration, there is still opportunity for groundwater recharge since rainfall occurs as storms on very few rain days and because the infiltration capacity of coarse alluvium and fissured rock is high. Soil development is thin or in pockets so there is little soil moisture retention. After heavy storms there is dramatic runoff from the steep and bare hills into the wadis and the surface water flow may last only for a few hours or days. However, on occasions floods may reach the sea or inland *sabkhas*. Most wastewater is disposed of through septic tanks. The collection and treatment systems for some 25% of the Muscat municipal population as well as for some other towns. Otherwise treatment depends upon plants of individual commercial and industrial businesses. The treatment and recycling of industrial water is practiced, at present, only in a limited way. Administratively the country has been divided into seven (7) regions, namely Musandam and Batinah to the north, Dhofar and Al Wusta to the south, Dhahira and Interior to the west and Sharqyiah to the east.

## RENEWABLE RESOURCES AVAILABILITY

The water resources assessment studies conducted by the Ministry of Regional Municipalities, Environment and Water Resources (MRME&WR) indicate that within a number of catchments there are areas where consumption now exceeds the available resource of the underlying aquifers MWR, 2000. The excess abstractions, derived from aquifer storage, are lowering groundwater levels, as well as inducing saline intrusion in coastal areas, figure 1.

It is evident from these studies that over abstraction is severe in Al Batinah, while it is moderately severe in Salalah, parts of the Interior, notably Umayri and Al Batha, and locally in Musandam. The consequential saline intrusion in Al Batinah has destroyed good agricultural land and water supplies, causing irreparable damage to the aquifer. In parts of Umayri and Al Masarrat, over-abstraction has caused *aflaj* to run dry while in the middle and lower reaches of Al Batha *aflaj* discharges show a steady decline. In Muscat the import of desalinated water provides a large addition to the natural recharge through leakage from the distribution system and infiltration from the disposal of wastewater. This augments aquifer storage and groundwater flow to the sea and can be expected to cause drainage problems in low-lying areas.

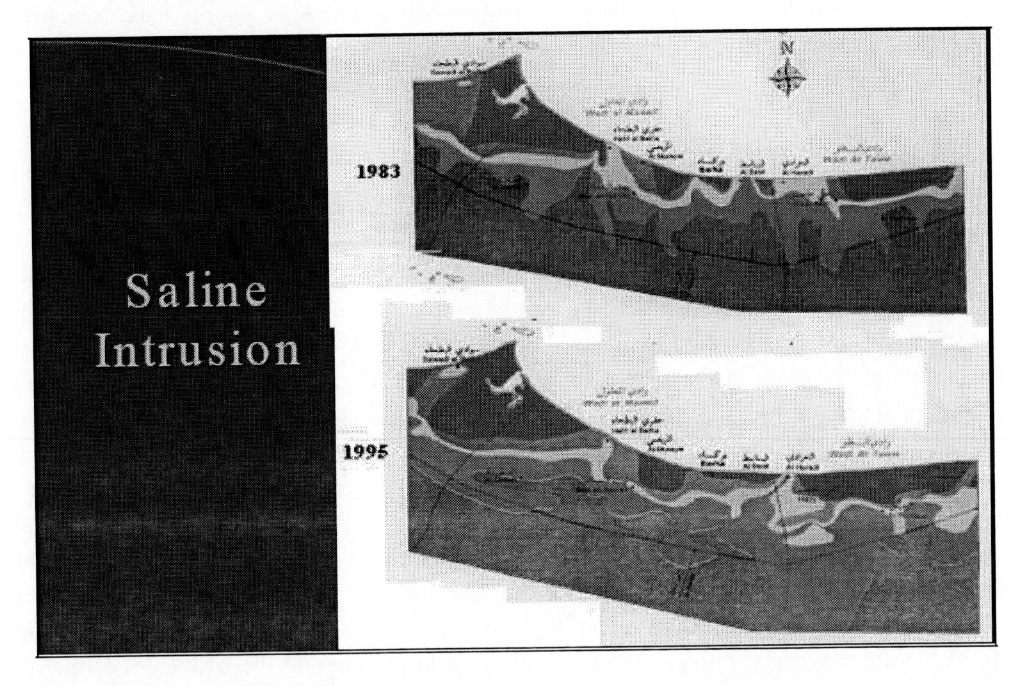


Figure 1. Saline Intrusions along the Al Batinah Coast.

The Ministry of Water Resources (MWR, 1995&2000) master plan of 2000 has indicated that recharge amounts to almost 1,300 $Mm^3$/year while the groundwater resource available for present use is of the order of 900 $Mm^3$/year. In the Musandam and North East Coastal Water Management Areas there are substantial surface and groundwater losses to the sea.

## The Study Area (Wadi Ma'awil)

Wadi Al Ma'awil catchment, located in the Batinah Region of the Sultanate of Oman, is part of the Barka Water Assessment Area as classified by the Water Resources Department of Ministry of Regional Municipalities, Environment and Water Resources. The current water situation in Wadi Ma'awil catchment where the groundwater levels are declining, the water quality is deteriorating and the saline intrusion is expected to continue in the shallow unconfined aquifer systems, has led to abandonment of some of the farms in the coastal plain, change of the agricultural settlements in the area, change of the demographic set up of the catchment which if allowed to continue will affect the social and economical structure of the area and eventually the country in general. To address the water situation in the study area and all other possible complications mentioned, an integrated water resources management plan (IWRMP) is required. It is planned to investigate the water situation of the area to formulate for the first time an integrated water resources management plan, which then can be simulated across the entire country. It should be noted that all of the previous investigations done in Wadi Ma'awil (and in the most of the catchments in Oman) were conducted to meet specific objective(s) i.e. exploration drilling, flood control, dam construction etc. The purpose of this study however; is to formulate an Integrated Water Resources Management Plan (IWRMP) that would provide a framework for achieving sustainable water resources development and management in Wadi Ma'awil with a planning horizon to 2020.

## The Need for an Integrated Water Resources Management Plan

Protection of the water resources has to be secured and the environmental degradation that is a consequence of over abstraction should be identified. The hydrological flows that are necessary for important environmental features should be restored or maintained. Desalination of seawater has become a significant contributor to national resources providing supplies for domestic use and industry, particularly in Muscat and for some other towns on the coast. Desalination of brackish groundwater currently plays a small role for supply in the Interior. Desalination is, however, relatively expensive, at least double the cost of using a potable groundwater resource. A relatively minor amount of treated wastewater is being used for municipal greening but greater consideration is needed of its beneficial reuse, either directly by agriculture or through recharge of aquifers. With the further planned development of water and wastewater systems there will be a much enhanced availability of treated effluent for these uses. The role of water is of great importance to the country's continuing socio-economic development. Since the population is expected to increase by more than 40% between 2000 and 2020, there is an urgent need for a plan that takes into account the needs of the water using sectors in Oman, as well as the relative scarcity and the optimum use of the renewable resources. Therefore, it is proposed to develop an Integrated Water Resources Management Plan (IWRMP) for Wadi Ma'awil, northern part of Oman as a pilot study area. Methodology and approach could be followed in other areas of Oman. An IWRMP expert system will be developed for the first time, customized to the arid region water problems, to facilitate the decision making of the IWRMP.

## Research Methodology

Detailed hydro geological assessment/management study is been conducted in Wadi Ma'awil using the available data from different government agencies (i.e. surface water, groundwater, hydro chemical, statistical data, etc). All existing hydrological, hydro geological, hydro-chemical, water demand, water supply and water use data was reviewed and analyzed to meet overall objectives. Data on well fields (municipal supply and major agricultural developments) within the study area was also reviewed.

The level of work and depth of analysis was conducted to provide and to enable:

1. An overall assessment of the catchment groundwater and surface water resource with appropriate maps, plans, cross-sections and aquifer details;
2. Water balances calculations;
3. Delineation and quantitative assessment of areas of significant deficit and surplus in water resources;
4. Establishment of conceptual hydro geological model;
5. Identification of key water resource issues;
6. Identification and evaluation of options to increase water resource availability; and
7. Development of an Expert System programme to be used in water resources management for arid regions.

## IWRMP-Expert System

Water resources development and management have been and still are dominant (Lilburne et al., 1998; Salman et al., 2001) but there is need for a shift towards a holistic approach to avoid fragmented and uncoordinated water management options (Rosegrant et al., 2000; Staudenrausch and Flugel, 2001). Expert system (ES) is both a process and a tool for solving problems that are too complex for humans alone, but usually too qualitative for only computers. As a tool, an ES consists of mathematical models, data, and point-and-click interfaces that connect decision-makers directly to the models and data they need to make informed, scientific decisions. An ES collects, organizes, processes information, and then translates the results into management plans that are comprehensive and justifiable. This ES uses the concept of IWRMP, defined as a set of different water management scenarios required to enhance the groundwater availability through either supply and/or demand management. The water balance components need to be calculated using classical hydro geological methodology which then fed to the ES. The ES being constructed under visual basic (VB) environment which will allow the system to operate within many different environments and concert with a variety of other software's (GIS, MAPINFO etc). The tool is designed according to the four step schema presented in figure 2, that involves; a) the database; b) the object model linked to mathematical models for water balance calculation; c) water management options model; and d) the user interface which allows for the definition of parameters related to the management option and its benefit cost ratio and social impact. Water management strategies or single interventions can be simulated under different scenarios, compared, and the decision maker or the analyst can formulate responses to mitigate water stress impacts with respect to their objectives.

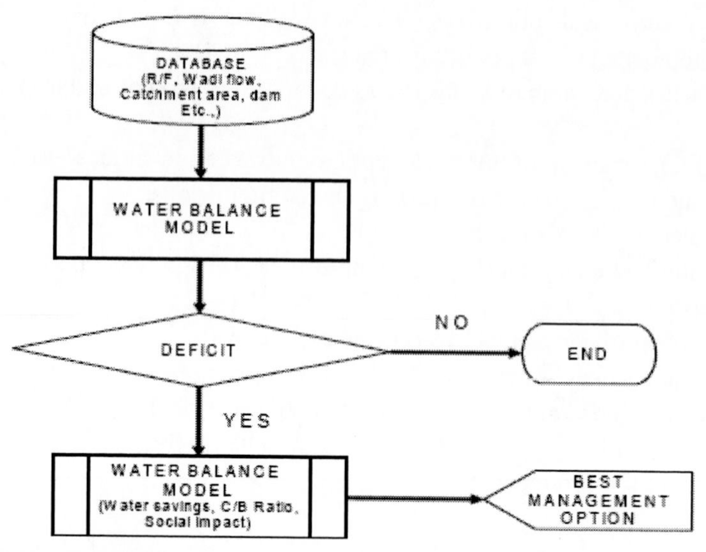


R/F: Rainfall, C/B: Cost Benefit.

Figure 2. The IWRMP-ES Design.

## WATER BALANCE MODULE IN IWRMP-ES

Water balance calculation is also carried out using the IWRMP-ES. The estimated hydrological/hydro geological information such as the average volume rainfall, Wadi flow and percent of recharge in the catchment need to be evaluated and entered in the IWRMP-ES. The catchment is subdivided based on hydrological characteristics (i.e. upper vs. lower catchment). The IWRMP-ES will then calculate the amount of recharge, evaporation, groundwater through flow and hence the amount of surplus or deficit in each sub-catchment for the base year as seen in figure 3.

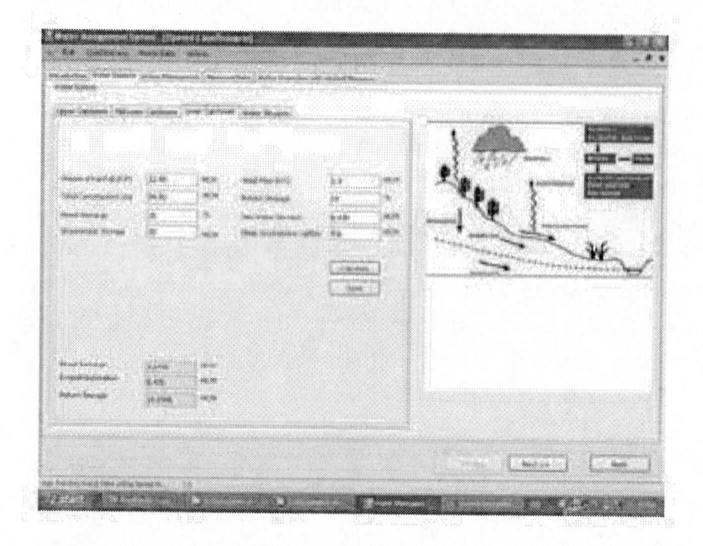

Figure 3. Water Balance Calculation Using IWRMP-ES.

In case of deficit in the water resources the IWRMP-ES would indicate that in red colour as shown in figure 4.

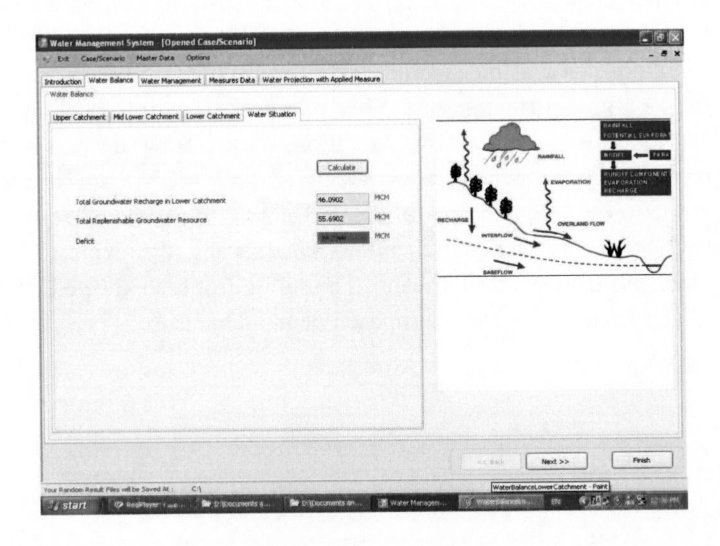


Figure 4. Water Situation, Deficit (Highlighted by the IWRMP-ES).

As it could be seen that using the IWRMP-ES as a tool to evaluate the water situation in the catchment has facilitated and streamed line the IWRMP approach. Figure 5 illustrates the changes in the balance if the remedial measures to manage water demand suggested above are implemented gradually between 2005 and 2015. The demand for domestic and industrial supply (about 4.39 MCM) is partially fulfilled by desalinated water (1.2 MCM) and the rest (3.19 MCM) through groundwater pumping to the end of 2005. After 2005, the rise in domestic demand attributed to population growth is to be completely met by desalinated water. Intervention in the agricultural sector is to begin in 2005/2006. Implementation of measures, including MIS and metering, is expected to yield savings of 4.5 and 11.00 MCM, respectively, by 2010.

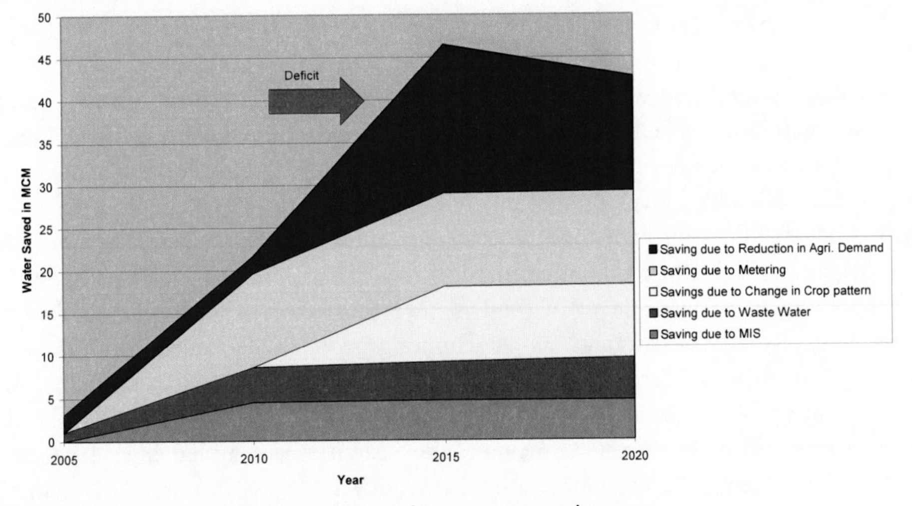


Figure 5. Integrated demand measures to meet future water requirement.

Further saving due to introduction of MIS and change in cropping pattern will be 4.634 and 8.866 MCM, respectively. All the measures taken together will bring down agriculture consumption to an average of 16,625 $m^3$/ha from 25,574 $m^3$/ha. Agriculture pumping is estimated to be 91.66 MCM in 1993 and projected to 53.65 MCM by 2020 on account of salinity and agriculture demand management. About 4.93 MCM of treated wastewater will also be available for use in irrigation and other purposes. Due to reduction of cropped area and modified agriculture pumping, the irrigation return flow decreases to 8.05 MCM correspondingly reducing the available resource to 54.57 MCM. Introduction of remedial measures and reduction in cropped area to about 3232 ha in 2015 will result in a surplus of about 4.92 MCM by 2015. This will help in re-saturating the depleted aquifers thereby preventing any increase in seawater intrusion. The saline front is expected to stabilize around the year 2015. Water savings due to the proposed demand measure is presented in figure 6.

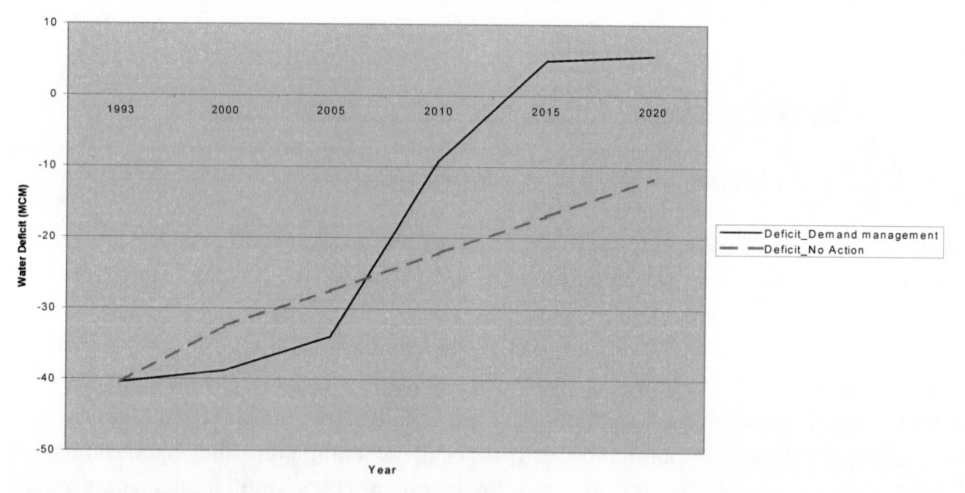


Figure 6. Water Savings Due to Demand Management.

## Socio- Cultural and Economic Study

The socio- cultural and economical analysis of an IWRMP Project aims to study the agricultural conditions current as well as future with and without the IWRMP. Integrated water resources management has proved to be an important strategy for water management and water scarcity at the level of the Wadi basin. Water development is crucial for the sustainable development through food security, particularly in the water scarce economies. Without irrigation, agriculture output required to feed the population is not possible. In oil-based economies, subsistence level of food is necessarily grown locally. The economy of Oman is basically oil based, and its economic growth is dependent on oil reserves. Considering that oil is not forever, it is therefore essential for sustainable economic growth that other resources should be used efficiently in the economy of Oman. Water is one of such resources, which are under threat. Growing water scarcities are a challenge to sustainable economic development. The catchment area the Wadi Al Ma'awil in the Al-Batinah region includes major portions of the three Wilayats: Barka, Wadi-Ma'awil, and Nakhal. The total population of these Wilayats comprises 15 % of the total population of Al Batinah and about 4.41 % of the country, as per the 2003 Census. Moreover, this area constitutes the 'agriculture bowl' of the Sultanate and is known to possess tremendous economic potential and therefore some priority ranking at least with regard to agriculture production was expected. Considering the socio-economic characteristics, the area of interest comprises between the coastal plains around Barka and interior around Nakhal. About two-third of the families have been living in the area for less than 25 years. About 28% of the population of the Sultanate lives in Al Batinah Region. Population has remained stable over the last decade. The population of the catchment area is about 9% of the governorate. Agriculture is the largest consumer of water all over the world and any increased share of water supply to agriculture should be determined as the residual available after the supply to non-agriculture purposes is met. In Oman, the share of Agriculture in GDP was as high as 34 per cent in 1967 but declined to 2.8 per cent in 1975 and thereafter has remained stable (MNE, 2004). In 2002, it was about 2% at current prices, and 3% at 1988 price. By contrast, the share of manufacturing sector is 45 per cent and that of tertiary (service) sector is 52 per cent. However, agriculture accounts for a major share (around 96%) of the total withdrawals of water.

## SYSTEM VALIDATION

The IWRMP-ES is used to validate the calculated data for Wadi Ma'awil. Using the water balance parameters the IWRMP-ES is used to calculate the water balance in the three different sub-catchments. The outcome of the calculations and the water situation in Wadi Ma'awil is shown in figure 7.

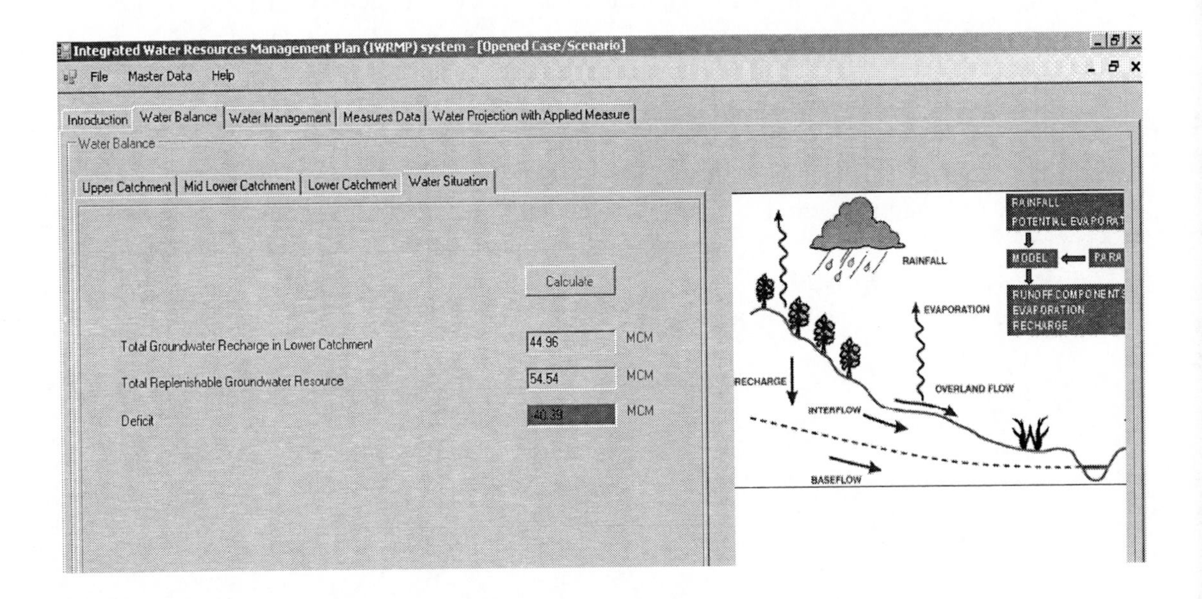


Figure 7. Water Situations as Calculated by the IWRMP-ES.

The results of the IWRMP-ES calculation, compared very well with the classical calculation done in the study. Using the water management module domestic, industrial, agricultural and livestock demands are calculated and projected to the year 2020.

## Conclusion

The review of international practice on Integrated Water Resources Management Planning (IWRMP) shows that the key to a successful water resources management is the implementation of IWRMP through people's participation in decision making. It is recognized that an integrated management of resources is the most effective way of dealing with current water management taking into consideration environmental as well as economical and social issues. Using the IWRMP-ES for Wadi Ma'awil has helped not only to calculate the water balance in the catchment and estimate the water situation but also to come up with a workable water management scenarios that economically feasible and socially acceptable.

## References

Lilburne, L; Watt, J. and Vincent, K. A Prototype DSS to Evaluate Irrigation Management plans. Computer and Electronics in Agriculture. 1998, vol 21, 195-205.

Ministry of National Economy (MNE). (2004). Statistical Year Book [http://www.moneoman.gov.om]. Oman.

Ministry of Water Resources. Data Compilation and Review, Northern Batinah: Oman, 1995.

Ministry of Water Resources. Barka Water Resources Assessment Report, Oman, 2000.

Rosegrant, M.W.; Ringler, C.; McKinney, D.C.; Cai, X, Keller, A. and Donoso, G. Integrated Economics- Hydrologic Water Modeling at the Basin Scale: the Maipo River Basin, Agricultural Economics, 2000. Vol 24, 33-46.

Salman, A.Z.; Al-Karablieh, E.K; and Fidher, F.M, 2001. An Inter-Seasonal Agricultural Water Allocation System (SAWAS), Agricultural Systems. 2001, vol 68, 233-252.

Staudenrausch H.; Flugel, Q.A. Development of an integrated water resources management system in southern African catchments. Physics and chemistry of the Earth, part B: Hydrology, Oceans and Atmosphere 2001, vol 26(7), 561-564.

In: Technologies and Management for Sustainable Biosystems ISBN: 978-1-60876-104-3
Editors: J. Nair, C. Furedy, C. Hoysala et al. 

*Chapter 11*

# UTILIZATION OF BIO GAS, SLURRY AND SLUDGE IN CHINA

***Li Kangmin***
Asian Pacific Regional Research & Training Centre for Integrated Fish Farming,
Freshwater Fishery Research Centre, Chinese Academy Fishery Sciences,
No.9 Shanshui West Road Wuxi 214081 China

## ABSTRACT

Biogas from anaerobic digester is an important source of renewable energy. The reasons for developing biogas plants are to treat the increasing organic wastes and waste water; to mitigate global warming because methane is a major greenhouse gas with a potential 22 times that of CO2. Biogas solves the energy supply in rural areas, where people traditionally forage for fuel wood in forest. It can also be used to co-generate electricity and heat, prolonging the active hours of the day and enabling the family to engage in social activities or to earn extra income. It solves sanitation problems by taking in human as well as animal manure, improving hygiene and environmental conditions in rural areas. Finally, anaerobic digester not only yields biogas, but also slurry and sludge rich in nutrients, minerals and biologically active compounds that form excellent fertilisers and bio insecticides for crops and/or provide fodder for animals and substrate for mushroom cultivation.

**Keywords**: Bio gas, Slurry, Sludge, Dry fermentation, Bio insecticide, Sanitation in rural areas.

## INTRODUCTION

China's energy structure indicates that biomass energy accounts for only a few percent. So biogas can not top the list of renewable energies [Li Junfeng, 2007]. However, it is still very important in China, especially in rural areas. Why? As is well known the population in China is 1.3 x $10^9$, among which 0.9 x $10^9$ live in rural areas.

Fermentation processes producing methane is a natural phenomenon. Biogas is a combustible mixture of gases produced by microorganisms, and its main constituents are methane ($CH_4$), which accounts for about 60%, carbon dioxide ($CO_2$), which accounts for about 35% and small amount of water vapour, hydrogen sulphide ($H_2S$), carbon monoxide (CO), and $N_2$. The composition of biogas produced in different anaerobic conditions might be different e.g. the methane content of biogas produced from night soil, chicken manure and waste water from slaughterhouse sometimes could reach 70% above, while that from stalk and straw of crops in general 55%. The concentration of $H_2S$ in biogas produced from chicken manure could reach 4,000mg/m$^3$ and from molasses alcohol waste water even higher at 10,000mg/m$^3$. Biogas is mainly used as fuel, like natural gas, while slurry and sludge can be used as excellent organic manure for crops. But there are other uses for biogas, bio slurry and bio sludge.

## THE BIOGAS TECHNOLOGY DEVELOPMENT IN CHINA

China is one of countries to use biogas technology earlier in the world. Luo Guorui invented and built an 8 m3 Guorui biogas tank in 1920. Chinese Guorui Biogas Digester Practical Lecture Notes, the first monograph on biogas in the world was published in 1935 [Bian Yousheng, 2000]. The second wave of biogas use in China originated in Wuchang in 1958 in a campaign to exploit the multiple functions of biogas production, which treated manure and improved sanitation in rural areas. The third wave of biogas use occurred between the late 1970s and early 1980s when the Chinese government considered biogas production an effective and rational use of natural resources in rural areas. Some 6 million digesters were set up in China, which became the biogas capital of the world, attracting many from the developing countries to learn from it. The 'China dome' digester became the standard construction to the present day [ESCAP, 1984], especially for small-scale domestic use. But many new types of rural household digesters have also been built on the basis of water pressure type and super small digester by dry fermentation. China's 2003-2010 National Rural Biogas Construction Plan was announced in 2003. From then on, a government subsidy of 1, 000 Yuan (about US$ 150) would be provided for each biogas digester. During the 10$^{th}$ Five Year Plan (2001-2005) the government invested 35 x 10$^9$ Yuan to promote eco-models with biogas as a key link. China has developed with great effort 2,200 large sized biogas engineering projects for wastes from intensive animal husbandry and poultry treating more than 60 million tons manures a year and built 137,000 digesters to purify sewage as well. In the 10$^{th}$ Five Year Plan China increased small sized biogas digester from 11 million to 22 million, which accounted for 15% of households in rural areas. In the 11th FYP China will continue to adhere to the strategic thinking that "Development is the absolute principle", at the same time it will lay emphasis on commanding the overall situation with the "Scientific outlook on development". By 2010, China would increase biogas-using households further from 31 million to 50 million, so the rate of user would reach over 30% households in rural areas and China would finish 4,000 large- scale biogas engineering projects.

## Why Develop Biogas Plants?

According to the government's Chinese Ecological White Paper issued in 2002, the total amount of livestock and poultry wastes generated in the country reached 2.485 x $10^9$ tonnes in 1995, some 3.9 times the total industrial solid wastes [Wu T. and Gao J., 2002]. These wastes are precious resources if used properly, but constitute major pollution when discharged into rivers and lakes. It is estimated that less than 10 percent of the wastewater in China is currently treated, and that 10 million ha of farmland are seriously polluted by wastewater and solid wastes as well. According to CAS China Natural Resources and Environment Data bank, if annual manure and night soil totally are used to produce methane, the theoretic production is about 130 x $10^9$ $m^3$ equivalent to standard coal 93 million tons [Yi A., 2003]. The COD of wastewater from a distiller often reaches 40 000 mg/litre while aerobic treatment only permits COD below 1 000 mg/litre, which means if aerobic fermentation is applied we have to dilute the wastewater 40 times. Fortunately with anaerobic digestion, 90 percent of the pollutants could be readily removed, thereby greatly reducing pollution to farmland, rivers and lakes.

Developing biogas plants mitigates global warming because methane is a major greenhouse gas, second to carbon dioxide in amount generated, but with a potential 22 times that of carbon dioxide [Ho MW. 2006]. The methane flux from exposed slurry is 3.92 – 7.76 mg/m2/hour, compared with 10.26 mg/m2/hour from compost in rice fields. The comparing experiments of different fertilizers showed that the methane emission from rice fields using bio slurry and sludge reduced by 50% than from rice fields using compost [ESCAP, 1984]. Methane mitigation decreases carbon emissions and can be traded as carbon credits under the Clean Development Mechanism of the Kyoto Protocol for climate change [Ho MW., 2005].

Developing biogas plants also solves the energy supply in rural areas, where people traditionally forage for fuel wood in forest. A 10 $m^3$ digester in rural areas can save 2 000 kg of fuel wood, which is equivalent to reforesting 0.26-4 ha land [Wu T. and Gao J., 2002]. Africa lost 64 million ha of forest between 1990 and 2005, more than any other continent, and fuel wood gathering was a major cause of forest depletion [Mygatt E. 2006]. Biogas methane provides fuel for cooking, not only saving the forests but also the women fetching and carrying heavy loads of fuel wood. Unlike firewood, biogas burns without smoke, thus saving women and children from respiratory distress and disease. Biogas can be used to generate electricity, prolonging the active hours of the day and enabling the family to engage in social or self-improvement activities or to earn extra income. Anaerobic digester solves sanitation problems by taking in human as well as animal manure, improving home and farm hygiene and the general environmental conditions in rural areas.

# Products from Anaerobic Fermentation and Their Utilization

Anaerobic digestion not only yields biogas but also bio-slurry and bio-sludge rich in nutrients, minerals and biologically active compounds that form excellent fertilisers for crops for food or fodder for animals and substrate for mushroom cultivation.

Nowadays a circular economy model of forest-grass-animals-biogas-fungus is promoted in central China. Biogas is one of key links of this agriculture circular economy, in which a farmer utilizes cow dung, stalk and straw by anaerobic fermentation to produce biogas and in turn to produce electricity, to reuse bio sludge to cultivate *Agaricus bisporus, Pleurotus* or wood ear and to plant rapid- growing poplar, among poplar rows he intercrops ryegrass as feeds for animals.

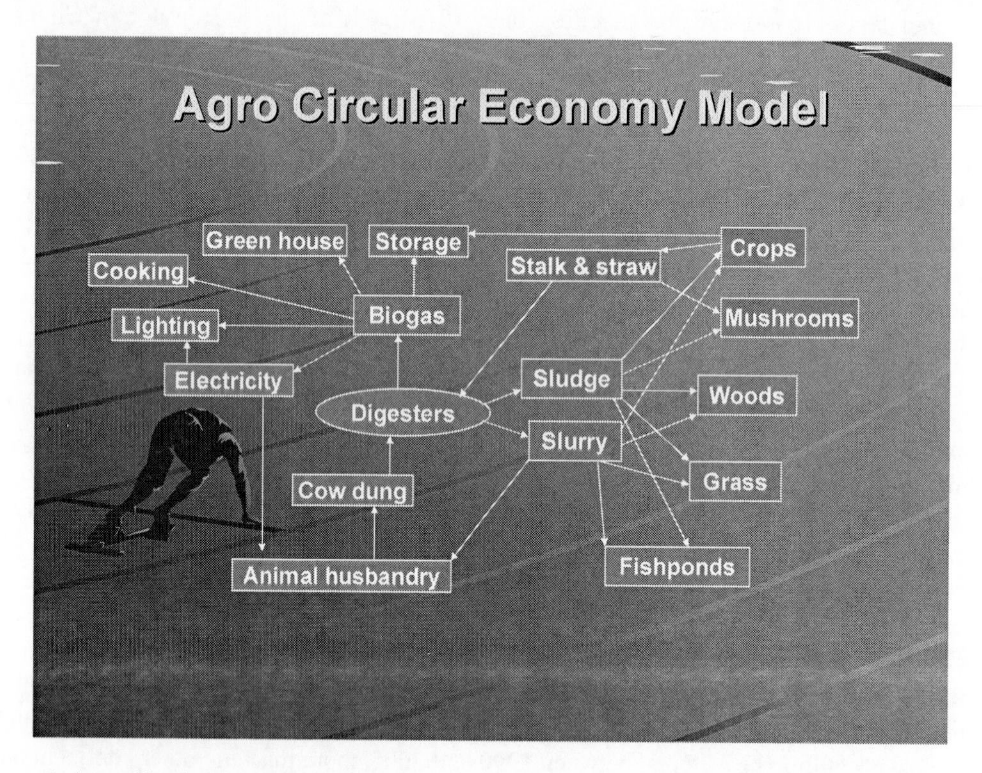


Figure 1. Agro Circular Economic Model.

## Utilization of Biogas

The utilization of biogas depends on its quality. Low quality biogas can be directly used for cooking and lighting, high quality biogas for co-generation of electricity and heat, which is especially feasible when it is used at or near the site of generation.

1 Biogas can be used in ovens and lamps or as gas fertilizer to heat greenhouses, increasing the carbon dioxide concentration to boost photosynthesis of plants and yields. In Shanxi Province a test showed the effect of CO2 on cucumber yield in a 333 m2 greenhouse, half for control. In test area three biogas lamps were lit from 6 to 8 o'clock for 2 hours the maximum concentration of CO2 reached $1600 \times 10^{-6}$ (volume fraction), the yield increased by 67.2%.

2 A biogas lamp gives both light and warmth to silkworm eggs increasing their rate of hatching as well as cocooning over the usual coal heating.

3 Biogas can be used to prolong storage of fruits and grains. An atmosphere of $CH_4$ and CO2 inhibits the metabolism of fruits or grains in storage, thereby reducing the formation of ethylene in fruits and grains. It also kills harmful insects, mould, and bacteria that cause diseases and rats as well [Bian Yousheng, 2000].
4 Biogas methane can also be used to make methanol, an organic solvent and important chemical for producing formaldehyde, chloromethane, organic glass, and compound fibre [Hun K. (ed.), 1990].
5 Upgraded biogas can not only be used to generate electricity and heat, but also as bio diesel for metropolitan transportation [West Start-Calstart Inc. 2004].

## Utilization of Bio Slurry: Liquid Soluble Substances (Small Amount of Undigested Material and a Variety of Metabolic Products)

Bio slurry in the digester is a treasure trove of valuable biological resources. These include major nutrients for crops such as N, P and K as well as large amount of metabolic products of microorganisms including vitamins, protein, various enzymes, trace elements and biologically active compounds such as amino acids, growth hormones, gibberellins, sugars, humic acid, unsaturated fatty acids and probiotics that may suppress the growth of pathogens, which benefit both plants and animals.

1 Slurry with complete nutrition is used as liquid fertilizer. It can promote the growth of fruit trees and veggie and reduce application of chemical fertilizers. The foliar spray comes from the digester in operation for 3 months with quick acting effectiveness and high utilization rate. In 24 hours 80% of the spray will be absorbed by leaves and nutrition could be obtained in time. Fruits taste better. If slurry is sprayed into peripheral ditches at the root zones of fruit trees, Different tree ages adopt different methods. Ditch expands to the outward every year for young trees. Ditches for fruit- bearing trees should appear radiating shape.
2 As solution for seed soaking, bio slurry can be used as a promoter in sowing season. Seeds soaked in bio slurry germinate better. The germinating rate can be raised by 15-20% and the seedlings grow earlier 1-2 days, being stronger. The method is to put 15-20 kg of seeds into a permeable plastic woven bag with some space in bag, tighten its mouth and put it into the outlet tank of a digester. Soaking time depends on species. Seeds are sun dried for 1-2 days before soaking. If they are conventional rice cultivar it needs about 24 hours; if they are hybrid rice cultivar intermittent soaking method is adopted i.e. soaking 8 h and cool dry 6 h for 3 times until seeds getting sufficient moisture and then rinse 3 times by clean water cool dry for germination. Sometimes seed colour might change it doesn't matter. If they are sweet potato seeds, 2 h; maize seeds, 12h; wheat seeds, 6-8 h; cotton seeds, 24 hours. Then take it out, rinse with clean water and dry in cool place for germinating and broadcasting.
3 As bio insecticide, bio slurry which contains some hormone and anti bacteria could raise resistance to adverse circumstances and could inhibit pests and diseases such as aphid, red spider, wheat scab, cotton anthracnose. The products will be free of chemical residues and it will not pollute ambient environment during the production

process. To control aphid of fruit trees or veggies, spray a mixture of 14 kg slurry + 0.5 kg 10% washing powder solution on the dosage of 35 kg/mu (525kg/ha). Practices indicate that the effectiveness is about 70% by one spray; if sprayed once more next day, the effectiveness could reach 96%. To control red and yellow spiders of citric fruits the higher the concentration of slurry the better the effectiveness. To control wheat scab which is a main wheat disease with high incidence and broad prevalence, slurry equals to a germicide on the dosage of 50-100kg per mu (750-1500kg/ha) when flowering and spray once again five days later the effectiveness reaches 80%. To control blight of water melon and grape powdery mildew, etc., seed soaking in slurry is applied for 8 hours and slurry is sprayed to water melon plants 3-5 times. Slurry and sludge are applied during the expansion period of water melon. The quality of water melon could be raised.

4 Slurry can be poured into fish ponds as nutrients, the proportion of poly-culture is filtering fish (silver carp, bighead carp, tilapia, and what have you) by 30%, omnivorous fish (common carp, crucian carp by 40~ 50%), herbivorous fish (grass carp) by 20~30%. Part of bio slurry is utilized as feed but most part of slurry used to enhance the biomass of phytoplankton and zooplankton as feed for filtering fish species, to improve aquatic ambience, to promote fish growth and quality. The dosage depends on the transparency of a fishpond, in general slurry dosage 4.5 t/ha once; sludge 2.25 t/ha once, less than 3 times a week. If the transparency is greater than 30 cm it indicates the biomass of zooplankton is greater than that of phytoplankton. Slurry could be applied every other day until the transparency comes back to 25-30 cm. The dosage is about 100-150 kg/mu (1.5-2.25 t/ha).

5 As feed for raising pigs, slurry can be fed to pigs as an additive to speed up the growth and shorten the rearing period by 25%, at the ratio of 1: 1 of slurry and feed, saving feeds by 15% or 150-200 kg feeds per head, reducing the cost. At first some slurry in a keg should be put in pigsty for pigs to smell and when they are getting used to the smell. Mix feed with slurry from small amount to more. The dosage of slurry depends on body weight of pigs, BW 25 kg below add 0.3 kg of slurry per head; BW 25-50 kg, add 0.6 kg of slurry; BW 50-100 kg, add 1 kg of slurry and BW 100 kg above, add 1.5 kg of slurry. Pigs fed with slurry love to sleep with the hair shiny and the skin smooth (Agriculture Broadcast World Program CCTV, 2007).

6 Slurry can be fed to broiler or egg layers with two methods: slurry as part of drinking water or slurry mixed with chicken feed as chicken feed. The dosage is 0.3 kg per hen which accounts for 26.79% of chicken feed. 90 days after broiler or layers fed with slurry gain weight by 34% than those fed without slurry. When fed to broiler and layers with the proportion of 3: 7 bio slurry to clean water, slurry come from the digesters using cow, chicken and pig manure increases the rate of laying egg by 14 %, 9 % and 7 %, respectively [Bian Yousheng, 2000]

## Utilization of Bio Sludge: Solid (Residue of Solid and Newly Produced Microbe Colonies)

1 As soil conditioner bio sludge can be applied to ameliorate soil in fields as basal manure by the dosage of 2,000 kg/mu (30t/ha) using a pump and then tiller the soil.

If bio sludge containing high level of humus, which accounts for 10-24%, be applied in three consecutive years the land would be ameliorated are required.

2 As raw material for fabricating nutritive cube for veggie, flowers and plants or nutritive soil for nursery. To fabricate 100 kg of nutritive soil, in general 30 kg slurry, 50 kg clay, 5 kg saw dust, 5 kg sand, 10% others including N, P, K 0.06 kg, trace elements 0.03 and soil germicide 0.01.

3 Sludge can be used as substrate for mushroom cultivation because its nutrition is complete, of loose texture, with good retention of moisture and moderate pH so it is a good substrate for mushroom growth such as *Agaricus bisporus, Pleurotus saja caju,* and Wood ear *Auricularia auricular-judae*. After harvesting, mushroom bed can be reused as feed for pig and cow while pig manure and cow dung in turn to be raw material for digesters to produce biogas. The carbon to nitrogen of the substrate for mushroom is 30:1 e.g. 100 m3 of the substrate need 5000 kg sludge, 1500 kg straw or stalk, 15 kg cotton husk, 60 kg gypsum and 25 kg lime. First shed straw or stalk into small pieces and put the first layer of shredded straw or stalk on the ground and spray slurry on it and then put the second layer of straw and spray the second layer of slurry up to the seven layers as a huge heap. Cover the heap for fermentation to 70 and then turn the heap. If the heap turns dry spray lime water to kill bacteria and after fermentation the substrate can be filled into mushroom bags (Agriculture Broadcast World Program CCTV, 2007).

4 For earthworm cultivation bio sludge can be used to culture earthworms which can be fed to chickens, duck, pig and fish. Earthworm is a healthy food for human. Earthworm faeces can activate soil containing humic acid 11%-68% to promote crops to absorb phosphorus and increase the yield of canola and cotton by 10%. Chickens fed earthworms will lay 15 to 30 % more eggs [Li Kangmin, 2005].

## BIOGAS AS THE BASIS OF ECO-ECONOMY

Biogas is at the centre of a burgeoning eco-economy in China. As animal husbandry goes intensive, there are many large or medium size livestock and poultry farms in the suburbs of cities. An example is Fushan farm in Hangzhou, with 32.47 ha paddy fields, 4 ha tea trees, 13.7 ha water shields and 7.3 ha fishponds. It also produces 30 000 laying hens, 150 000 broilers, and 8 000 pigs a year, with 15 tonnes of solid waste and 70 tonnes of wastewater discharged daily, a huge amount of pollution. But using biogas digesters to deal with the pig and poultry wastes, biogas energy becomes available for processing tea and heating the chicken coop, and there's fodder for fish and pigs and fertilisers for tea trees and the paddy fields, Also, the large amounts of water in slurry could be reused to wash away wastes in hog houses as a water-saving measure. So no pollution is exported to surrounding areas [Yi A., 2003] [Cai C, et al., 2005]. This eco-farm has now moved to the outskirts far from the city because of its malodour, however, it is possible to use a combination of multiple microorganisms to deodorize animal manure.

Northern China has cold winters but sufficient sunshine. Digesters do not operate below 10°C, and pigs raised in winter eat but do not get fattened. People also lack fresh vegetables in winter. All these problems could be solved in a Four-in-One eco-model that includes vegetable planting, pig-raising, a digester underneath the pig shed and a toilet adjoining the pig shed in a big greenhouse. The pigs grow well with manure flowing into the digester together with human excreta. The digester works well because the temperature could be kept above 10°C and it greatly improves the living conditions of farmers. The digester provides biogas as energy, slurry and sludge as fertilisers, and the pigs produce carbon dioxide to enrich the greenhouse to produce plenty of quality vegetables.

In southern China, a Five-in-One Eco model incorporates pigs, digester, fruit orchard, light trap, and fishponds. The pig manure flows into a digester to be fermented. Biogas is harvested to provide energy for cooking and lighting. Bio slurry and sludge are used as fertilizers for the fruit gardens and as feed for animals. The light-trap hangs above the fishpond to attract and kill pests, which become additional feed for fish. This model is practised especially in Guangxi Province in southern China, where a yellow sticky board (a kind of fly paper) is hung in the orchard for additional pest control [Li H, et al., 2006].

Nowadays a circular economy model of Forest-Grass-Animals-Biogas-Fungus is promoted in central China (figure 1). Biogas is one of key links of this agriculture circular economy, in which a farmer utilizes cow dung, stalk and straw by anaerobic fermentation to produce biogas and in turn to produce electricity, to reuse bio sludge to cultivate *Agaricus bisporus*, *Pleurotus* or wood ear and to plant rapid- growing poplar, among poplar rows he intercrops ryegrass as feeds for animals.

## CONSTRAINTS AND PROSPECTS

China is unleashing a campaign of Building New Socialist Villages in rural areas in the 11$^{th}$ FYP (2006-2010). The guidelines are to develop production and to clean the environment, to encourage innovation so as to save resources. Developing biogas fits in well with this programme.

The major constraint in China is lack of technical capacity for running and maintaining biogas digesters in rural areas. Now a new breed of biogas farmer workers has appeared. Up to the end of 2005, Shanxi Province held 40 biogas technical training courses and trained 6 000 farmers, 4 037 of which gained National Biogas Professional Technician Certificate [News from People Daily, 12 Feb 2006]. In order to popularise the biogas technology, a training course is held twice a month lasting 7 days and costing 600 Yuan per person (280 Yuan by correspondence). After completion of the course, participants would be given a corresponding professional technical grade certificate, which is recognised by the state [Readers Service Department, 2003].

Another constraint is water shortage. It is difficult to build water- pressure type digester in dry and semi-dry areas because it needs 6-7 m3 of water at initial stage when we build a 10 m3 digester and it needs to add water after slurry flows out when operating. In an extreme dry area e.g. Minqin County Gansu Province we need to build 10,000 digesters, the total water volume needed would be 60,000-70,000 m3. It is impossible to afford so much with poor precipitation over there. In North Western Region annual precipitation is only 50 mm but annual evaporation 2,000mm. Thus, dry fermentation should be promoted over there. It needs policy support from the government.

Using biogas as a transportation fuel is potentially worth about 10 times using the biogas for heating and even more for power generation. Hence reducing the cost of biogas upgrading is a high priority [WestStart-Calstart, Inc 2004]. To maximise the benefits of biogas as energy, China also needs not only advanced technologies to upgrade and compress methane and to build new engines that use methane effectively and efficiently, but also policies to promote biogas utilization. In pre-treatment of raw materials micro organisms can play an important role in deodorization.

The prospects for biogas as renewable energy are excellent, provided the constraints can be removed. Anaerobic digesters can also treat crop and food wastes as well as algal blooms and water hyacinths that clog up polluted rivers and turn them into biogas energy and common cord grass *Spartina anlgica*, which is an invasive grass from England in coastal areas, etc.

## REFERENCES


"Agriculture Broadcast World" Program CCTV7, 2007. Utilization of slurry and sludge, Series Disc produced by Central Agriculture Broadcast TV School 2007.

Bian Yousheng, 2000. The Treatment and Reutilization of Wastes in Ecological Agriculture, 1st Edition, Beijing, Chemical Industry Publisher, May 2000. P287-308

CAI C, Zhu X, Shou Y and Huang W., 2005. Fushan Farm – an example of integrating biogas use, papers provided by the company, 2005.

Chinese Environment Protection White Paper (1996-2005), 2006, published on June 5, 2006 from China Network, Beijing, 2006.

ESCAP (UN Economic and Social Commission for Asia and the Pacific), 1984. Guide for biogas development in Asian Pacific Region,

Ho MW. 2005. Biogas bonanza for Third World development, *J. of Science in Society*/ 27, p29, 2005.

Ho MW.2006. Dream Farm 2 – Story so far. *J. of Science in Society.* 31, p40-43, 2006.

Hun Kuhn (ed.), 1990. Organic Chemistry, 2nd edition, Beijing, High Education Press, 1990.

Li Junfeng, 2007. Speech of Deputy Director of Energy Research Institute of NDRC in China, In: *Power & Alternative Energy Summit 2007, Beijing*

Li Huashou, Jiang Chunxiao, Li Gangsen, Li Kede & Nie Chengrong, 2006 Development of integrated livestock-biogas-crop eco-agriculture models in Guangxi Province, In: *Proceedings of International Conference on Circular Economy and Regional Sustainable Development,* 1-4 November 2005, Hangzhou, China, 2006.

Li Kangmin, 2005. "Vermiculture Industry in Circular Economy", International Organisation for Biotechnology and Bioengineering, *IOBB E-Seminar*-14, 2005,

Mygatt Elizabeth 2006 World forests continue to shrink, EPI Eco-Economy Indicators, News from CCTV7 Aug 19, 2007.

Readers' Service Department, 2003. *J. of Village Know-all, Nanchang,* 2003 (8) p4

West Start-Calstart Inc. 2004. Swedish Biogas Industry Education Tour 2004: *Observations & Findings.*

Wu T. and Gao J., 2002. To re-integrate livestock and poultry with planting in Lake Taihu Basin (in Chinese), *J. of Jiangnan Forum,* Wuxi, 2002.

Yi Aizhong, 2003. Biogas kindles the hope of rural areas in China in Village Know-all 2003 (8)

Zheng Erlu 2007. Introduction to super small digester (S-2 model), China Biogas Network.

In: Technologies and Management for Sustainable Biosystems ISBN: 978-1-60876-104-3
Editors: J. Nair, C. Furedy, C. Hoysala et al. 

*Chapter 12*

# WASTEWATER TREATMENT USING NATURAL SYSTEMS: THE INDIAN EXPERIENCE

***Manoj K. M. Chaturvedi* and Shyam R. Asolekar*****

Centre for Environmental Science and Engineering,
Indian Institute of Technology Bombay,
Mumbai 400 076, India

## ABSTRACT

Inadequate infrastructure for rural and urban sanitation coupled with improper wastewater management practices, including disposal of untreated or partially treated wastewaters into the natural water courses, have deteriorated the water quality of almost all the aquatic resources in India. This chapter analyses secondary data on status of wastewater management in India and highlights significance of natural aquatic systems for treatment of sewage in urban and rural communities. Although, the number of wastewater treatment plants has increased over the years in urban India, this increase is not adequate to keep pace with escalating generation of wastewater. The conventional mechanised systems turn out to be rather expensive in terms of both, the installation as well as operation and maintenance costs. It is argued here that the newer solutions should be such that the peri-urban and small communities should be able to own and operate their wastewater treatment plants. Interestingly, in the recent past, communities seem to accept the natural treatment systems that are capable of giving adequate treatment to wastewaters in conjunction with supplementing fish and nutrition to the food baskets of the fishing communities engaged in managing the systems as well as by generating adequate water for irrigation of farms and agro-forests. Above all, the engineered natural treatment systems blend well with the agricultural, peri-urban, and rural ecosystems.

**Keywords:** Natural treatment systems; Waste stabilization ponds; Constructed wetlands; Duckweed ponds; and Sewage-fed aquaculture.

---

* Corresponding Author: Global Tech, Dubai, United Arab Emirates (Current Affiliation);
Contact Details: mkmchaturvedi@gmail.com ; +971-50-3951 421 (Cell)

** Professor & Head, CESE, IIT Bombay; Mumbai, India;
Contact Details: asolekar@iitb.ac.in ; +91-22-2576 7851 (Phone) +91-22-2572 3480 (Fax)

# 1. Introduction

Urbanization, which is proceeding at an accelerated speed around the world, has posed several new problems before urban residents. Inadequate water supply and poor water quality have been taken as serious contemporary concerns by many municipalities, industries, and farmers. Communities are thirsty for potable, irrigation as well as process waters.

Globally, around 2.4 billion people in rural and urban areas are devoid of adequate sanitation facility and it is expected that in coming 20 years an additional two billion people will be added, especially in developing countries, demanding adequate sanitation (Langergraber and Muellegger, 2005). India is one among those developing countries facing problems related to water and sanitation.

On one hand, there is an escalating demand for water for domestic, agricultural, as well as industrial purposes. On the other hand, the available water is getting deteriorated as a result of disposal of domestic and industrial effluents into watercourses. Disposal of partially treated and mostly untreated effluents into rivers and lakes and runoff from urban and agricultural areas are the two main reasons responsible for deterioration of drinking water resources in India.

In this chapter, an attempt has been made to articulate the status of sewage management in India followed by classification of natural aquatic systems for wastewater treatment. An effort has also been made to introduce the various natural treatment systems being applied throughout the country including waste stabilization ponds, constructed wetlands, duckweed ponds, and sewage-fed-aquaculture.

# 2. Status of Sewage Management in India

A large volume of wastewater continues to be discharged into natural watercourses leading to pollution of the coastal zones and drinking water reservoirs in India (Asolekar, 2001). Disposal of partially treated and mostly untreated effluents into rivers and lakes and runoff from urban and agricultural areas are the two main reasons responsible for deterioration of drinking water resources. In addition, excessive withdrawal of water for agricultural and municipal utilities as well as use of rivers and lakes for religious and social practices, and perpetual droughts limits the capacity of river for dilution of wastes (Asolekar, 2002).

In spite of the unprecedented laws and policies in India including Policy Statement for the Abatement of Pollution (1992), the National Conservation Strategy (1992) and the Policy Statement on Environment and Development (1992) the National Water Policy of (2002), and the National Environmental Policy (2005), there has been a steady deterioration of all the environmental sub-systems during the past five decades. Existing regulations and the approach of regulatory agencies appear to be inadequate to address the pollution caused by domestic sewage emissions and industrial wastewaters (Asolekar, 2002 and 2005).

Kantawala (2001) has reported the status of sewage treatment in Indian Metros (population more than one million) in year 2000. Out of the 23 Metros, 19 have sewerage covering 75% of the population while the remaining 4 have about 50% coverage. Further, out of 9,275 million litres per day (MLD) of the total wastewater generated by the 23 metros, only 2,923 MLD (i.e. 31%) is treated while the rest is disposed off untreated. The ultimate disposal in four cities is into rivers/lakes, in two other cities into sea/creeks and in the rest it is partly used for agriculture and partly disposed off into rivers. In 12 metros there is some level of organized sewage farming under the control of the state government or the local body. The total wastewater generated by the 23 Metros is 9,275 MLD, out of which about 58% is generated by Mumbai, Delhi, Calcutta and Chennai. Mumbai generates the maximum, 2,456 MLD, while Madurai generates the least, 48 MLD. The situation in Class I cities (population between 100,000 to 1 million) is not much different either! A comparison of the data of year 2000 (Kantawala, 2001) presented in the preceding paragraph with the recent data reported by Bhardwaj (2005) paints rather grim picture of wastewater treatment in India (figures 1 and 2).

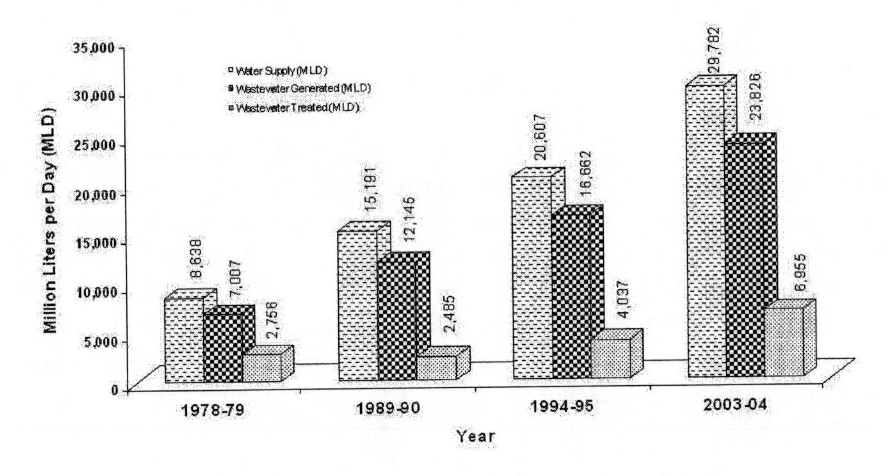


Figure 1. Water Supply and Wastewater Management in Class I Cities and Metros.

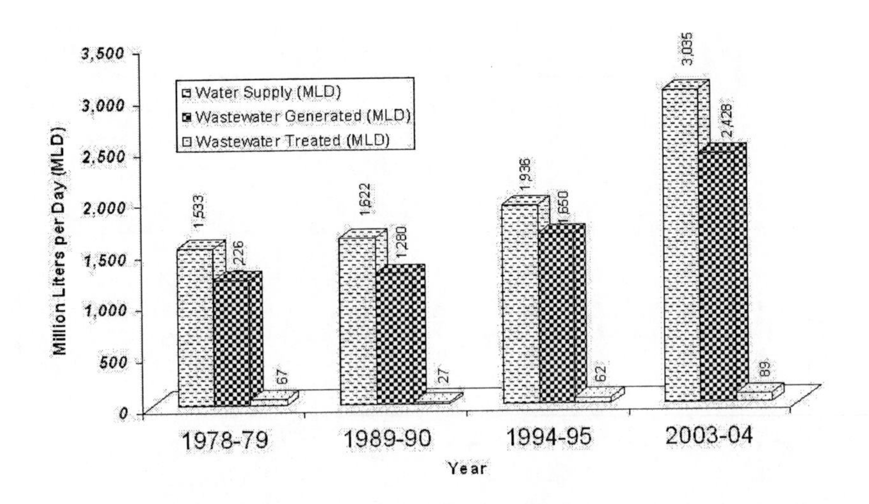


Figure 2. Water Supply and Wastewater Management in Class II Cities.

The conventional-mechanized wastewater treatment plants, involving collection and pumping of sewage at a centralized location and giving the primary and secondary treatment, require large capital investments and demand high operation and maintenance costs as well. Therefore, these treatment plants do not seem to be preferred in India. In contrast, natural systems for wastewater treatment use the "nature's principles". Such systems usually require relatively much smaller operation and maintenance costs. Natural systems can also give the quality of water comparable with the mechanized and energy intensive treatment systems but their biodegradation rates are rather slow and footprint areas are large! The natural systems can sometimes yield treated water having lower count of harmful pathogens and thus eliminate dependence on chemical disinfection provided the detention periods are long. For example, waste stabilization ponds (WSPs) are favored in rural communities because they are economical; produce adequate reductions in BOD, nutrients, and greater pathogen reduction.

## 3. Classification of Aquatic Natural Treatment Systems

As shown in figure 3, aquatic natural systems can broadly be classified into two classes: [I] intrinsic and [II] engineered systems. The intrinsic systems can be further subdivided into two divisions, namely: self-supporting and stressed systems. A self-supporting system typically allows degradation of pollution without altering its own mechanisms and processes; for example, rivers and lakes polishing traces of biodegradable organic matter or treated sewage with the help of plants and microorganisms present in the system. It must be noted that the natural systems also concurrently process other biological mass arriving to the system via other natural biogeochemical cycling routes through the routine humification of natural organic matter.

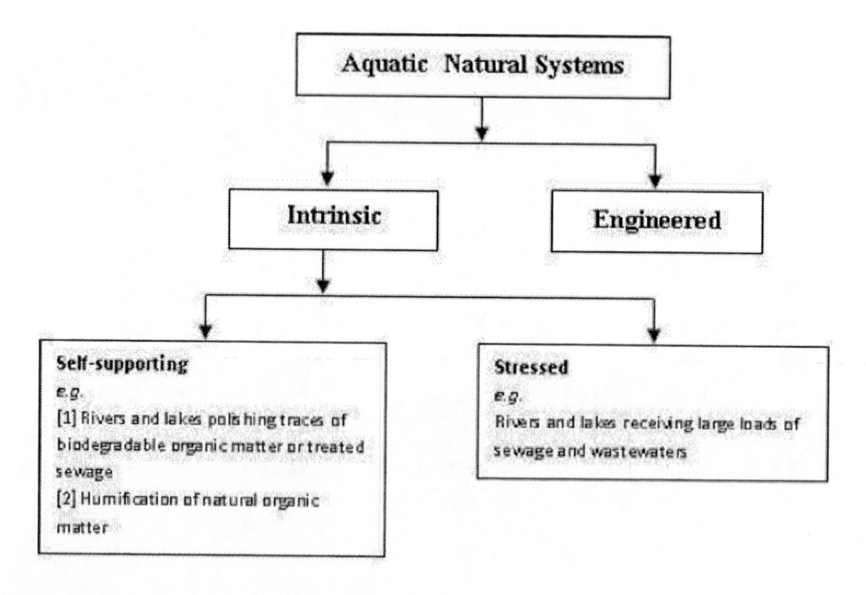


Figure 3. Broad classifications of natural aquatic systems.

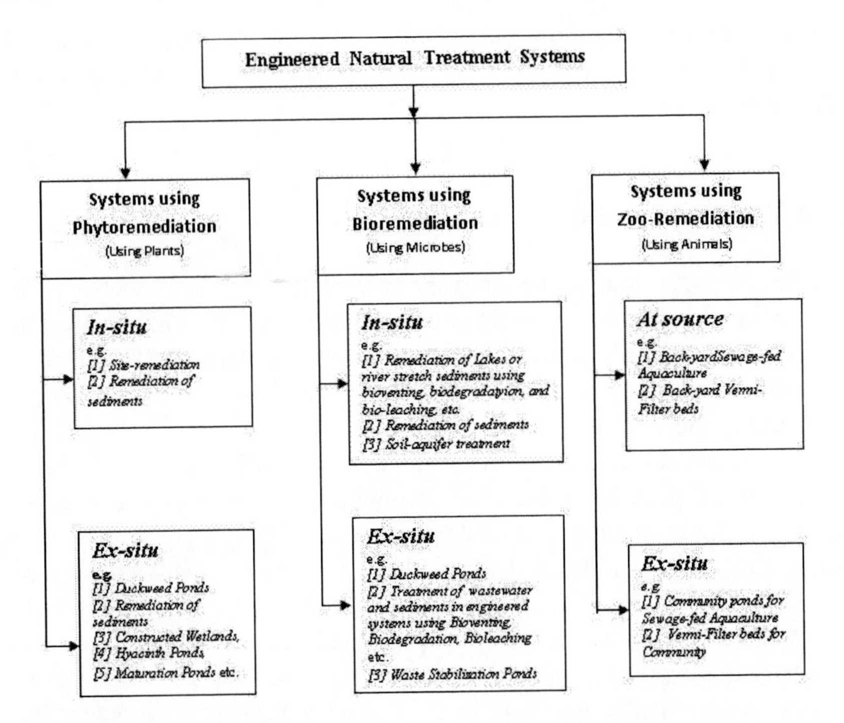


Figure 4. Classification of eco-technologies in engineered natural systems.

The stressed systems are usually characterized by their display of inability to cope with and degrade rather large amounts of pollution subjected to them (for example, rivers and lakes receiving large loads of sewage and wastewaters from urban or peri-urban communities). Engineered natural treatment systems essentially involve expansion of natural processes that occur in ecosystem as, both, inorganic and organic constituents cycle through different biotic and abiotic environmental compartments (*see figure* 4).

Aiming at understating significance of aquatic natural systems for wastewater management in Indian cities and rural areas, several sewage treatment principles (STPs) based on natural systems across India were visited and secondary data were collected by interviewing the operating staff of the respective STPs as well as by utilizing the literature, log books, and progress reports supplied by respective personnel.

It is important to note that all the facilities based on natural systems were observed to be producing effluent to a level which could easily be utilized for reuse and recycle purposes except in few cases such as STPs at Delhi and Panihati where the operating bodies have typically neglected maintenance of the system, which has lead to undesirable performance.

## 4. WASTE STABILIZATION PONDS

Waste stabilization ponds (WSPs) are the simplest of all waste treatment techniques available for sewered wastewaters. Their advantages stem from their extreme simplicity and reliability of operation. The principal mechanisms include sedimentation, nitrification, denitrification, and enhanced biodegradation by using physical components of the systems including oxygen transfer at air-water interface as well as solar radiation. In a typical waste stabilization pond ecosystem, the principal abiotic components are oxygen, carbon dioxide, water, sunlight, and nutrients while the biotic components include bacteria, protozoa, and a variety of other organisms.

### 4.1. Usage of Waste Stabilization Ponds

Configuration of the WSPs would vary according to the situations including characteristics of the wastewater to be treated as well as objective of the treatment (for example treatment and disposal versus treatment and reuse). Typical WSPs for treatment of sewage consist of anaerobic ponds followed by facultative pond followed by maturation ponds or aquaculture ponds.

In recent years, like other natural treatment systems, WSPs too have been gaining increasing acceptance, especially in the small communities, where population densities are very low and options for reuse of treated wastewaters are available. For example, in the State of West Bengal, almost all small towns have WSPs in combination with aquaculture ponds. Some of the working WSPs are as big in the treatment capacity as the conventional sewage treatment plants and are being successfully applied to treat municipal wastewater up to 9-10 MLD and in some cases even more than that. Possibly, the first small installation of WSP was a pond for Madras University campus built in 1957 (Arcievala, 1970). Subsequently, several WSPs were installed around the country. Performance data of some WSPs, currently working in India is given in table 1.

**Table 1. Performance of some waste stabilization ponds currently working in India**

| Location, State (code)[#] | Hydraulic Loading Rate (MLD)* | Performance (% Removal) | | | Ultimate use of treated wastewater |
|---|---|---|---|---|---|
| | | COD | BOD | Pathogen | |
| Miraj, Maharashtra | 9.20 | 70 | 64 | 38 | Irrigation |
| Karad, Maharashtra | 9.00 | 50 | 53 | 93 | Irrigation |
| Bandipur, W. Bengal | 1.71 | - | 90 | 99 | Pisciculture and Irrigation |
| Kalyani, W. Bengal | 6.00 | 85 | 90 | 99 | Pisciculture and Irrigation |
| Natagarh,W.Bengal | 3.00 | 85 | 90 | 99 | Pisciculture and Irrigation |

* 1 MLD = $10^6$ L $d^{-1}$ (Million Liter per Day).

### 4.2. Case Study: WSPs of the STP at Kalyani, West Bengal

The sewage treatment plant at Kalyani receives about 17 MLD of domestic sewage from about 70,000 inhabitants of the Kalyani township in West Bengal. Of the total of 17 MLD sewage, 11 MLD is treated through conventional systems before discharge into the river Ganga. Only about six MLD is treated through natural system using a three-step waste stabilization pond system consisting of two anaerobic ponds in the first step, two facultative ponds in the second step followed by four stocking ponds in the third step. Performance of the WSPs of STP at Kalyani is given in table 2.

**Table 2. Performance of waste stabilization ponds at Kalyani in the State of West Bengal**

| Parameters | Value |
|---|---|
| Flow | 6 MLD |
| $BOD_{inlet}$ | 135 (mg/L) |
| $BOD_{outlet}$ | 10 (mg/L) |
| $COD_{inlet}$ | 220 (mg/L) |
| $COD_{outlet}$ | 15 (mg/L) |
| $SS_{inlet}$ | 220 (mg/L) |
| $SS_{outlet}$ | 40 (mg/L) |
| $pH_{inlet}$ | 7.2 |
| $pH_{outlet}$ | 7.5 |

The facultative pond of the waste stabilization system was found to be most dynamic in the treatment process and it accounts for 22-69% of the treatment in the unit. The next subsystem of maturation ponds (*i.e.* fish growing ponds) happens to be primarily responsible for converting organic wastes into fish biomass.

## 5. Duckweed Ponds

Duckweed refers to family of floating aquatic plants with the scientific name *Lemnaceae*, which are monocots (like grasses and palms) and are divided into five genera: *Lemna*, *Spirodela*, *Wolffia*, *Landoltia*, and *Wolffiella*. Duckweed species are the smallest of all flowering plants. Their structural and functional features have been simplified by natural selection to only those necessary to survive in a given aquatic environment. A typical Indian duckweed plant consists of a pair of leaf-like flat structure (called as fronds), which function like leaf as well as stem and rootlets. Some species may have relatively more developed hair like rootlets. In either case, the rootlets are mere extension of stem and are not in true sense the roots (Chaturvedi *et. al.*, 2004 and Arceivala and Asolekar, 2006). Species of the genus *Spirodela* have the largest fronds, measuring as much as 20 mm across, while those of *Wolffia* species are 2 mm or shorter in diameter. *Lemna* species are of intermediate size at 6 to 8 mm. Compared with most other plants, duckweed fronds have little fiber (<5% w/w) because they do not need structural tissue to support leaves or stems. Duckweed is capable of producing more protein on an average and therefore has got great potential as a feedstock for aquaculture.

### 5.1. Usage of Duckweed Ponds

Although the use of duckweed systems in sewage-fed fisheries has been explored at lab-scale or pilot-scale, the status of awareness about the technology is still in its infancy. So far, only a few demonstration projects or pilot-scale systems have come up, around the world except some full scale plants in some parts of South-East Asia including Vietnam, China, and Bangladesh.

In the Indian subcontinent, there are not many examples of duckweed ponds for sewage treatment for large-scale treatment; though duckweed grows in abundance in natural ponds and lakes, particularly in the eastern parts of India and contributes to removal of pollutants in its natural setting. More research through field trials is needed in order to refine the sizing of the ponds used and to determine the correct inoculum of plant species to achieve a desirable effluent quality. Based on the personal interviews of the personnel at sites during field visits, performance of some demonstration projects employing duckweed ponds are listed in **table 3.**

**Table 3. Performance of some duckweed ponds currently working in India**

| Location and Capacity | Duckweed Sp. Used | Fish produced | Performance (% removal) | | |
|---|---|---|---|---|---|
| | | | BOD | COD | Pathogen |
| # Bhubaneswar, Orissa (1MLD) | Spirodella, Wolffia, Azzola, and Lemna | Catla, Rohu, Mrigal, Silver carp. Grass carp, Magur, and Freshwater Prawn | 75-90 | 70-80 | 95-99 |
| # # Halisahar, West Bengal (0.7 MLD) | Spirodella, Wolffia, and Lemna | Catla, Rohu, Mrigal, Silver carp. Grass carp, Mangur, | 85-90 | 70-85 | 97-99 |
| # # # Wazirabad, Delhi (1 to 3 MLD) | Lemna | Catla, Rohu, Mrigal, Silver carp and some local genera. | 66-80 | 65-70 | 99.3 -99.8 |

# A demonstration site developed by Central Institute of Freshwater Aquaculture (CIFA)
## A demonstration site developed by the M/s Sulabha International
### A demonstration site developed by the M/s Sulabha International.

## 5.2. Case Study: Duckweed Ponds of the STP at Wazirabad, New Delhi

A full scale sewage treatment plant at the Nehru Vihar suburban community of the New Delhi Metropolitan Area in the region of Wazirabad, happens to be an excellent example of application of duckweed pond technology for treatment of domestic sewage. About 25% of the flow (1 to 3 MLD) is subjected to duckweed ponds and the remaining goes to WSPs (6 to 8 MLD). M/s Sulabh International, New Delhi, is operating the duckweed pond system.

The raw domestic sewage is given a primary treatment in a settling pond and the primary treated sewage is fed to duckweed pond. The floating duckweed pond is periodically transferred to aquaculture/pisciculture pond. This system produces fish/shrimp as well high quality treated water suitable for irrigation.

Initially, in a bid to develop and demonstrate the low cost treatment technology, the Central Pollution Control Board, Government of India sponsored a research project on duckweed based wastewater treatment to M/s Sulabh International, New Delhi. During the project span the plant performed very well. The project was conducted for three years. Locally available genera, Spirodela, Lemna and Wolffia were used for the study after acclimatization.

The duckweed ponds were operated at different flow rates giving hydraulic retention times from 5.4 to 22 days. The BOD reduction in the system was recorded to be between 66%

and 80%. It was observed that on an average with 10.62 g $m^{-3}$ $d^{-1}$ volumetric organic loading and 106.2 kg $ha^{-1}d^{-1}$ surface organic loading, the city sewage could be treated effectively. A 30-50% reduction in phosphate and 56 to 80% reduction in ammonical nitrogen were observed. Bacteriological analysis in influent and treated effluent indicated removal of faecal coliform above range of 99% at hydraulic detention times of 6.4 -14.2 d.

The favorable temperature regime for adequate growth of duckweed reported to be in range of 20°C -35°C. Harvested duckweed is being fed to fish in a separate fishpond. The yield of fish has been recorded to be approximately 6 MT $h^{-1}$ in a period of 11 months. The financial gain through the sale of fish was reported to be Rs.88,000, Rs.66,000, Rs.40,000 per year for one MLD of sewage, sullage and secondary treated sewage respectively, which come to about 50% of the operational cost.

# 6. Constructed Wetlands

Constructed wetlands (CWs) are used extensively to treat domestic (Billore *et al,* 1995; Kadlec and Knight, 1996; Srinivasan et al., 2000; Ayaz and Akca, 2001) and industrial wastewaters (Hammer, 1989; Billore *et al.* 2001). They have also been applied to passive treatment of diffuse pollution including mine wastewater drainage (Hammer, 1989; Kadlec and Knight, 1996; Mungur *et al.*, 1997; Robinson et al., 1999; Jing *et al.*, 2001), and highway runoff following storm events (McNeill and Olley, 1998). Besides, wetlands, being a model ecosystem, can serve as wildlife habitats and can be perceived as natural recreational areas for the local community.

Configuration of CWs would vary according to the situation including characteristics of the wastewater to be treated as well as objective of the treatment (for *e.g.* treatment and disposal *versus* treatment and reuse). Further, they can vary according to the flow pattern selected *i.e.* horizontal flow, vertical flow, surface flow, and sub-surface flow. In any case, typical CWs for treatment of sewage comprise of one settling tank followed by the wetland bed.

## 6.1. Usage of Constructed Wetlands

Constructed wetlands appear to perform all of the biochemical transformations of wastewater constituents that take place in conventional, energy intensive processes including activated sludge process, septic tanks, and other forms of land treatments. In last four five years dozens of CWs based sewage treatment plants (providing secondary treatment) and industrial treatment plants (providing tertiary or polishing treatment) have come up across the country. However, their applications have been limited to the decentralized treatment facilities (for example CWs systems at Ravindranagar Colony at Ujjain and University Campus, Ujjain) and relatively low hydraulic load ranging from 0.2 to 100 $m^3$/d. Given the appropriate climatic condition of the country, CWs may be successfully established with plant species acclimated to the tropical environment and able to be harvested for use in secondary functions like fuel production. Performance data of some existing constructed wetlands in India is given in table 4.

**Table 4. Performance of some Constructed Wetlands currently working in India**

| Location and Capacity | Wastewater Type | Performance (% removal) | | | | References |
|---|---|---|---|---|---|---|
| | | BOD | TSS | P | N | |
| Uni. Campus, Ujjain Madhya Pradesh (13 $m^3$/d) | Municipal Sewage | 65 | 78 | 58 | 36 | Billore *et al.* (1999) |
| Ravindra Nagar, Ujjain, Madhya Pradesh (40 $m^3$/d) | Municipal Sewage | 67 | 74 | - | 71 | Billore (2006) |
| Barwah Distillary Madhya Pradesh (10 $m^3$/d) | Industrial Wastewater (Distillery) | 85 | 40 | 80 | 65 | Billore *et al.* (2001) |
| Ekant Park, Bhopal Madhya Pradesh (70 $m^3$/d) | Municipal Sewage | 71 | 78 | - | 80 | Billore (2006) |
| Chennai, Tamilnadu (5 – 100 $m^3$/d) | Industrial Wastewater (Tannery) | 50-62 | 25-52 | - | - | Emmanuel (2000) |
| Pune, Maharashtra (0.2 – 40 $m^3$/d) | Industrial Wastewater (Chemical, Automobile) | 83-90 | 90 | - | - | Gokhale (2000) |

## 6.2. Case Study: CWs of the STP at Ujjain, Madhya Pradesh

A constructed wetland system in the Ravindranagar residential neighborhood in urban Ujjain in Central India was installed to treat sewage, which has now eliminated this source of pollution to the Kshipra River (Billore, 2006). This gravity-based system has an entrance at one end through which the wastewater flows in and an outlet at the opposite end for collection of the treated water. It works with the help of a perennial reed (*Phragemitis karka*), which is found in abundance in Indian plains around water streams and swampy areas. The dense root planted in the gravel bed supports thousands of microorganisms, protozoa, metazoan and a complete food chain system that kill harmful microbes and decontaminate the pollutants in the wastewater. The hollow grass acts as an oxygen pump and through its root canal diffuses and oxygenates the wastewater.

Average treatment performance of the CW system recorded removal efficiencies of 78% for $NH_4$-N, TSS; 58-65% for P, BOD and TKN . Effluent dissolved oxygen levels increased to 34% indicating existence of aerobic conditions in the rooted-gravel bed(Billore, 2006). The SF system has apparently has yielded two concrete benefits: (a) very cost-effective treatment technology and (b) Removal efficiency above 50% for BOD, $NH_4$-N, TKN, and P.

# 7. Sewage-Fed-Aquaculture

The sewage fed aquaculture (SFA) system has been practiced since the beginning of 20th Century in different parts of India. SFA, mainly being applied by the State of West Bengal in India, has two-fold advantage: firstly, it provides a cost effective treatment of domestic sewage up to a certain extent and secondly, it provides an alternative of free nutrients to fishes being reared in such ponds. However, the possibility of potential health hazards posed to the fishermen involved in the aquaculture by the pathogens in the ponds cannot be overlooked and deserves further investigation.

The Kolkata Municipal Corporation area generates roughly 600 MLD sewage. The wastewater is led by underground sewers to the pumping stations in the eastern limit of the city, and then pumped into open channels. The responsibility of the Kolkata Municipal Corporation ends with the reaching of the wastewater to the outfall channels. Thereafter, the East Kolkata Wetlands manage the sewage. A detention period of 20-40 days brings about bio-degradation of the organic compounds in sewage. Organic loading rate on these fishponds appears to vary between 20-70 kg BOD $h^{-1}$ $d^{-1}$. There are networks of channels that are used to supply untreated sewage and to drain out the treated wastewater. The cumulative efficiency in reducing the B.O.D. (a measure of biodegradable organic pollution in wastewater) was found to be above 80% and that in reducing the coliform bacteria is 99.99% on an average. In fact, the sewage fed fishery ponds act as solar reactors. A dense population of plankton traps solar energy. The fishes consume planktons. While the planktons play a highly significant role in degrading the organic matter in the wastewater, at times excessive growth of planktons poses problems for pond management.

It is at this critical phase of the ecological process that the fishes play an important role by grazing on the plankton. The two fold role played by the fishes is indeed crucial – they maintain proper balance of the plankton population in the pond and also convert the available nutrients in the wastewater into readily consumable form (*viz.* animal protein as fish) for the humans. This complex ecological process has been adopted by the fish farmers of the East Kolkata Wetlands, who have reportedly developed a great mastery of these resource recovery activities and thus have become an excellent example of self-sustaining wastewater treatment system through community intervention.

It is clear that empowerment of the communities on the fringes of urban and per-urban localities through improving their participation in the management of such treatment systems on one hand and raising their awareness towards potential economic benefit *vis-à-vis* health impacts will continue to play the significant role in improving adoption of natural treatment technology in general SFA in particular.

## 8. Significance for India

As stated earlier, Wastewater treatment capability of India is far from the desired capacity because conventional mechanized treatment systems are beyond the reach of most of the rural and urban communities in India. Since they are typically expensive to build, operate, and maintain. . In the face of this imminent crisis, application of engineered natural treatment systems (NTSs) seems to be more promising for developing economies like India and has been considered as potential wastewater treatment alternative.

Although land may be a limiting factor in dense urban areas (Class I cities), natural aquatic systems are potentially well suited to smaller communities (Class-II & III and for our 5 lakh Villages) where municipal land surrounding schools, hospitals, hotels and rural areas is not in short supply. If, for the sake of simplicity, capital investment costs are taken into account, conventional activated sludge process costs Rs. 600-700 per capita; assuming each person contributes average 180 L volume and 50 g of $BOD_L$ every day, whereas the cost for stabilization ponds could be Rs. 150-200 per capita.

The constructed wetlands, duckweed ponds, and sewage-fed-aquaculture systems are very similar to waste stabilization ponds from a maintenance and operational perspective, therefore the indicative cost for natural systems can be taken as say Rs. 200 per capita /day. Thus, the difference in the cost is about Rs.500 per capita which could be applied towards purchase of land. At what price is land "reasonable" to buy? While it is a tricky question to answer, based on the above example; the maximum price one can pay will be Rs. 250 /$m^2$ because the natural systems will be Rs. 500 per capita cheaper. In fact, the break-even price of land could even be higher if the comparison of O&M costs is applied towards land purchase! In addition to their cost effectiveness, the natural treatments systems have enormous potential for application in India, as its climate is conducive for higher biological activity and productivity, hence can harness better performance of these systems.

It appears that the environmental issues associated with the natural treatment systems and alternate competing conventional wastewater treatment practices have been poorly understood by most of the agricultural and rural development professionals and policy makers in India and even throughout the world to some extent. The authors of this study have initiated comprehensive serious studies in various natural treatment technologies because of their potential for contribution towards elimination of poverty without altering the contiguous ecosystem as well as without affecting the health of fishermen and personnel engaged in management of these treatment systems.

## 9. CONCLUSION

India is facing a grave crisis with respect to water supply in urban as well as rural areas for potable, domestic, agricultural, and industrial purposes while the available water resources are getting polluted due to disposal of partially treated and untreated domestic and industrial effluents. It is evident from the recent statistics on wastewater management in India that out of about 26,000 million litres per day (MLD) of wastewater reportedly collected cumulatively in two mega cities (population above 5 million), 11 large metro cities (population between 2 to 5 million), 26 small metro cities (population between 1 to 2 million), 384 Class I cities (population between 100,000 to 1 million) and 498 Class II cities (population between 50,000 and 100,000), which houses more than 70% of India's 500 million urban population, merely 27% of urban wastewater has access to some kind of treatment. The reason behind the inadequate domestic wastewater treatment across the country is due to the costs associated with conventional mechanized systems. Electrical power, too, is not available in most of the places and costs of operation and maintenance are not affordable. About 30% population in urban India (about 150 million population) and 500 million populations in rural India continue to depend on septic tanks and land disposal of sewage and sullage.

The reason behind the inadequate domestic wastewater treatment across the country is due to the costs associated with conventional mechanized systems. Electrical power, too, is not available in most of the places and costs of operation and maintenance are not affordable. About 30% population in urban India (about 150 million population) and 500 million populations in rural India continue to depend on septic tanks and land disposal of sewage and sullage. Engineered natural treatment systems (NTSs) appear to be more promising for developing economies like India and have greater potential to become wastewater treatment alternative especially in the small communities where reuse of treated wastewater would be welcome.

In sum, the engineered natural treatment systems have a great potential for application in India. By the virtue of their simplicity and ability of generating employment for fisherman and rural agricultural workers deriving irrigation water from them. In urban localities land may be a limiting factor where conventional mechanized systems may has edge over the engineered natural treatment systems. However, engineered natural treatment systems are potentially well suited to smaller communities where municipal land surrounding schools, hospitals, hotels and rural areas is not in short supply. In addition, such treatment systems blend well with the agricultural, peri-urban, and rural ecosystems.

## ACKNOWLEDGEMENTS

The authors acknowledge partial funding from the Swedish International Development Cooperation Agency (Sida) and European Union (EU) for pursuing research in natural treatment systems.

## REFERENCES

Arceivala, S. J. and Asolekar, S. R., *Wastewater Treatment for Pollution Control and Reuse*, Tata McGraw-Hill Publishing Company Limited, New Delhi, 2006.

Asolekar, S. R., Enabling policies and technologies for reuse of treated domestic wastewater in India. In: proceedings of the first workshop entitled: *Reuse of treated wastewater and sludge for agriculture in South Asia*, co-organized by the SASTAC-GWP and IWWA-Pune-Centre in Pune, India. December 7-8; 2001.

Asolekar, S. R. and Gopichandran, R., *Preventive environmental management– an Indian perspective*, Foundation Books Pvt. Ltd., New Delhi, 2005 (the Indian associate of Cambridge University Press, UK).

Asolekar, S. R., Greening of industries and communities. In: *Rio to Johannesburg and Beyond* (Lead India edition), Orient Longman, Hyderabad, India, pp 125-166; 2002.

Ayaz, S. C. and Akça, L., Treatment of wastewater by natural systems. *Environment International*, 2001, 26(3), 89-195.

Bhardwaj, R. M., Status of wastewater generation and treatment in India. In proceedings of *IWG-Env international work session on water statistics*, Vienna, June 20-22; 2005.

Billore, S. K., Head of Institute of Environment Management and Plant Sciences, Vikram University, Ujjain. *Personal communications,* October 2006.

Billore, S. K., Bharadia, R., and Kumar, A., Potential removal of particulate matter and nitrogen through roots of water hyacinth in a tropical natural wetland. *Current Science*, 1998, 74(2), 154-56.

Billore, S. K., Singh, N., Ram, H. K., Shrama, J. K., Singh, V. P., Nelson, R. M., and Das, P., Treatment of molasses based distillery effluent in a constructed wetland in central India. *Water Science and Technology*, 2001, 44(11), 441-448.

Chaturvedi, M. K. M., Langote, S.D., and Asolekar, S. R., Sustainable treatment of small-community wastewater: significance of duckweed-fed-fishponds. In Proceedings of *International Conference on Wastewater Treatment for Nutrient Removal and Reuse*, jointly organized by Asian Institute of Technology and Technical University of Denmark (DTU) in cooperation with International Water Association at Asian Institute of Technology, Bangkok, Thailand, January 26-29: 2004.

Emmanuel, K. V., Application of Constructed Wetland in Tertiary Industrial Effluent Treatment – UNIDO's Experience. In: Proceeding of *International Conference on Constructed Wetlands: Wastewater Treatment in Tropical and Sub-tropical Region*, organized by Indo German Projects at Centre for Environmental Studies (CES-GTZ), Anna University, Chennai, India and Central Pollution Control Board (CPCB-GTZ) Delhi India on December 11-13, 2000.

Gokhale, A., Experience with Sewage and Industrial Effluent Treatment using Indian Reed Bed Systewms. In: Proceeding of *International Conference on Constructed Wetlands: Wastewater Treatment in Tropical and Sub-tropical Region*, organized by Indo German Projects at Centre for Environmental Studies (CES-GTZ), Anna University, Chennai, India and Central Pollution Control Board (CPCB-GTZ) Delhi India on December 11-13, 2000.

Hammer, D. A., *Constructed Wetlands for Wastewater Treatment*. Lewis Publisher, Chelsea, pp319-351; 1989.

Jing, S.R., Lin, Y.F., Lee, D.Y., and Wang, T.W., Nutrient Removal from Polluted River Water by Using Constructed Wetlands, *Bioresource Technology*, 2001, 76(2), 131-135.

Kadlec, R. H. and Knight, R.L., *Treatment Wetlands*, Lewis Publishers, Boca Raton, New York, London, Tokyo, 1996.

Kantawala, D., Urban wastewater management in India: present status. In: *Roundtable on urban wastewater management: technological and financial issues*, organized by the Indian Environmental Association, Mumbai on February 2; 2001

Langergraber, G. and Muellegger, E., Ecological Sanitation- a way to solve global sanitation problems? *Environment International,* 2005, 31(3), 433– 444.

Lim, P. E., Wong, T. F., and Lim, D. V., Oxygen demand, nitrogen and copper removal by free-water-surface and subsurface-flow constructed wetlands under tropical conditions, *Environment International*, 2001, 26(5-6), 425-431.

McNeill and Olley, S. The effects of motorway runoff on watercourses in south-west Scotland. *J. CIWEM*, 1998, 12, 433-439.

Mungur, A.S., Shutes, R. B. E., Revitt, D. M., and House, M. A., An Assessment of Metal Removal by Laboratory Scale Wetland. *Water Science and Technology*, 1997, 35(5), 125-133.

Robinson, H., Harris G., Carville M., Barr, M., and Last, S., The use of engineered reed bed system to treat leachate at monument hill landfill site, Southern England. In: Mulamoottil, G. McBean, E.A., and F. Rovers, eds. *Constructed Wetlands for the Treatment of Landfill Leachates*. Lewis Publishers, Boca Raton, FL7. pp 71-79; 1999.

Srinivasan, N., Weaver, R.W., Lesikar B.J., and Persyn, R. A. Improvement of domestic wastewater quality by subsurface flow constructed wetlands, *Bioresource Technology*, 2000, 75(1),19-25.

## VITAE OF AUTHORS

**Dr. Manoj K M Chaturvedi**

**Dr. Manoj K. M. Chaturvedi**, currently working at *Global Tech*, Dubai, was affiliated to *Centre for Environmental Science and Engineering (CESE), Indian Institute of Technology-Bombay (IITB)* during April 2001 – Oct 2007 as senior project staff and research scholar. He was recruited at the CESE, IITB as a research fellow on a Sida-sponsored project entitled "Significance of Pond Systems in Treatment of Domestic wastewater in India" and has worked on several other sponsored and consultancy projects funded by various international and national agencies including but not limited to Central Pollution Control Board, New Delhi, Ministry of Environment and Forests, Government of India, Government of Rajasthan, Government of Maharashtra, European Union, and the World Bank as well as several industrial units. Earlier, Dr. Chaturvedi has provided his services as a project fellow to the Indian Institute of Environment Management (IIEM), where he was assigned to work on a World Bank aided project entitled "Delineation and dissemination of green corporate accounting procedures-with case studies in medium sized Textile, Pharmaceutical, and Chemical industries".

He has more than 25 technical papers in National and International conferences to his credit and has been a recipient of 'The Best Poster-Paper Award at the EnviroVision 2003 (5$^{th}$ Annual Conference of Indian Environmental Association entitled "Advances in Environment Management and Technology" held at Mumbai, during May 29-31, 2003. In addition to the above, he had been active member of the Indian Environmental Association (IEA) and was involved in a number of initiatives intended to broaden the scope of the IEA and was notably involved in organizing conferences and skill-building workshops on various environmental management issues, environmental policies and law.

**Dr. Shyam R. Asolekar**

**Dr. Shyam R. Asolekar** is **Professor and Head** at the Centre for Environmental Science and Engineering at the Indian Institute of Technology, Bombay. He is author of two books, two patents, policy documents, chapters of books and several research papers in international and national journals. Dr. Asolekar has been a *member* of the "Dahanu Taluka Environmental Protection Authority" (since 1997) as well as the "Expert Committee on Conversion of Municipal Solid Wastes to Energy" (since July 2005), both constituted by the Honorable Supreme Court of India. Recently, the Honorable Planning Commission of India has recruited him in the "Working Group on Rivers, Lakes and Aquifers". The Govt. of Maharashtra appointed him recently in the "expert committee" entrusted with recommending clearance for the *High-rise Buildings* after investigating environmental suitability of the tower-projects having more than 20 floors and taller than 70 meter heights. Earlier, the Govt. of Maharashtra had appointed him in the "Fact-finding Committee" to probe the causes of the collapse of civic infrastructure in the aftermath of the deluge in Mumbai between July 26 and 29, 2005. In December 2005, the Ministry of Environment and Forests (MoEF), Govt. of India invited him to participate in the deliberations of "Consultative Group on Chemicals (CGC)" set up to discuss India's strategy at the *second session of the Joint ILO/IMO/Basel Convention Working Group on Ship Scrapping held at the UNO, Geneva* during 12-14 December, 2005. In addition, Dr. Asolekar is currently advising the MoEF on the projects undertaken by the National River Conservation Directorate (NRCD), Hazardous Substances Management (HSM) Division, and Environmental Information System (ENVIS). The Honorable Bombay High Court, Govt. of Goa, Govt. of Maharashtra, Gujarat Pollution Control Board, and Maharashtra Pollution Control Board also routinely call him upon for seeking his advice on environmental matters. Dr. Asolekar's one of the interests lies in teaching undergraduate and post-graduate students as well industrial and regulatory professionals. Some of the courses recently developed by him include: *Environmental Change and Sustainable Development, Municipal and Industrial Wastewater Management and Reuse, Environmental Law and Policy* for post-graduates, and *Environmental Studies* for the fourth-year undergraduate engineering students. His areas of research have been *Treatment, Recycle and Reuse of Industrial and Municipal Wastewaters, Environmental Policy, Environmental Management Plans and Systems, Preventive Environmental Management, Eco-Industrial Networking, Global Environmental Issues and Phase out of ODSs, Treatment of Hazardous, Biomedical, and Solid Wastes,* and *Application of Remotely Sensed Data for Monitoring of Environmental Systems.*

Dr. Asolekar has been a recipient of the award for "Outstanding Ph.D. Thesis" (1991) given by the American Chemical Society. In 1994, he has been a recipient of the Rockefeller Foundation *Fellowship for Leadership in Environment and Development*. He received two awards (excellent article publication and best lecture delivery) from Indian Water Works Association in 1998. He also received "Dr. Patwardhan Award for Innovative Technology Development" instituted by IIT-Bombay (1999). He was awarded the "Prof. R.C. Singh Medal" by the Institution of Engineers (India) for his publication in 2000. Dr. Asolekar was awarded the *Leadership for Environment and Development (LEAD) – India Society* fellowship for authoring a policy paper in the context of the World Summit on Sustainable Development (WSSD), held during August 26 - September 4, 2002 in Johannesburg, South Africa. He has been included as a member of the steering committee of "Network for Reuse of City Effluents for Agriculture" under the Global Water Partnership initiative in South Asia. He was also a partner in the "Asian Regional Research Programme on Environmental Technology (ARRPET)" (2001-2004) funded by the Sida *i.e.* Swedish Development Co-operation Agency. Until recently, he was the *President* of the Indian Environmental Association (IEA) for 2003-2005 term.

In: Technologies and Management for Sustainable Biosystems ISBN: 978-1-60876-104-3
Editors: J. Nair, C. Furedy, C. Hoysala et al. 

*Chapter 13*

# REHABILITATION OF DEGRADED ECOSYSTEMS IN DRYLANDS OF SOUTHERN PAKISTAN: COMMUNITY-LED INNOVATIVE INTERVENTIONS FOR GETTING OPTIMUM BIO-PRODUCTION FROM WASTELANDS

***Sahibzada Irfanullah Khan***
Farm Forestry Support Project (Intercooperation-Pakistan)

## ABSTRACT

Drylands in southern Pakistan are home to communities living in poverty and depending on livestock rearing for livelihood. The subsistence agriculture is losing its importance due to uncertain rainfall and very low productivity. Due to increasing population of livestock, the pressure on silvo-pastures is increasing resulting in degradation of natural resources and loss of soil fertility, a fact that adversely affects the livelihood of communities. The Farm Forestry Support Project (FFSP) of the Intercooperation-Pakistan (funded by Swiss Agency for Development & Cooperation), initiated rehabilitation work in 2003 in extreme dry region of Karak using the silvo-pastoral system with hillside ditches and sand dune stabilization techniques. The objective was to recover vegetation and increase productivity of the area with minimum cost and hence support livelihoods. The activity was carried out with participation of civil society organizations and farmers' associations.

The results recorded in 2008 showed a profuse plant growth in terms of trees, shrubs and grasses with a potential to provide timber, fuel wood and fodder for livestock. Maximum harvesting of rainwater and conservation of moisture also resulted in growth of natural grasses and shrubs. Within a short period of 5 years, plant growth in height and diameter of 6 meters and 20 centimeters respectively was recorded. The average vegetation cover of 45% and increase in soil organic mater and nitrogen content was also recorded. All this happened with a minimum cost of US$ 82 per hectare. The rejuvenation of wells in few cases was an additional positive affect of the activity. On the other hand, an annual income of US$ 735 per hectare from *Saccharum spontaneum* planted in sand dunes was a real benefit to farmers against the other land-uses in dry sand dunes.

**Keywords:** Drylands, Poverty, Silvopastures, Degradation, Livelihoods, Rehabilitation, Hillside Ditches, Sand Dune Stabilization, Water Harvesting, Vegetation Cover, Soil Fertility, Costs.

## BACKGROUND

Drylands are generally defined as arid, semi-arid or dry sub-humid lands receiving less than 500 mm annual rainfall with an aridity index between 0.05 and 0.65 (the aridity index is the ratio Average Annual Precipitation / Potential Evapotranspiration) [UNCCD/UNEP, 1999/7]. There are more than 3 billion people living in drylands globally that cover 40% of earth's surface [Robin, P.W., 2002]. In climatic terms, drylands are defined as lands receiving less than 500 mm of annual rainfall. In Pakistan, the situation is severe with 75% of the country's area receiving less than 250 mm of annual rainfall [Pakistan Meteorological Department, 1998]. Most parts of Sind and Balochistan, and Southern parts of Punjab and North West Frontier Province (NWFP) are falling within this dry zone.

Over 30 million people in Pakistan live in drylands. Their livelihoods depend heavily on the natural resource base in form of provision of food for human beings, fodder for livestock, fuel for cooking and heating, and water for drinking. Some scanty income from the sale of medicinal plants and herbs, livestock and dairy products, and wildlife also add to the meager earnings [Fischler M., et al. 2006].

The poor in these ecologically fragile marginal lands are increasingly locked into patterns of natural resource degradation. Due to the low production and regeneration potential, drylands are not able to support an ever-increasing population of human beings and livestock. Most of the silvo-pastoral ecosystems in drylands are degraded due to overstocking beyond their carrying capacity, whereas rainfed croplands are increasingly being abandoned due to prolonged drought periods. These adverse factors are continuously undermining the livelihoods of poor farming families.

## THE STUDY AREA

This chapter relates to joint activities of the FFSP Project (SDC, IC), local NGOs and rural community organizations in Karak, one of 22 districts in the North West Frontier Province of Pakistan. District Karak is situated in southern region of NWFP, covering an area of 3372 square kilometers [Govt. of Pak., Bureau of Statistics, 1996]. Total population of Karak is 430,000 heads [Govt. of Pak., Pakistan Census Organization, 1998].

The area comes under tropical and sub-tropical climatic zone, characterized by arid and semiarid conditions. It can be divided into three distinct geographical divisions: the dry hilly zone in north, sandy desert in southwest and sandy-loam plains in the eastern part. The northern hilly zone is famous for mining of various minerals like salt and gypsum. The south-western desert is characterized by shifting sand dunes, dry and hot winds, and subsistence cultivation of gram, mustard, groundnut and wheat. The eastern region is famous for a number of agricultural crops (millets, wheat, and maize) and vegetables (chilies, okra, egg-plant, tomato) mainly because of availability of some irrigation water. As a whole, 19% area is under cultivation out of which water is available for 2% area only [Govt. of Pak., Agri. Census Organization, 2000].

People in this area live on subsistence agriculture, livestock rearing and minor trade of daily use commodities. Literacy rate is surprisingly high (above 50%) as compared to prevailing rate in Pakistan (44%) [Govt. of Pak., Pakistan Census Organization, 1998]. Due to harsh living conditions and limited opportunities on land, people prefer to join civil and armed services out of the area. The remittances they send back to their families is thus an important source of living.

## THE DRYLAND ECOSYSTEMS

The interplay between human beings, land resources, climatic conditions, natural vegetation and livestock constitute the ecosystem in most of the drylands in Pakistan. In all these, the climatic factors and availability of land for productive practices are limiting factors. Again, in most of the cases vast tracts of land are available but production systems are limited to only a few patches because of climatic conditions that limit the availability of water [See figure 1: The Drylands in Karak].

In the area under consideration, mean maximum temperature can reach to 46°C in summer (May to September). The mean minimum temperature in winter months (November to February) goes down to 3° C. The extreme arid conditions prevailing in major part of Karak limit agriculture to a profit-less rather under-paying activity. Subsistence agriculture is totally dependant on rainfall that is sporadic, uncertain and does not exceed 350 millimeters per annum [Govt. of Pak., District Census Report of Karak, 1998]. Livestock rearing (mainly goats and sheep) is thus adopted as major source of livelihood that supports the family in terms of nutrition and income from sale of animals, wool and milk.

These limitations lead towards a silvo-pastoral way of living where natural vegetation plays deciding role in the sustenance of the system. Sporadic grasses, shrubs and stunted trees are all what is required for grazing herds. The local tree vegetation in this area include *Acacia modesta, Prosopis cineraria, Capparis aphylla, Prosopis glandulosa,Tamarix aphylla, Zizyphus mauritiana, Olea ferruginea* and *Tecoma undulate*. Some of the important shrub species include *Zizyphus numularia, Vitex negandu, Saccharum munja, Callygonum polygonoides, Callotropis procera,* and *Nannorrhops ritchiana.* Among grasses, Chrysopogon spp., Cenchrus spp., and Cynodon dactylon are important. Whereas *Salsola foetida, Withania spp.,* and *Erva javanica* are common herbs. The natural forest is limited to only 2% of the total area on distant hills, comprising mainly *Acacia modesta* and *Olea ferruginea* [Govt. of Pak., Bureau of Statistics, 1996].

Figure 1. The drylands in Karak, Pakistan.

Availability of water for drinking purpose is also not certain. The water table is as low as 500 feet and it costs high to drill and pump the water out. There were some natural springs in the hills that were providing drinking water to communities but dried out in recent droughts (1992, 1998, 2002).

## Statement of the Problem

Most of the people in Karak live below poverty line. Their livelihood is dependant on rainfed subsistence agriculture and livestock. The livestock is then dependant on natural vegetation in the form of low trees, shrubs and grasses. However, due to increasing drought conditions and scarcity of rainfall, the agriculture is not more a productive activity and croplands are increasingly abandoned. To fill up this gap in livelihood, the number of livestock per household is increasing with time. This exerts great pressure on natural vegetation of the area that gets grazed more intensively and more frequently. This leads to the degradation of ecosystem and depletion of natural vegetation. The scanty rainfall condition, hot weather and sustained grazing pressure restricts recovery potential of natural vegetation. The phenomenon thus adds to desertification that compounds the problem of poverty and makes communities utterly vulnerable to the situation.

The net effect of the problems stated above is observed in the form of increase in poverty and vulnerability of the poor. The droughts leave negative effects on their capacity to survive. In the efforts to survive, they become heavily indebted, their health is badly affected and most of them migrate to urban areas.

## THE REHABILITATION OF DEGRADED ECOSYSTEMS

Keeping in view the importance of natural vegetation and the support it does provide to local livelihoods, the FFSP started the dryland management and rehabilitation program in 2003 in District Karak. The purpose was to rejuvenate the productive capacity of degraded lands so that the support these lands were providing to livelihoods previously could be restored.

The FFSP has already been working in this region since 2000 as a support to farmers in rainfed regions of NWFP to promote their farm forestry related initiatives by providing an enabling environment. The project is funded by SDC and executed by the Intercooperation-Pakistan Delegation Office, Peshawar.

Detailed surveys were conducted in the region to identify problems and suggest rehabilitation measures. The surveys were jointly conducted by technical experts from FFSP, local NGOs and community members in different villages. Based on results, the activities were suggested and thoroughly discussed in the communities for feasibility, and specifying roles and responsibilities. Following rehabilitating measures were adopted on selected sites to see the outcome.

### A. The Development of Silvopastures

Vast tracks of land were lying as wastelands because of degradation by over-grazing and water shortage. The degradation was evident in form of depletion of vegetation cover, loss of soil fertility and soil erosion. These wastelands were producing very little to support livestock (mainly sheep and goats) grazing. Within the Karak region, 5 sights were selected for applying the "Hillside Ditch" technique to recover fertility of soil, productive potential of land and hence vegetation cover [Farm Forestry Support Project, 2008, Peshawar].

Figure 2. Lay-out of hillside ditches.

The intervention was planned after carrying out detailed analysis of previous and present land-uses on the site, the vegetation and socioeconomic parameters of inhabitants. These lands were previously providing a number of goods and services to the human and livestock population in past. For example, the scattered and stunted trees were a source of fuel wood and timber for households, and bushes and grasses were browsed by animals to fill their stomach. However, these types of outputs were no more available due to depletion of vegetation.

***Suitability***

The Hillside Ditches were especially designed for these silvo-pastoral lands. Keeping in view the gentle sloping topography of the sites, interventions were so designed where machinery (tractors) could be used to reduce labor cost [See figure 2: Layout of hillside ditches].

***Description***

Continuous ditches along the contour line having plant pits at regular interval were excavated. The ditches were 66 centimeters wide and 30 centimeters deep, with excavated soil from ditch placed on downhill side making continuous ridge of 30 centimeters.

The soil excavated from plant pits was placed within the ditch on one side of plant pit to impound water. Spacing of ditches and plant pits was kept as 7 meters and 5 meters respectively [See figure 3: Design of hillside ditches]. The size of the ditches and spacing of plants and ditches was fixed keeping in view annual rainfall received in the the area.

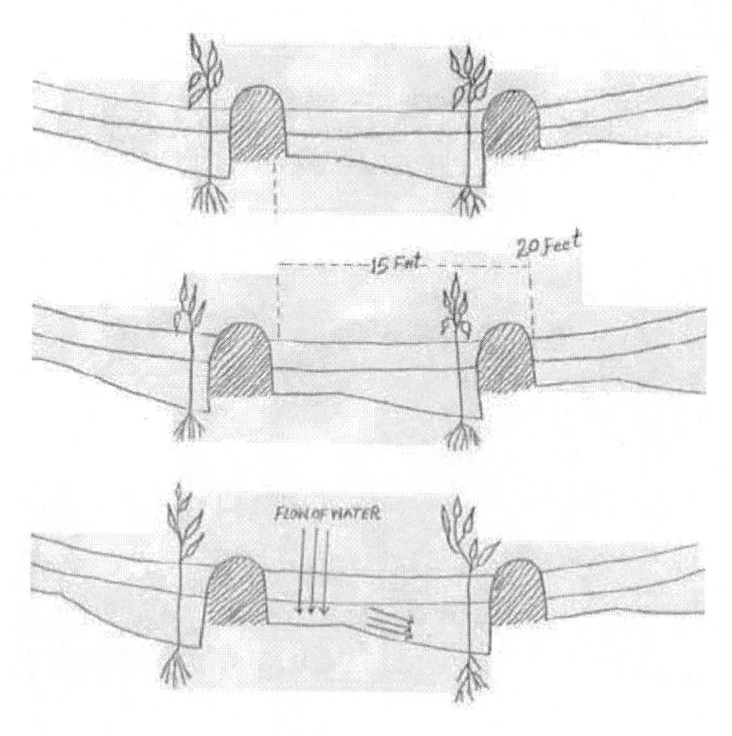


Figure 3. Design of hillside ditches.

The plant pits were planted with tree species that were fast growing and having fodder value. The inter-spaces between plants were sown with seeds of grasses and fodder shrubs to have maximum utilization of space. The species used on different sites included *Acacia albida, Dalbergia sissoo, Acacia nilotica, Melia azadarich,* and *Acacia victoria* in trees, *Dodoneae viscose* and *Acacia modesta* in shrubs, and *Sorgham almum* and *Cenchrus ciliaris* in grasses.

***Instruments Used***

The hillside ditches were excavated with the help of a tractor driven "Ditcher" specially designed for the purpose to reduce cost [See figure 5: Ditcher, specialized for ditch making]. The ditcher that was fabricated in a local workshop, consisted a modified form of mould plough commonly used by farmers in hilly areas for cultivating hard gravelly soils. The front two blades were replaced with strong chisels and the rear blades by enlarging its length to 1 meter and depth to 0.6 meter.

For reducing the cost of manually excavated plant pits within ditches, a pit excavator was designed and used [See figure 4: Pit maker for making pits in ditches]. The front blade commonly used with tractor was modified to have a top width of 1 meter and bottom width of 0.6 meter. The pit excavator was fitted with the tractor in front to excavate pits in hillside ditches. The pit excavator was so used that it produced a gently increasing slope towards the planting point from the middle of the space between two plant pits.

Figure 4. Ditcher; specialized for ditch making.

Figure 5. Pit maker for making pits in ditches.

***Functioning***

Function of the whole arrangement of ditches and pits was to take maximum benefits of rain water in arid zones by making maximum rain water available for plant growth for prolonged period. This was with the purpose to eliminate high establishment costs in arid zones involving labor in plantation and manual watering at frequent intervals.

By keeping the depths of ditch and pits within the ditch as 30 centimeters both plus the 30 centimeter high ridge, a 90 centimeter deep and 66 centimeter wide space at each planting site was available for storing run-off water coming from up-slope side. By keeping the space between ditches and plant pits as 7 meters and 5 meters respectively, rain water falling on 35 square meters land surface on uphill space was collected at each planting point.

***Effectiveness***

The maximum on-site conservation of rainwater and its utilization for plant growth was the major effect visible on these sites. In an area arid to a limit that could not support the slow-growing vegetation, produced fast growing trees and obtained profuse growth of shrubs and grasses within a few years of time.

According to the data collected from different sites, the average survival rate of trees planted was 40%, the average number of trees growing per hectare becoming 218 [See figure 6: 3 years old plants of *Acacia nilotica*]. This number was manifold more than the number of trees growing on these types of lands without treatment (i.e. 14 trees per hectare) [Pakistan Forest Institute and FFSP, 2005, Peshawar]. The height and diameter growth rate on these sites recorded was also considerably higher. Maximum diameter and height growths were recorded in case of *Acacia albida* as 20 centimeters and 6 meters respectively, followed by *Acacia nilotica* as 15 centimeters and 5 meters respectively. See table 1 below for growth data.

**Table 1. Growth data for trees, shrubs & grasses in Hillside Ditches [Dr. Khattak, J., K., 2008]**

| S/No | Parameter | Species | Data recorded |
|---|---|---|---|
| 1 | Average diameter | *Acacia albida* | 20 centimeters |
| | | *Acacia nilotica* | 15 centimeters |
| 2 | Average height | *Acacia albida* | 6 meters |
| | | *Acacia nilotica* | 5 meters |
| 3 | Av. no. of trees surviving / hectare | Overall | 218 numbers |
| 4 | Average vegetation soil cover | Overall | 45 % |

Due to retention of run-off and percolation of run-off water into soil on the site, a profuse growth of local annual and perennial grasses was recorded, in addition to the *Sorgham almum* and *Cenchrus ciliaris* that was sown during plantation activity. The average soil cover on these soils recorded was 45%, considerably high over normal cover on these degraded lands (10- 15% on the average). These grasses and shrubs were of high value as a feed for local goats and sheep. The farmers were advised not to allow animals for grazing in initial 2 years. They could, however cut grasses and stall-feed their animals during these 2 years.

The activity also contributed to the overall fertility status of soil. The laboratory analysis of soil samples taken at three sites each from treated and controlled plots showed a higher organic mater content and total nitrogen concentration in treated plots. A slight increase in phosphorus content and decrease in lime content could also be attributed to the treatment of site. No significant change in the electrical conductivity, pH and potassium content was however recorded. The treatment period of 4-5 years was too less to demonstrate any significant change in soil properties, except the content of organic mater that was recorded higher in treated plots (see table 2 below).

Figure 6. 3 years old plants of *Acacia nilotica*.

**Table 2. Soil properties in treated and controlled plots [Irfanullah, S., 2008, Peshawar]**

| S.No. | Parameters | Control plot | Treated plot |
|---|---|---|---|
| 1 | Organic mater (%) | 0.65 | 1.01 |
| 2 | Total nitrogen (%) | 0.13 | 0.20 |
| 3 | Phosphorus (mg / kg) | 3.05 | 3.14 |
| 4 | Potassium (mg / kg) | 155.13 | 114.1 |
| 5 | Electrical conductivity (d S / m) | 0.10 | 0.13 |
| 6 | Lime content (%) | 6.96 | 6.75 |
| 7 | pH (1:5) | 8.29 | 8.38 |

In addition to increase in on-site productivity and soil fertility, the activity also contributed to the re-charge of ground water in down the slope areas. According to information provided by local community, 2 wells that dried out due to prolonged drought were rejuvenated near to the activity sites.

### *Cost Analysis*

Due to use of specialized instruments and machinery, the cost was very low for applying hillside ditch technique to the development of silvo-pastures. The total cost including use of machinery, planting stock, seeds, and labor was calculated as US$ 82 per hectare (see table 3). The usual cost per hectare plantation activity by the Forest Department was Rs. 19,800 or US$ 330 that was considerably higher than the cost on using hillside ditches [Govt. of Pak., 2007]. The additional benefit of this silvo-pasture development was that it re-established the whole vegetation cover as compared to ordinary plantation work by the department that considered only trees.

**Table 3. Cost analysis of silvopastures development per 1 hectare of land**

| Activity | Cost Description | Rate (Rs) | Amount (Rs) | Amount (US$) |
|---|---|---|---|---|
| Preparation of hillside ditches with tractor and ditcher | 3 Hours | 300 | 900 | Total cost = US $ 82<br>@ PK Rs. 60/$ |
| Preparation of pits with tractor and pit blade | 2.5 Hour | 300 | 750 | |
| Planting stock | 540 Plants | 2/plant | 1,080 | |
| Planting with first watering | 540 Plants | 2/plant | 1,080 | |
| Restocking (30%), including cost of plants and planting | 160 Plants | 4/plant | 640 | |
| Grass seed | 3 Kg | 50/kg | 150 | |
| Seed of shrubs | 2 Kg | 100/Kg | 200 | |
| Sowing of shrubs/grasses seeds | 1 Labor day | 100/Day | 100 | |
| Total Cost | | | 4,900 | |

## B. Sand Dune Stabilization

Considerable portion in south-western part of Karak region is comprised of sandy desert. This is part of a greater "Thal" desert that is stretched on southern parts of NWFP and Punjab. The Thal desert is home to extreme poverty due to shortage of food and income sources. Gram, mustard and wheat are cultivated in sandy area but return very little or no yield due to prolonged droughts. The shifting sand dunes also cause hurdle to crop cultivation and infrastructure.

FFSP conducted consultation with farmers and proposed the introduction of *Saccharum spontaneum* (locally called as *Kana*), in sand dunes for the purpose of stabilizing sand dunes, acting as wind-breaks for crop lands, and contributing to household income in the form of proceeds from sale of its stalks and leaves. The *Saccharum* plant was found most suitable for sandy land as it did withstand prolonged droughts, lesser cost involved in its establishment and high return for its marketable products [See figure 7: *Saccharum spontaneum* in sand dunes].

Local communities selected 9 sites for demonstration of *Kana* [FFSP, 2008, Peshawar]. *Kana* suckers were obtained from an adjacent district at the cost of Rs. 7 per sucker and planted at a spacing of 5 meter x 3 meter in straight lines. Total cost per hectare of *Kana* establishment on sandy land including the cost of suckers and labour was Rs. 5,000 (US$ US$ 83). The average annual return from *Kana* site was Rs. 44,100 (US$ 735) that was profitably comparable with other land uses available for sand dunes, except wheat (see table 4 below).

**Table 4. Annual cost/benefit per hectare for various crops of sand dunes in Karak [Agri. Research Station, Ahmadwala, Karak, 2008]**

| Cost/Benefit | Canola (Rs) | Gram (Rs) | Mustard (Rs) | *Kana* (Rs) |
|---|---|---|---|---|
| Annual Cost | 6,052 | 9,139 | 10,003 | - |
| Annual Income | 14,795 | 53,097 | 74,055 | 44,100 |
| Net profit (Rs.) | 8,743 | 43,958 | 64,052 | 44,100 |
| Net profit (US$) | 146 | 732 | 1,067 | 735 |

The investment cost for *Kana* was only one time as this was a perennial plant. It was cut each year and sprouted again. Both the long stalks and leaves were sold in market (these were used for furniture making, as roofing material, sunscreens and making of decoration items). The outstanding characteristic of *Kana* was that its production did not depend on rainfall and even did well in prolonged droughts when all other crops failed.

Figure 7. *Saccharum spontaneum* in sand dunes.

## CHALLENGES AND COPING STRATEGIES

More than 60 percent land in Karak is not producing any agricultural crops, and hence is treated as wasteland where free and unrestricted herding and grazing of animals is practiced. Due to no or lesser productivity of economic goods, the use rights for livestock grazing are not reserved. Free, unrestricted and extensive grazing of animals is thus practiced by local communities, even by those who don't own any land and totally depend on their livestock.

The rehabilitation measures however demand care of the land and protection from grazing for initial two years to provide relief to the recovering vegetation. Due to silvopastoral practices that have become a way of life, it is difficult for land owners to abandon grazing on their land. It is due to this reason that communities usually demand for fencing the area or keeping watchmen to protect the site which enormously increase the establishment cost of the activity.

Without attending to the protection parameters, activity in some places have resulted in no conspicuous results after the planted seedlings and shrubs were completely clean washed by roaming herds of goats and sheep.

On the other hand, it is a common concept among local people that investing on silvopastures is a profit-less venture. Failures due to water shortage in past and the lack of protection from free grazing animals have further strengthened this perception. The already marginalized communities therefore find it very difficult to invest on pasture development.

The interventions in silvo-pasture development have proved significant in overcoming the water shortage and rejuvenating the vegetation for the benefit of human beings and livestock. The cost of these activities is also very low and within the bearing capacity of farmers. These facts and results need to be spread wide through extension and mobilization of communities at regional level. The mater of free livestock grazing should be dealt with at regional and not at local level. Communities should be facilitated to reach a mutual consensus for protecting sites under treatment and keeping their animals grazing in other areas. A controlled grazing system in which area is divided into blocks, keeping one block under protection on rotational basis, may also be one of the options.

## ACKNOWLEDGEMENTS

The cooperation extended by local farmers, especially those who offered their land for interventions and invested in planting cost and labor, and those other community members who cooperated in the execution of the activities, is highly adorable. The commitment and efforts of local NGOs (*Khwendo Kor* and *Yaraan* in Karak) are highly appreciable. These organizations and their role are crucially important in the sustainability of the activity on long-term basis.

Technical expertise and support provided by Dr. B.H.Shah in the initial phases of the activity was very much helpful in designing tailor-made interventions for sites in the field. The author is highly thankful to management of FFSP, the IC-Pakistan Delegation Office and the SDC in providing funding resources, logistic support and facilitation in the execution of the activities in Karak District.

## REFERENCES

Agriculture Research Station, 2004. Agriculture statistics for drylands of Karak. *Ahmadwala, District Karak.*

Civil Secretariat FATA, 2007. Development of forestry sector resources for carbon sequestration in FTA (PC-I).

Fischler, M., et al, 2006. Dryland management: A perspective for livelihood improvement in rural areas, Experiences from Pakistan. Intercooperation – Pakistan, Peshawar.

Forest Department NWFP, 2003. Schedule of rates (unit cost estimates) for Malakand Forest Circle.

Government of Pakistan, 1998. Census Report of District Karak. Provincial Census Organization, Peshawar.

Government of Pakistan, 1998. Census Report of Pakistan. Pakistan Census Organization, Islamabad.

Government of NWFP, 1996. NWFP Development Statistics. Bureau of Statistics Planning, Environment and Development Department, Peshawar.

Government of Pakistan, 2000. Pakistan Agricultural Census Report. Pakistan Census Organization, Islamabad.

Intercooperation-Pakistan, 2004. Annual Progress Report 2003, Farm Forestry Support Project, Peshawar.

Intercooperation-Pakistan, 2005. Phase I Report (2002-2004), Farm Forestry Support Project, Peshawar.

Intercooperation-Pakistan, 2008. Yearly Plan of Operation 2008, Farm Forestry Support Project, Peshawar.

Irfanullah, S., 2008. Dryland management and rehabilitation: A case study from Karak District, Pakistan. *Studies in Indian Economy.* Volume 3. 108-114.

Khattak, Jamal K., 2008. Laboratory analysis of soil samples from dryland management sites in Karak. Department of soil and environmental sciences, NWFP University of Agriculture, Peshawar.

Meteorological Department, 2006. Annual Normal Rainfall Map.

Pakistan Forest Institute, Farm Forestry Support Project, 2005. Tree Growth on Farmlands and Wastelands of Karak, Haripur and Kurram Agency.

Pakistan Meteorological Department, 1998. *Envi. Profile of Pakistan,* 1998.

Robin, P. W., et al, 2002. An ecosystem approach to drylands: Building support for new development policies, Information Policy Brief No. 1. World Resources Institute, Washington, DC.

Shah, B.H., 2006. Field manual of the role of water harvesting for dryland management in Pakistan. Intercooperation - Pakistan, Peshawar.

United Nations Convention to Combat Desertification (UNCCD), 1999. United Nations Convention to Combat Desertification in Those Countries Experiencing Serious Drought and/or Desertification, Particularly in Africa. UNCCD, France.

United Nations Environment Program (UNEP), 1997. World Atlas of Desertification, 2nd edition. Edited by N. Middleton and D. Thomas. UNEP, London.

# SECTION THREE: CONSTRUCTED WETLANDS

In: Technologies and Management for Sustainable Biosystems ISBN: 978-1-60876-104-3
Editors: J. Nair, C. Furedy, C. Hoysala et al. 

*Chapter 14*

# PERFORMANCE EVALUATION OF A FULL-SCALE CONSTRUCTED WETLAND SYSTEM PROVIDING SECONDARY AND TERTIARY TREATMENT OF MUNICIPAL WASTEWATER – AN AUSTRALIAN CASE STUDY

***Kathy Meney, Ljiljana Pantelic and Kathryn Hardcastle***
Syrinx Environmental PL, 12 Monger St, Perth, Western Australia 6000

## ABSTRACT

This chapter reports on the performance of a fully operating multi-component constructed wetland system treating the municipal wastewater of 1000pp in the main township of King Island, Australia. The system combines a pre-treatment unit with a series of vegetated surface and subsurface biofilters, with a low specific treatment area. This is the only Australian example of a full-scale wetland used for the treatment of raw municipal wastewater to tertiary level standards.

A range of parameters were analysed over an 18-month monitoring period. Results showed that the CWS was highly efficient in removal of BOD (93%), TSS (87%), TN (81%), $NH_4$-N (88%), O&G (90%), but less efficient in removal of TP (32%). The system achieved 4.7 log units reduction of faecal coliforms (>99.9%), 5.11 log units of *E.coli* (>99.9%), 3.6 log and 2 log units of *Giardia* and *Cryptosporidium*, respectively (>99.1%), 6.23 log units of coliphages (>99.9), and up to 1.5 log units of enteroviruses and adenovirus (>96%). All contaminants were reduced below limits set for non-potable water reuse without further treatment.

**Keywords:** Constructed wetlands, wastewater treatment, case study.

# 1. INTRODUCTION

Wastewater treatment and possible options for water reuse are a pressing issue worldwide, especially in small and remote communities that must rely on low cost technologies. Many coastal towns in Australia have reticulated sewage networks, but discharge effluent directly to the ocean with no treatment. There is a move towards decommissioning of direct ocean discharge both to relieve environmental and human health impacts, and in recognition of the value of reusing wastewater to offset declining rain-fed storages. An appropriate technology for the treatment of sewage in small-medium sized communities is constructed wetland systems (CWS) (Hench et al. 2003; Luederitz et al. 2001; Luederitz and Gerlach 2002). However published data on the performance of full-scale systems are still scarce, and their use for treatment for raw sewage, as opposed to polishing of secondary treated water, is still uncommon generally and particularly in Australia. The size requirement of the CWS, combined with perceived health risks are often constraints to the adoption of this technology in more densely populated areas. To overcome barriers to their uptake, data are needed at full-scale to optimise design and demonstrate proof-of-performance.

In general, CWS perform extremely well in the removal of a variety of physico-chemical pollutants such as biochemical oxygen demand (BOD), total suspended solids (TSS) and nutrients (Ayaz and Akça, 2001; Luederitz et al. 2001; Kadlec and Knight, 1996; Muasya Nyengy'a and Wishitemi, 2001). Pollutant removal is achieved through a combination of physical, chemical and biological processes such as sedimentation, filtration, complexation, microbial degradation, plant uptake etc. In addition to removing physical and chemical pollutants, there is a growing body of evidence that indicates a high efficiency of constructed wetlands in the removal of pathogens from domestic wastewater. Most microbiological studies have focused on the fate of key faecal indicator organisms (e.g. total or faecal coliform, *E.coli* and faecal streptococci) and have shown their reduction in treatment wetlands of up to 95-99% (Gersberg, 1987a; Green et al. 1997; Perkins and Hunter, 2001; Vymazal, 2005). More recent investigations have demonstrated the capacity of wetlands for advanced removal of a wide spectrum of other pathogenic microorganisms including viruses and protozoa (Hagendorf et al. 2005; Hench et al. 2003; Quinonez-Diaz et al. 2001; Sleytr et al. 2007; Thurston et al. 2001). While the performance record of CWS's is good, optimal sizing of such systems appropriate to the climate and loading patterns is still a current research focus.

Most wetland performance data reported to date are results obtained from either small experimental or pilot-scale CWS connected to existing treatment facilities, in which both wastewater flows and contaminant loads can be controlled and climatic influences ignored. In contrast, there is very limited data regarding the performance of full-scale operational CWS's treating municipal wastewater (for exceptions see Behrends et al. 2007; Hagendorf et al. 2005; Luederitz et al. 2001). In Australia, the use of CWS for wastewater treatment is not common since regulatory approvals require proof of performance at the operational scale.

This chapter reports on the performance of a CWS which has been designed to provide secondary and tertiary level treatment of sewage from approximately 1,000 residents and seasonal visitors to the Township of Currie on King Island, Australia. A CWS was considered appropriate for this township since it could be co-located downgrade of the existing sewer network, could add biodiversity and educational value to a degraded coastal dune area, including the use of local rare and restricted plant species, and could allow later reuse of water for non-potable uses. The integration of wastewater treatment with these other objectives also enabled sharing of costs between different government funding schemes and budgets (natural resource protection, infrastructure, coastal protection). In consideration of the potential risks to public health (from discharge, reuse and recreational access), emphasis was given to determining the capacity of the CWS for removal of common pathogenic microorganisms (bacteria, protozoa and viruses).

# 2. Materials and Methods

## 2.1. Site Description

Currie is the main township on King Island which is situated at the western entrance to Bass Strait midway between Victoria and mainland Tasmania in Australia (Lat. 39° 55.8' S, Long.143° 50.5' E) (figure 1). The township is a small community with a resident population of approximately 850, escalating to about 1000 in the tourist season. Prior to construction of the CWS, up to 275 kL per day of untreated wastewater was discharged to the Southern Ocean. The CWS is located in coastal dunes on the southern side of the Currie Township, near the existing ocean outfall and approximately 1.3 km from the town centre.

## 2.2. Treatment System Description and Design Parameters

The CWS combines a pre-screening unit with a horizontal flow system with both surface flow (SF) and subsurface flow (SSF) components (figure 2). There are seven successive treatment stages:

1. Pre-treatment - Balance tank fitted with an oversized screen to remove large waste and balance flows; Huber ROTAMAT® Mini Complete Plant (Minicop) screen used for separation of fine (>3 mm) non-degradable solids.
2. Primary treatment - Facultative aeration pond (Main Inlet) 1.5 m deep with a venturi aeration device used for removal of suspended solids (TSS), oil and grease (O&G) and biological oxygen demand (BOD).
3. Secondary treatment - Open water sedimentation pond with vegetated filter strips (Zone 1);
4. Tertiary treatment - SF wetland (Zone 2) - includes an open water area (Zone 2A) and a vegetated area (Zone 2B); SSF vegetated wetland (Zone 3); SF vegetated wetland (Zone 4 and Outlet)

The system currently receives an average daily influent flow of 275 $m^3$. The size and loading characteristics of each component is given in table 1. The total system area is approximately 11,500 $m^2$ and the theoretical detention time is 22-23 days in average rainfall years. The specific treatment area is 42 $m^2/m^3$/day. The entire system is gravity fed and fully lined. The SSF biofilter (Zone 3) is filled with local bluestone gravel. The system was commissioned in November 2005 and was fully operational in January 2006.

Vegetated zones are planted with local species as follows: *Schoenoplectus validus* ((Vahl) A.Love & D.Love), *Eleocharis sphacelatus* (R.Br.), *Eleocharis acuta* (R.Br.), *Juncus kraussii* (Hochst.), *Juncus procerum* (E.Mey.), *Ficinia nodosa* ((Rottb.) Goetgh., Muasya & D.A.Simpson), *Carex fascicularis* (Sol. ex Boott), *Carex appressa* (R.Br.) and *Triglochin procerum* (RBr.). The latter submergent aquatic was planted only in Zone 4.

## 2.3. Physical Analysis

Analysis of physical water parameters was done for the main wetland components excluding Zone 3 which is subsurface. Temperature and pH were measured *in situ* using an Ecoscan meter. Dissolved oxygen (DO) was measured *in situ* using a TPS (model WP82Y). All physical data were collected on a weekly or monthly basis for 18 months. Samples were taken at a minimum of three locations, including the outlet, within each zone. Data for each zone were averaged for comparison.

## 2.4. Chemical Analysis

Water samples (2000 mL) were collected at outlets of individual system components: For Z2 and Outlet, samples were taken at two locations, one at the midpoint and one at the outlet, while for other zones they were taken only at the outlet; data were averaged for comparison.

Samples of raw sewage (RS) were not collected after commissioning of the wetland, so previously recorded data were used (minimum 12 samples over a 2-year period). In all sampling events, a dipping tube was used to collect water from the entire water column (surface to bottom). Collected samples were transferred into plastic bottles and kept on ice until freighted to the laboratory the same day. All samples were analysed for BOD, TSS, $NO_3$-N, $NO_2$-N, $NH_4$-N, TP and O&G using standard methods (Clesceri et al. 1998; US EPA, 1993). Sample analysis was carried out by Ecowise Environmental Laboratory (Melbourne) which holds National Association of Testing Authorities (NATA) accreditation for the specified analyses.

## 2.5. Microbiological Analysis

Sampling for microbiological indicators was undertaken using standard procedures and were analysed in the Ecowise Environmental Laboratory (Melbourne). For faecal coliforms (FC) water samples (250 mL) were collected at outlets of individual system components, except for Z2 and Outlet where an additional sample was also collected at the midpoint. FC were sampled on 10 occasions, while sampling for other microorganisms was undertaken on two occasions. For analysis of *Eschericia coli* (EC) coliphages, *Giardia* and *Cryptosporidium*, samples were collected at the balance tank and at outlets of Z1, Z2, Z3 and Outlet, while analysis of viruses was done on samples from the system inlet and outlet only.

Each sample was taken using a dipping tube to collect a sample that was integrated through the full water column. Sample volumes collected for analysis were: 500 mL per bacteria sample, 10 L per protozoa and 10 L per virus sample. Presence and concentrations of faecal coliforms, *E. coli*, salmonella, *Giardia* and *Cryptosporidium* were determined by enumeration; of enteroviruses, adenovirus, and reovirus by presence/absence and enumeration; and of hepatitis A virus, rotavirus and norovirus by presence/absence.

### 2.6. Statistical Analysis

Data obtained from individual system components were averaged for comparison. Linear regression analysis was used to determine the relationship between bacterial removal and levels of TSS and BOD. The student t-test was used for comparison of individual zones for physical and chemical parameters.

## 3. Results and Discussion

### 3.1. Removal of Chemical Contaminants

The CWS was highly efficient in the removal of the full range of chemical contaminants (table 2), and achieved much higher unit rate removals than typical in the literature for the relatively high hydraulic loading rates and low specific treatment area (e.g. Luederitz et al. 2001). In particular, the SSF wetland, which had a HLR of 25.5 cm/day, achieved a 2.1 – 2.5 times higher removal efficiency of all parameters except O&G.

Concentrations of all pollutants in treated effluent, with the exception of TP, were significantly ($p<0.001$) lower than in the influent. The system achieved an average 31% removal of TP, initially showing high removal (>90%) presumably due to plant uptake (Zhang et al. 2007) then declining in the following months. This declining trend has been reported by others (Behrends et al. 2007; Luederitz et al. 2001).

The progressive removal of contaminants through the system (figure 3) shows the first two components (Main Inlet and Z1) removed most of the BOD, TSS and O&G. The vegetated component (Z1) was more efficient than the facultative pond, which may reflect plant-associated differences, or may simply reflect the longer hydraulic retention time (HRT) in Z1 (9.7-10 days) compared to the Main Inlet (2.2-2.4 days).

The SSF biofilter (Z3) showed a removal efficiency of 53% of influent BOD and 65% of influent TSS, which while lower than that reported by Luederitz et al. (2001), was impressive given the low detention time through this system (<2 days compared with ~10 days in the Luederitz study). The Outlet did not achieve additional removal of BOD or O&G. Since the Outlet also showed an increase in TSS compared to Z3, it is likely that the performance of this exposed zone was affected by constant wind turbulence exacerbated by only partial vegetation coverage at this stage.

In terms of TN and $NH_4$-N removal efficiency, Z1 (SF) and Z3 (SSF) were the most efficient (figure 3) and the subsurface component most efficient per unit area (table 2). While observed reductions of TN in SF and SSF wetlands are generally in good agreement with previously reported performances of CWS, $NH_4$-N removal efficiency was approximately 15% higher in the case of SSF and up to 30% higher in the case of SF than in previously reported performances of constructed wetlands (Huang et al. 2000; Kaseva et al. 2004; Kadlec and Knight, 1996).

As expected, reduction in $NH_4$-N levels was accompanied by the production of its degradation products nitrates and nitrites ($NO_x$) (table 3). These were effectively and progressively denitrified in anaerobic parts of the system, most effectively in Z3 (table 3). Whilst average DO levels are typical of aerobic conditions, the range in all system components indicate sufficiently low oxygen conditions occur both temporally and spatially to promote nitrate reduction.

Phosphorus was progressively removed through the CWS, most effectively in the Main Inlet and the SSF biofilter (Z3) (table 2). Total removal was at the lower range reported for CWS (e.g. Kadlec and Knight, 1996), possibly reflecting very low adsorption and precipitation to sand and media, still low humic development due to the age of the system and continuous rather than intermittent loading (which limits storage of organic P), and finally the slow growth of vegetation and hence P-uptake at the Outlet. Since the system has limited P-adsorption capacity, it is also likely that plant uptake efficiency, which should theoretically be higher (Lantzke et al. 1999), is constrained by N:P ratios, which decreased progressively throughout the system.

Temperature was relatively constant throughout the CWS (average value ~17.5 °C) and did not have a significant effect on overall system performance (data not shown). Levels of DO in the effluent decreased steadily in the first stage of the system (from Main Inlet to Z1), due to the consumption of $O_2$ required for the oxidation of organics. Slightly lower temperatures in later components partly contributed to the higher DO levels observed in the vegetated system components (Z2 and to a lesser extent Outlet, correlation of -0.68). Average pH values of wastewater passing through the CWS were relatively constant (7.4-7.6), with no significant differences between system components and no significant annual variations in effluent pH.

### 3.2. Removal of Microbial Contaminants

The wetland system was highly effective in FC removal with an average efficiency of >99.9% which resulted in a 4.72 log reduction of the influent FC concentration (figure 4). This is in good agreement with previously reported data (summarised in Kadlec and Knight, 1996). The performance of the SSF wetland (Z3), which had low bacterial loading and a low hydraulic residence time (<2 d) was similar to other systems with these characteristics (Hagendorf et al. 2005; Vymazal, 2005).

While generally accepted as a broad bacterial indicator, FC cannot be used as a specific indicator of human faecal bacterial contamination or as an indicator of pollution (Gersberg et al. 1987a ; Kadlec and Knight 1996). Therefore, a snap-shot study (two sampling events) was undertaken in which the fate of a broader community of enteric microorganisms in the CWS was monitored. The CWS reduced the influent concentration of EC by 5.11 log units demonstrating an overall removal efficiency of 99.99% (figure 4). This is in line with the best published results recorded for a similar multi-component wetland system (Hagendorf et al. 2005). The dynamic of EC removal in the CWS was very similar to that of FC (figure 4). *Salmonella* was not detected in influent wastewater.

*Giardia* cysts were present in the influent at an average concentration of 2,500 cysts $L^{-1}$ and were reduced by 3.6 log units resulting in an average effluent concentration of 0.65 cysts $L^{-1}$ (figure 4). *Cryptosporidium* oocysts were reduced by 2 log units, from an influent concentration of ~ 140 oocysts $L^{-1}$ to a concentration of 1.2 oocysts $L^{-1}$ in the discharge effluent (figure 4). The efficiency of CWS in removal of these pathogenic protozoa was 99.99% and 99.13% for *Giardia* and *Cryptosporidium*, respectively. Similar removal rates of protozoa have been reported by others for constructed wetlands, however these focus on secondary treated water, not raw wastewater, and are pilot scale experiments (Al-Herrawy et al. 2005; Thurston et al. 2001).

The CWS was particularly effective (removal efficiency 99.99%) in removal of FRNA coliphages from wastewater (figure 4). The influent concentration of these viral indicators of faecal contamination (~$1.7x10^6$ pfu/100 mL) was reduced by 6.23 log units at the end of the system and virus particles could not be detected in the final effluent. Although generally comparable, these removal efficiencies are somewhat lower than previously reported in other wetland systems (Gersberg et al. 1987b; Hench et al. 2003; Thurston et al. 2001) most likely due to the very low influent concentrations.

Adenovirus accounted for >96% of the total enteric viruses detected (influent concentration 735 pfu/L) while enteroviruses represented <4% of the total virus population (influent concentration 29 pfu/L) (table 4). The CWS removed both viruses with an efficiency of 96.6%, reducing their concentration by up to 1.5 log units. Norovirus (genotype I and II) was detected in wastewater entering the system but not in the final effluent. Hepatitis A virus, reoviruses and rotaviruses were not detected in influent wastewater.

Linear regression analysis showed a significant positive effect ($p<0.01$) of TSS on FC and EC removal. Pathogen removal in wetlands is the combined result of simultaneously occurring physical, chemical and biological removal processes, including attachment to suspended solids followed by sedimentation or filtration through wetland vegetation (Karim et al. 2004). In addition, the lower coliphage removal efficiencies recorded in Z2 and Z3 could, at least in part, be due to the greatly reduced TSS levels in the effluent passing through these system components. The correlation between virus removal and removal of TSS is well known (Kadlec and Knight, 1996).

## 4. CONCLUSION

This chapter focuses on the performance evaluation of a SF and SSF wetland system used to treat raw wastewater from the remote island township of Currie at King Island in Australia. Our chapter reports on all three major contaminant classes (physical, chemical, microbiological), which are key parameters for assessing environmental and public health risks associated with discharge or reuse of treated wastewater . The CWS proved to be highly efficient in the removal of the range of pollutants during an 18-month monitoring period. Nitrogen species, BOD, TSS and O&G were reduced by 81-93%. TP was the least efficiently removed at 31%. Generally pathogens were removed at >99.9% efficiency.

The SSF was more efficient per unit area for removal of all parameters than SF components (e.g. 3.8 $mg/m^3$ for TN in the SSF compared with 1.3 $mg/m^3$ for the SF). A specific treatment area of $42m^2/m^3$/day was found to be adequate to achieve the high removal efficiencies, making this design a cost-effective and space-efficient system. Further space gains could be achieved by increasing the SSF proportion, given it showed 2-2.5 times better performance than the SF components.

Data on pathogen removal are particularly useful given this is one of a few full scale systems treating raw wastewater. This chapter demonstrates the capacity of passive systems to effectively remove the suite of both indicator organisms and other more specific pathogens such as enteric viruses and *Giardia* and *Cryptosporidium*. The data show that the reduction of microorganisms in wastewater was positively correlated with the turbidity of the water entering the system components.

The treated effluent is in compliance with the national Australian standards for non-potable water recycling (NRMMC 2006). At present, the King Island Council is considering using this water for irrigation of the adjacent golf course. A longer-term monitoring program is needed to verify removal efficiencies and wetland system stability during continuous and prolonged system operation, as well as more data on the role of the system in dealing with organic compounds. Application of this system elsewhere should consider additional phosphorous removal methods, where discharge and/or reuse applications are to phosphorous-sensitive environments.

## Acknowledgements

The authors would like to acknowledge the support and assistance of Michelle Drew, formally of Syrinx Melbourne, for her early monitoring work at the Currie Wetlands, Michael Brown and the King Island Council Outdoor staff for their assistance in monitoring and good management of the Currie Wetlands.

## References

Al-Herrawy, A; Elowa, S & Morsy, E. Fate of Cryptosporidium during wastewater treatment via constructed wetland systems. *Int. J. Environ. Studies*. 2005;62 (3):293-300.

Ayaz, S Ç & Akça, L. Treatment of wastewater by natural systems. *Environ. International.* 2001;26(3):189-195.

Behrends, LL; Bailey, E; Jansen, P; Houke, L; Smith, S. Integrated constructed wetland systems: design, operation, and performance of low-cost decentralized wastewater treatment systems. *Water Sci. Technol.* 2007;55(7),155–161.

Clesceri, LS; Greenberg, AE; and Eaton, AD. Standard methods for the examination of water and wastewater. 1998. American Public Health Association.

Environmental Protection Agency USA.. Methods for chemical analysis of water and waste. Cincinnati, Ohio, USA 1983.

Gersberg, RM; Brenner, R; Lyon, SR; Elkins, BV. Survival of bacteria and viruses in municipal wastewaters applied to artificial wetlands. In: KR Reddy and WH Smith, Editors. *Aquatic Plants for Water Treatment and Resource Recovery.* Orlando, Florida: Magnolia Publishing 1987a:237-245.

Gersberg, RM; Lyon, SR; Brenner, R; Elkins, BV. Fate of viruses in artificial wetlands. *Appl. Environ. Microbiol.* 1987b;53(4):731 – 736.

Green, MB; Griffin, P; Seabridge, JK; Dhobie, D. Removal of bacteria in subsurface flow wetlands. *Water Sci. Technol.* 1997;35(5):109-116.

Hagendorf, U; Diehl, K; Feuerpfeil, I; Hummel, A; Lopez-Pila, J; Szewzyk, R. Microbiological investigations for sanitary assessment of wastewater treated in constructed wetlands. *Water Res.* 2005;39:4849–4858.

Hench, KR; Bissonnette, GK; Sexstone, AJ; Coleman, JG; Garbutt, K; Skousen, JG. Fate of physical, chemical, and microbial contaminants in domestic wastewater following treatment by small constructed wetlands. *Water Res.* 2003;37:921-927.

Huang, J, Reneau Jr, RB; Hagedorn, C. Nitrogen removal in constructed wetlands employed to treat domestic wastewater. *Water Res.* 2000;34(9):2582-2588.

Kadlec RH & Knight RL. Treatment Wetlands. New York: Lewis Publishers; 1996.

Karim, MR; Manshadi, FD; Karpiscak, MM; Gerba, CP. The persistence and removal of enteric pathogens in constructed wetlands. *Water Res.* 2004;38:1831–1837.

Kaseva, ME. Performance of a sub-surface flow constructed wetland in polishing pre-treated wastewater – a tropical case study. *Water Res.* 2004;38:681-687.

Lantzke, IR; Mitchell, DS; Heritage, AD; Sharma, KP. A model of factors controlling orthophosphate removal in planted vertical flow wetlands. *Ecol. Eng.* 1999;12:93-105.

Luederitz, V; Eckert, E; Lange-Weber, M; Lange, A; Gersberg, RM. Nutrient removal efficiency and resource economics of vertical flow and horizontal flow constructed wetlands. *Ecol. Eng.* 2001;18(2):157-172.

Luederitz, V& Gerlach, F. Phosphorus removal in different constructed wetlands. *Acta Biotechnol.* 2002;22(1-2):91-99.

Muasya Nzengy'a D & Wishitemi BEL. The performance of constructed wetlands for wastewater treatment: A case study of Splash wetland in Nairobi, Kenya. *Hydrological Processes.* 2001;15(17):3239-3247.

Natural Resource Management Ministerial Council (NRMMC), National Water Quality Management Strategy. Australian Guidelines for Water Recycling: *Managing Health and Environmental Risks.* (Phase1). 2006.

Perkins, J & Hunter, C. Removal of enteric bacteria in a surface flow constructed wetland in Yorkshire, England. *Water Res.* 2000;34(6):1941-1947.

Quinonez-Diaz, MJ; Karpiscak, MM; Ellman, ED; Gerba CP. Removal of pathogenic and indicator microorganisms by a constructed wetland receiving untreated domestic wastewater. *J. Environ. Sci. Health. A.* 2001;36(7):1311-1320.

Sleytr, K; Tietza, A; Langergraber, G; Haberla R.. Investigation of bacterial removal during the filtration process in constructed wetlands. Science of the Total Environ. Contaminants in Natural and Constructed Wetlands: Pollutant Dynamics and Control - Wetland Pollution and Control Special Issue, 2007;380(1-3):173-180

Thurston, JA; Gerba, CP; Foster, KE; Karpiscak, MM. Fate of indicator microorganisms, Giardia and Cryptosporidium in subsurface flow constructed wetlands. *Water Res.* 2001;35(6):1547-1551.

Vymazal, J. Removal of enteric bacteria in constructed treatment wetlands with emergent macrophytes: A review. *J. Environ. Sci. Health A.* 2005;40(6&7):1355-1367.

Zhang, Z; Rengel Z; Meney, K. Growth and resource allocation of Canna indica and Schoenoplectus validus as affected by interspecific competition and nutrient availability. *Hydrobiologia.* 2007;589:235–248.

In: Technologies and Management for Sustainable Biosystems ISBN: 978-1-60876-104-3
Editors: J. Nair, C. Furedy, C. Hoysala et al. 

*Chapter 15*

# EFFECT OF EXTERNAL CARBON SOURCES ON NITRATE REMOVAL IN CONSTRUCTED WETLANDS TREATING INDUSTRIAL WASTEWATER: WOODCHIPS AND ETHANOL ADDITION

***S. Domingos**[*1]**, K. Boehler**[1]**, S. Felstead**[2]**,***
***S. Dallas**[1]** and G. Ho**[1]*

[1] Environmental Technology Centre, Murdoch University, WA, Australia.
[2] CSBP Ltd., Kwinana, WA, Australia

## ABSTRACT

The present chapter assessed the effect of ethanol and woodchips addition on nitrate removal in free water surface/vertical flow wetland microcosms operated at a 6 day hydraulic retention time, one received increasing ethanol concentrations (COD varying from 58 to 336mg/L) and the other received 2140g (9.3 kg/m$^2$) of dry woodchips. After the addition of COD both system had increased percentage removal of nitrate. COD:$NO_3^-$ ratios applied here with external carbon (16:1 minimum) were higher than the experimental 7:1 reported in the literature for complete denitrification. Excess COD, however, was successfully removed. Parallel to the wetland microcosm, the COD released from woodchips was measured by placing 100g woodchips in 1L of water, COD was measured and the 1L water batch changed weekly. After 65 days the 100g of woodchips released a total of 2262mg COD. These trials were preliminary to a large scale constructed wetland receiving up to 2,000 m$^3$/day of industrial wastewater with high nitrate and low COD. Woodchips, as a low cost biological waste product, can be considered as an alternative to expensive ethanol. Alternatively, the feasibility of using high COD wastewaters from nearby industries is being assessed.

**Keywords**: Denitrification; COD; external carbon source; ethanol; woodchips.

---

* Corresponding author: Mr Domingos. Address: ETC - MURDOCH UNIVERSITY South Street, Murdoch WA 6150 Australia. Email: S.Domingos@murdoch.edu.au

## Introduction

Nitrate nitrogen is an important parameter to be measured in water and wastewater. When released to lakes, rivers and coastal areas it constitutes a main risk for eutrofication and depreciated water quality. In anaerobic conditions denitrifying bacteria reduce nitrate to nitrogen gas, this process called denitrification, can be illustrated in a simplified form by equation 1:

$$5C + 4NO_3^- + 2H2O \rightarrow 2N_2 + 4HCO_3^- + CO_2 \quad (1.0)$$

when nitrogen is present in nitrate form, nitrogen removal via denitrification is generally rapid and complete when organic matter is available; however, denitrification is affected by several parameters, predominantly by a carbon source, anaerobic conditions, pH and temperature (Crites and Tchobanoglous, 1998).

Carbon is usually indirectly measured as Chemical Oxygen Demand (COD). The theoretical stoichiometric COD/N ratio calculated by Carrera *et al.* (2003) when using ethanol as C for denitrification was 4.2 g COD : g N. However, experimentally, the COD/N ratio for complete denitrification was found to be 7.1 ± 0.8 g COD : g N. Their experiment showed that there was a loss of about 39% of the COD added which was consumed by oxidation (aerobic respiration) and not by denitrification. Gersberg *et al.* (1983, 1984) also illustrated that the addition of COD should be higher then the theoretical ethanol/nitrogen ratios required for denitrification due to losses of the carbon fraction to aerobic decomposition. When using organic matter such as plant litter or woodchips the resistance to degradation of the lignin fraction must also be observed.

This paper summarises the experimental work performed at the Environmental Technology Centre/Murdoch University (ETC) from September 2007 to January 2008. The work tested ethanol and woodchips as possible carbon sources for a future denitrifying free water surface/vertical flow (FWS/VF) constructed wetland receiving high concentrations of nitrate to be built at CSBP Ltd. CSBP Ltd is a major fertiliser and chemical manufacturer in Western Australia.

## Material and Methods

### Water Analysis

Influent and effluent samples were collected and analysed according to the following methods: ammonia was determined as ammonium nitrogen by an ion selective electrode (I.S.E) according to APHA (2005) and nitrate was determined spectrophotometrically according to APHA (1998). COD was analysed by the colorimetric determination (potassium dichromate) method with a HACH test kit.

## COD Release from Woodchips

To determine the COD ranges of woodchips in water for their potential use as carbon source the following experiments were conducted. 100.0g of dry woodchips was put into a container and filled with 1.0L of tap-water. After storage, between 7 and 15 days, the water was drained and a sample taken. The container was again filled up with 1.0L fresh tap-water. During the 9 week sampling period the trials were run in triplicates. The samples were fixed and stored in the freezer at -15°C and later analysed. A variation of this experiment was later conducted using CSBP wastewater instead of tap-water, during this second trial shorter retention times were used (4-9 days).

## Experimental Wetland System Description and Operation

The arrangement of 200L plastic drums is shown in figure 1. Approx. 10cm of medium sized (10-14mm) gravel was placed on the bottom, just enough to cover the outlet pipe to prevent clogging. A 50cm layer of beach sand (porosity 0.3) was used as the main medium for the wetland. The surface was planted with *Schoenoplectus validus*. Once planted the systems were fed with wastewater from mid October to the end of November 2007 to allow bacterial establishment.

*Vertical Flow (VF)* - Drums A and C worked as VF wetlands, shown in figure 2. They were batch loaded with 10L of wastewater from CSBP each. Every day the water was drained completely and stored in a separate tank and the wetland was filled up with new wastewater. The intention was to convert nearly all ammonia to nitrate to get higher nitrate concentrations compared to the raw wastewater. Due to evaporation, plant uptake and evapotranspiration the effluent volume was 30% lower than the influent. The effluent from drums A and C is hereafter named A/C blend.

Free water surface/vertical flow (FWS/VF) –ethanol and woodchips addition

Drums B, D, E and F were operated as FWS/VF systems (figure 2). Batch loaded with 10L, every 3rd day with A/C blend. The retention time was therefore 6 days, with the water remaining 3 days in the surface and 3 days subsurface. In drums E and F ethanol was added to the influent in different volumes as a carbon source for 5 weeks. Drum D received 2140 g of woodchips as an alternative carbon source. Drum B was found to be faulty and therefore was not used.

Wastewater was pumped from the containment pond at CSBP Ltd, Kwinana into 200L drums, transported and stored at ETC with an average storage time of 14 days. All experimental wetland cells were located at the ETC- Murdoch University.

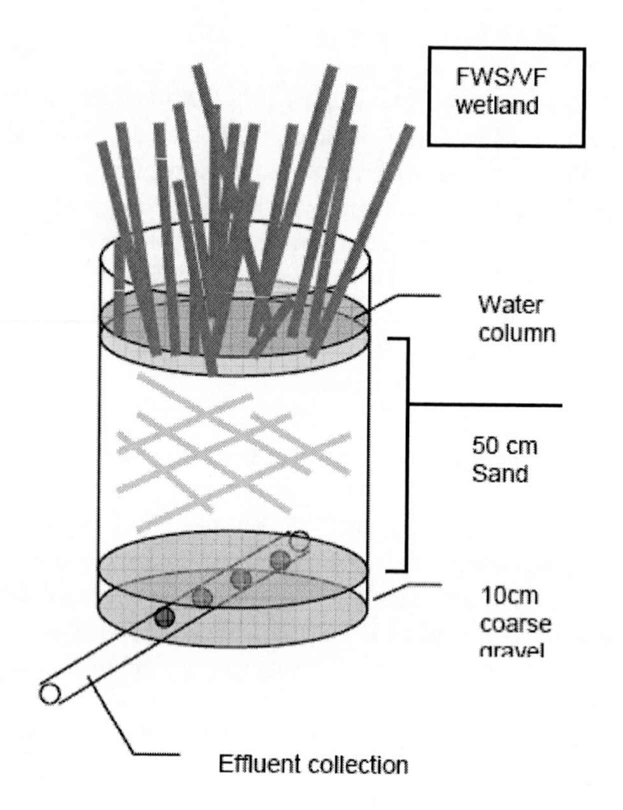


Figure 1. Wetland layout.

Figure 2. Experimental setup. A,C – VF. D – FWS/VF (woodchips). E,F - FWS/VF (ethanol); B – faulty, not used.

# RESULTS AND DISCUSSION

## COD Release from Woodchips in Water

After sitting in the water for a week the woodchips turned the water brown. As the water was changed weekly and the woodchips reused this effect became less intense. The results of the COD analysis showed the same trend. With every reuse of the woodchips the level of COD released into the water was lower, starting at 958mg/L (tapwater) and 858 (CSBP wastewater), decreasing very fast over the next few days, and then slower after 40 days of use.

For the tap water experiment, after 50 days the release of COD decreased to levels of 70 - 90 mg/L/week (figure 3). For the CSBP wastewater experiment, after 20 days, the COD released dropped to 205 mg/L. In figure 4 (tap water) the released COD was summed over the time of storage, which makes it possible to calculate the required amount of woodchips for a specific COD level over a certain time. At 65 days a plateau is evident at the cumulative graph (figure 3) with no significant increase of COD taking place. This may be the time to replace the woodchips. This result however should be used just as guidance as COD release will vary depending on the type, coarseness, and age of the woodchips.

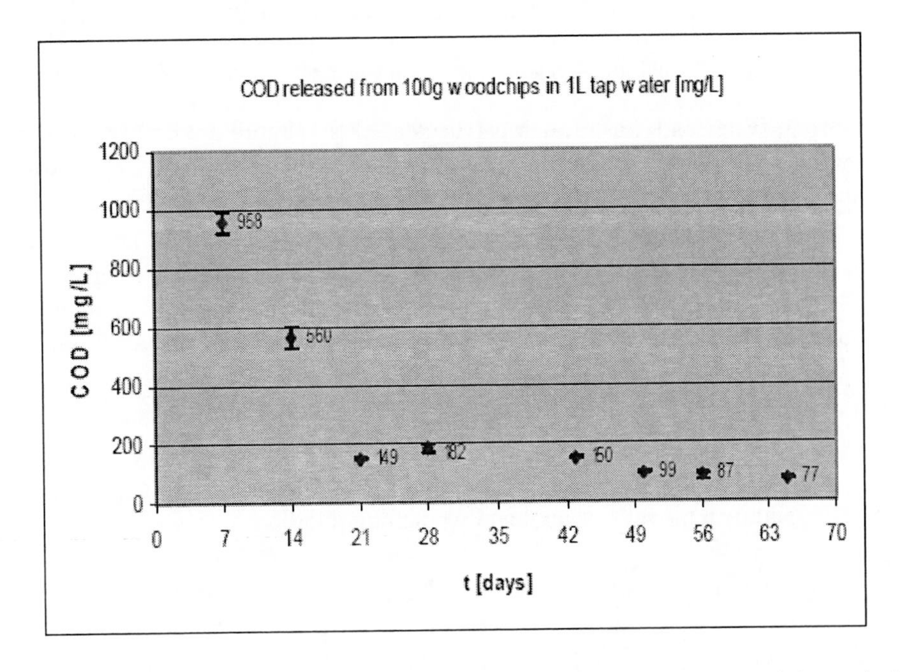


Figure 3. Average COD released from 100g woodchips in 1L tap water over one week intervals. Water exchanged weekly.

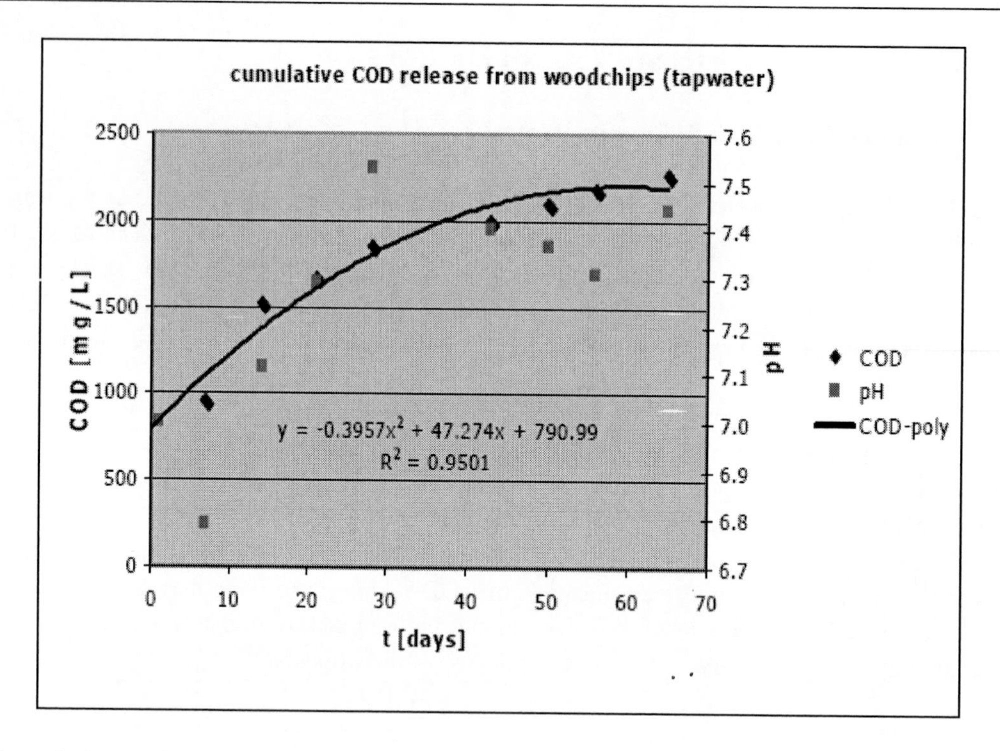


Figure 4. Accumulated COD of 100g woodchips in 1L tap water and pH variation.

For both tap water and wastewater pH decreased to a more acidic level and returned towards neutral again with each sequential reusing of the woodchips. For the wastewater experiment the initial COD of CSBP wastewater (as pumped from the containment pond) was analysed. In order to obtain the COD released from the woodchips, the initial wastewater COD was subtracted from the COD measured after the experiment. The average COD values from 3 CSBP samples were 59.8, 43.8, 49.0 mg/L. Considering the rule of thumb COD:N ratio of 5:1 this would be enough to fully denitrify only up to 11 mg/L nitrate. The experiment showed that woodchips can be used as a slow release COD substrate. A great proportion of COD is released in the first two weeks and then slowly released over the next two months.

FWS/VF – Nitrate removal - ethanol and woodchips addition

The performance of the wetlands dosed with ethanol and woodchips are presented in table 1. Due to unexpected nitrate removal by the VF system (low nitrate in the A/C blend), inflow nitrate concentrations were low for the FWS/VF cells before the addition of carbon. Once this problem became apparent, the addition of potassium nitrate into the A/C blend helped increasing nitrate levels in the inflow, alternatively, the use of raw wastewater from CSBP spiked with potassium nitrate also helped.

**Table 1. Performance of FWS/VF in removing nitrate when dosed with ethanol or woodchips. Inflow and outflow values shown are means ± standard deviation**

| | Cells E + F | | | Cell D | |
|---|---|---|---|---|---|
| | Before ethanol | Ethanol (cycle 4-8) | Ethanol (cycle 9-12) | Before woodchips | Woodchips |
| Inflow $NO_3$-N[mg/L] | 2.0±0.4 | 10.9±1.1 | 20.9±2.6 | 2.0±0.4 | 14.8±2.8 |
| Outflow$NO_3$-N [mg/L] | 1.1±0.2 | 1.9±0.4 | 2.0±0.2 | 1.4±0.17 | 2.3±0.3 |
| Removal [%] | 46 | 82 | 90 | 31 | 84 |
| Inflow COD [mg/L] | - | 247 | 336 | - | 1800-400 * |
| COD:$NO_3$-N | - | 22:1 | 16:1 | - | ≥27:1 * |

* Estimated based on the COD release, woodchip experiment.

*Ethanol-* Before ethanol addition nitrate removal was in the order of 46%. After adding ethanol to the system nitrate removal steadily increased reaching 90% removal by the end of the experiment. Because of improper mixing of the ethanol with the wastewater and erroneous sampling method used during the first cycles (1-3) the results obtained from this period are biased and must not be considered. For cycles 4-8 and 9-12 the mixing and sampling methods were corrected. During cycles 4-8 the average inflow COD:$NO_3^-$ ratio was 22:1, this resulted in approximately 82% nitrate removal. The following cycles (9 -12) had a lower COD:$NO_3^-$ ratio of 16:1, however, nitrate removal increased to 90%. This is contrary to what was expected as a higher COD: $NO_3^-$ ratio generally results in a higher nitrate removal. In terms of outflow nitrate concentration there was no difference between the two sets of trials. A few ideas arose from these results:

- COD: $NO_3^-$ ratios higher than 7:1 (literature) did not affect denitrification;
- outflow nitrate concentrations of 1 to 2 mg/L could be expected as a background level for FWS/VF wetlands, even when the system is performing best;
- System maturation may also have contributed to improved treatment performance.

Nitrate influent and effluent values can be seen in figure 5. Increased inflow and outflow nitrate concentrations towards the end of the experiment are noticeable.

*Woodchips-* Wetland cell D had on average 31% nitrate removal before the addition of COD in the form of woodchips. The addition of 2140g of woodchips increased removal to 84%. In terms of mass/area the application of woodchips was 9.3 kg/m$^2$. The influent COD was measured, but the amount of COD released by the woodchips within the wetland could just be estimated based on the previous experiment. When compared to the woodchip experiment mentioned earlier the quantity applied here (214g/L – from 2140g and 10L batches) is twice as much as the one used in the experiment (100g/L). The experiment showed that after 23 days under water and 6 day retention time the COD released to the wastewater was 205 mg/L. We could expect approximately twice as much COD being released in the water from each batch at cell D (~400mg/L after 23 days). It has been estimated that from the day woodchips were placed in cell D, COD:$NO_3^-$ ratio was maintained at a minimum of 27:1.

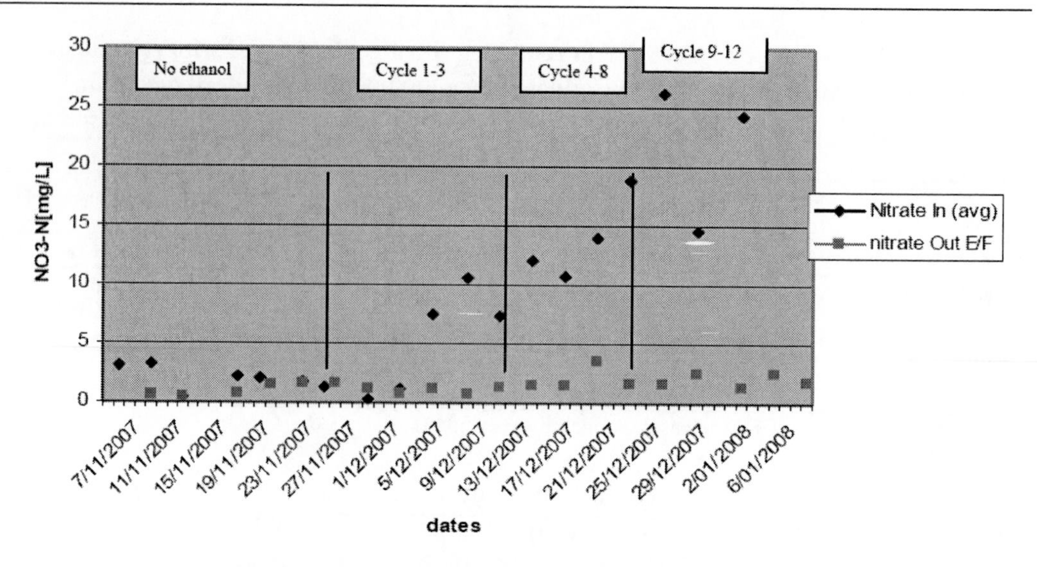


Figure 5. Nitrate removal in FWS/VF cells E/F where ethanol was added.

Inflow and outflow nitrate concentrations for wetland cell D can be seen in figure 6, influent nitrate concentrations were raised concomitantly when woodchips were introduced, outflow nitrate levels however remained low, although there was more COD available outflow nitrate could not be lowered below 1.4 mg/L. Nitrate removal increased after the addition of woodchips and ethanol into the systems. In both cases COD: $NO_3^-$ ratios were maintained at levels higher than those previously reported in the literature. In terms of final effluent quality the addition of an external carbon source did not play a role in decreasing nitrate concentrations beyond the 1 and 2 mg/L values. These effluent concentrations were also achieved without carbon, but in the period before carbon was added nitrate influent values were always low and never above 8 mg/L.

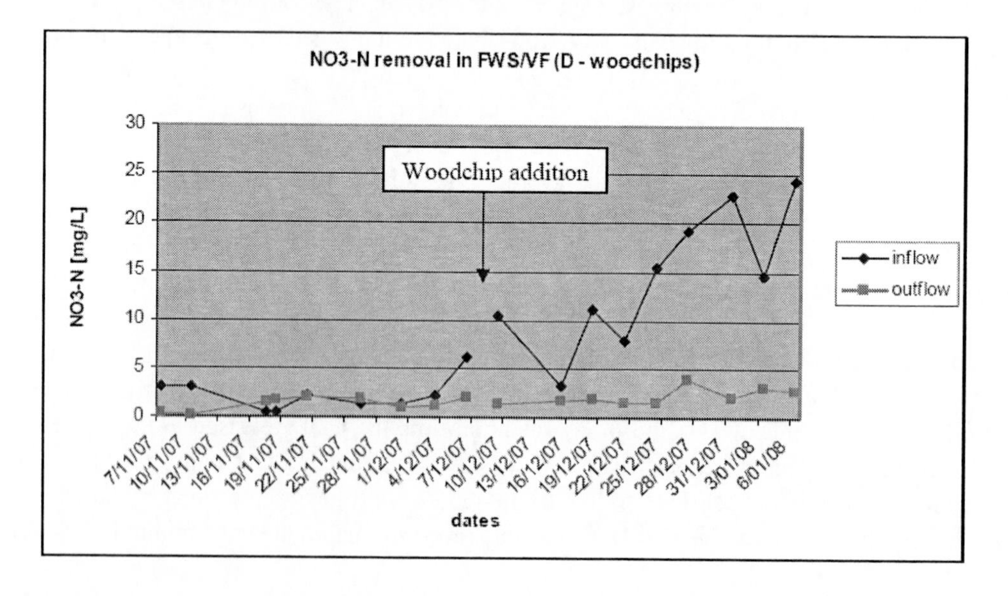


Figure 6. Nitrate removal in FWS/VF where woodchips were added.

The consumption of carbon in cells E+F brought COD values closer to the COD of CSBP wastewater (as pumped from containment pond) (figure 7). The same pattern was observed for cell D, the effluent COD after woodchip addition was on average 76.6mg/L (±11) while the influent COD of CSBP wastewater for the same period was 58.7mg/L (±17). The experiments showed that COD was successfully removed within the FWS/VF wetlands, nonetheless, final effluent COD from the future denitrifying wetland cells is an important parameter to be measured as it should not vary much from the actual COD values of the effluent being discharged, otherwise, COD discharge regulation/license will need to be reviewed. In many wastewater treatment systems the rule of thumb for the COD:N ratio is 5:1 at the denitrifying stage. Table 2 shows different carbon sources used in various denitrification studies.

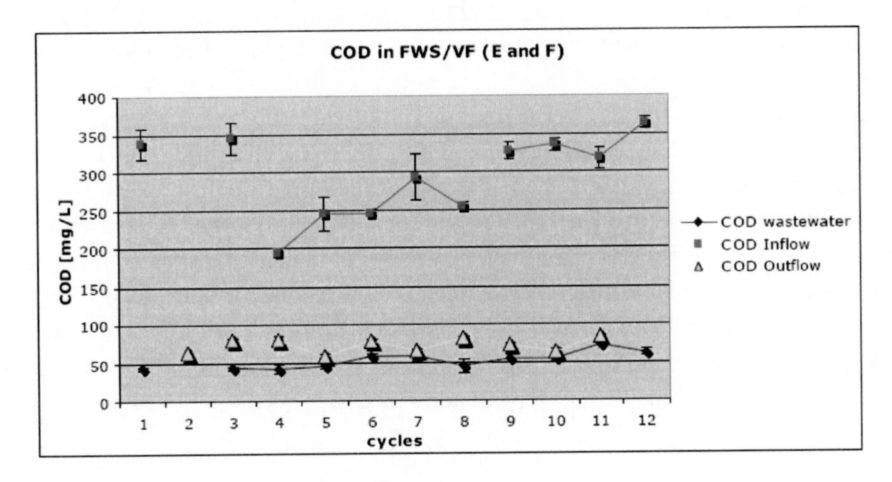


Figure 7. COD in FWS/VF cells E/F dosed with ethanol.

**Table 2. Different carbon sources, C:N ratios and nitrate removal reported in various studies**

| Reference | Carbon source | Ratios | $NO_3^-$ - N influent concentration | $NO_3^-$ - N removal |
|---|---|---|---|---|
| Davidsson and Stahl (2000) | Organic Matter (organic soils) Glucose | C : N = 6.6 : 1 | 200μM $KNO_3$ | ≤73% |
| Ingersoll and Baker (1998) | Organic Matter (chopped cattails) | C:N= 5:1 (C : NO3-N) | 30mg/L | >80% |
| Constantin and Fick (1997) | Ethanol Acetic acid | C:N = 1.38 : 1 1.46 : 1 (mol : mol) | 7.1g /L | - |
| Carrera et al. (2003) | Ethanol | COD: N mass ratio 4.2:1 (stoichiometric) 7.1 ± 0.8 : 1 (experimental) | - | - |
| McAdam and Judd (2007) | Ethanol | C: N mass ratio < 1.52 : 1 | 14.7mg/L | 92% |
| Gabaldon et al. (2007) | Methanol | COD:NO3-N mass ratio 3.31:1 | 140-210 mg/L | >90% |
| Lin et al. (2002) | Macrophytes and Fructose | COD:NO3-N mass ratio ≤6.2:1 | 21 -47mg/L | >90% |

## Conclusion and Future Studies

Although preliminary this research showed that:

Nitrate removal increased when ethanol and woodchips were added to the wetlands.

Both ethanol and woodchips are suitable carbon sources for enhancing denitrification in FWS/VF wetlands.

Woodchips (100g/L) released COD in sufficient quantity to support denitrification for at least two months in a FWS/VF system.

Excess COD was successfully removed by the FWS/VF wetlands.

The COD:$NO_3^-$-N ratio of 7:1 suggested by the literature should be followed for the future carbon dosing of theFWS/VF wetland cell at CSBP.

Even at the best performance final nitrate concentrations of 1.0 – 3.0mg/L can be expected for future FWS/VF wetlands at CSBP (HRT = 6 days) considering a 7:1 minimum COD:$NO_3^-$-N ratio.

Because the control (without external carbon) and treatment (with carbon) conditions did not occur simultaneously but consecutively, under different environmental conditions and subject to much lower influent nitrate concentrations when the carbon source was not present, the effect of the carbon source on the performance of the systems is just an indication. Comparisons between the periods prior to carbon addition and after carbon addition should be made only very carefully due to this experimental limitation. Further studies on low cost carbon sources for denitrification such as wastewater from a local soft-drink industry are being conducted under an improved experimental design where the control and treatment wetlands run in parallel subject to the same conditions and influent nitrate concentrations.

## Acknowledgements

This work would not have been possible without the contribution Mark Germain (CSBP Ltd.) and the team at CSBP water laboratory for the water analysis.

## References

APHA. (1998). Standard Methods for the Examination of Water and Wastewater; 20th Edition; American Public Health Association, Washington, D.C.

APHA. (2005). Standard Methods for the Examination of Water and Wastewater; 21st Edition; American Public Health Association, Washington, D.C.

Carrera, J., Vicent, T. and Lafuente, J., (2003). Effect of influent COD/N ratio on biological nitrogen removal (BNR) from high-strength ammonium industrial wastewater. *Process Biochemistry. 39*. 2035-2041

Constantin, H. and Fick, M. Vandoervre-lès-Nancy. (1997). Influences of c-sources on the denitrification rate of a high-nitrate concentrated industrial wastewater. *Water Research*. 31(3). 583-589

Crites, R., Tchobanoglous, G. (1998). Small and decentralized wastewater management systems. McGraw-Hill College, Blacklick, Ohio, U.S.A.

Davidsson, T.E. and Stahl, M. (2000). The influence of organic carbon on nitrogen transformation in five wetland soils. *Soil Sci. Soc. Am. J.* 64.1129-1136

Gabaldón, C., Izquierdo, M., Martínez-Soria, V., Marzal, P., Penya-roja, J.-M. and Alvarez-Hornos, F.J. (2007). Biological nitrate removal from wastewater of a metal-finishing industry. *Journal of Hazardous Materials.* 148. 485-490

Gersberg, R.M., Elkins, B.V. and Goldman.C.R. (1983). Nitrogen removal in artificial wetlands..*Water Resource.* 17(9): 1009-1014

Gersberg, R.M., Elkins, B.V. and Goldman,C.R. (1984). Use of artificial wetlands to remove nitrogen from wastewater. *J. Water Pollut. Control Fed.* 56(2).152-156

Ingersoll, T. L. and Baker, L.A., (1998). Nitrate removal in wetland microcosms. *Water Research.* 32(3). 677-684

Lin, Y.-F., Jing, S.-R., Wang, T.-W. and Lee, D.-Y. (2002). Effects of macrophytes and external carbon sources on nitrate removal from groundwater in constructed wetlands. *Environmental Pollution.* 119. 413-420.

McAdam, E.J. and Judd, S.J. (2007). Denitrification from drinking water using a membrane bioreactor: Chemical and biochemical feasibility. *Water Research.* 41(18). *4242-4250.*

# SECTION FOUR: SOLID WASTE MANAGEMENT

In: Technologies and Management for Sustainable Biosystems ISBN: 978-1-60876-104-3
Editors: J. Nair, C. Furedy, C. Hoysala et al. 

*Chapter 16*

# INTEGRATED BIOLOGICAL SYSTEM FOR POULTRY WASTE MANAGEMENT

***Harrie Hofstede****

SPARTEL PTY LTD. Environmental Solutions for a Sustainable World;
PO Box 1097 West Leederville 6901 Western Australia, Western Australia

## ABSTRACT

This chapter outlines the results of a demonstration of an innovative and sustainable approach to resource recovery of chicken litter and mortalities; the project aimed to produce a processed poultry litter product 'on farm' that is compliant with the Western Australian Health (Poultry Manure) Regulations 2001 (90% fly reduction), maintains high nitrogen levels, is free of pathogens and has low odour.

The project aimed to demonstrate the following objectives:

- Demonstrate the viability for on-farm poultry litter processing using the FABCOM® technology;
- Demonstrate the reduction in fly breeding potential of processed litter and compliance with Health Regulations;
- Demonstrate the improved nitrogen content of the compost; and
- Determine the minimum processing time in the FABCOM® system that is required to reduce fly breeding in mixed poultry litter to within compliant levels.
- Impact on phosphate mobility
- Impact on odour emission of the product
- Pathogens level in end product

The following technical issues associated with the use of poultry litter have been addressed and solved as part of the project:

**Health Regulations:** The processed litter showed 90 - 98 % fly reduction, i.e. 100% Compliance with Health Regulations.

* h.hofstede@spartel.com.au; www.spartel.com.au; ph +61 8 9200 6295; fax +61 8 9200 629; Mob: +61 414 37 66 99

**Odour Emission:** > 80% odour emission reduction in treated poultry litter compared to raw poultry litter.

**Pathogen Reduction:** Pathogen levels were reduced to non detectable and/or compliant with the compost standard (AS 4454).

**Nitrogen Conservation**: Nitrogen retention in the litter was improved significantly compared to conventional litter processing.

**Phosphate Mobility:** Phosphate leachability was reduced by 90% during the process resulting in lower Phosphate leaching potential

**Keywords:** Poultry litter, Mortality, Odourless, Horticulture, Sustainability, Health Regulations.

## 1. Introduction

In 2001 the Health (Poultry Manure) Regulations 2001 were promulgated by the Western Australian State Government restricting the direct application of raw chicken manure to agricultural land, notably horticulture in the interest of public health. The ban was driven by a desire to reduce the incidence of flies, particularly stable flies in areas where poultry litter was used on farmland.

It had been found that during summer where raw poultry litter had been applied followed by irrigation up to one million flies were bred per acre (ref). This resulted in significant nuisance and possible increased pathogen infections rates to the regional population.

This has resulted in a need to change poultry litter management which previously was being applied to land as a low cost and effective fertiliser. This project was initiated to provide a viable and more sustainable poultry management system that will provide benefits to all stakeholders, notably chicken and vegetable growers, but also Local and State Government environmental and planning agencies.

The WA poultry industry generates 108,000 tonnes of poultry litter per annum.

The benefits associated with utilizing poultry litter as an organic fertilizer include;

- Low cost nutrient source,
- Reduced reliance on irrigation,
- Replacement of chemical fertilizer inputs,
- Reduced nutrient leaching and enrichment of ground and surface water,
- Improved sustainability of horticultural production.

The project was initiated by the author and was funded by vegetable growers, WA State Government, Spartel P/L and the Federal Government through Horticulture Australia. The project further received a grant from the 'Federal National Landcare Program' to facilitate implementation of the concept into horticulture to promote sustainable land use practices in horticulture as well as providing farmers with a tool to reduce irrigation water requirements.

## ON FARM LITTER PROCESSING

This project has demonstrated an on-farm composting technology that was developed as a means of making the use of poultry litter based organic fertilisers sustainable in horticulture. Figure 1 demonstrates the role of this technology as a way of ensuring that the nutrients and organic matter stored in chicken litter are not wasted through costly and inefficient alternative disposal methods. The on farm system approach was preferred over centralised processing due to the following benefits:

- Dispersed truck traffic avoiding excessive concentrated traffic for which centralised processing plants are notorious
- Reduced transport requirements since centralised processing requires double transport from litter source (poultry farms) to the processing plant and from the plant to the product user (vegetable farms)
- Ability for vegetable farmers to also process vegetable waste in the plant to recover organic matter and nutrients, and also sanitise this organic waste from phyto-pathogens

The on farm system can also be applied on poultry farms where other benefits apply:

- Reduced litter volume to transport of site (30%)
- Sanitation of poultry litter prior to being released into the community
- Ability to process and sanitise poultry mortalities on farm as well as a management option for on farm bird flu mortalities
- Supplementary income for poultry farmers
- Improved environmental and social performance of poultry farms due to reduced odour emission

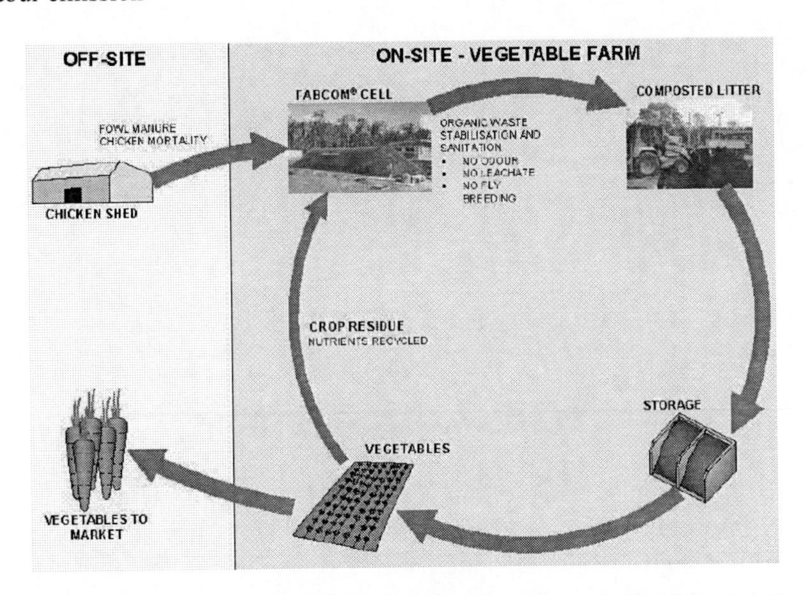


Figure 1. On-Farm processing of poultry litter and the Organic Matter and Nutrient Flow in the system.

## FABCOM TECHNOLOGY

The FABCOM® System applied in this demonstration is a globally patented manure processing technology developed in Western Australia.

The FABCOM system consist of the following components:

1. FABCOM® Aerobic Cells; Stage I
2. FABPro® process control software
3. Aerobic Curing Cells Stage II
4. integrated Biofilter

The project has proven to deliver a cost effective and environmentally acceptable means of processing organic waste in Western Australia and as such is making a substantial contribution to the development of best practice waste management in horticulture and chicken farming. The system has proven to be versatile in that is able to turn a range of organic waste materials into high quality compost products. Several alternative uses for the technology are also being explored as a result of advances made during the course of this project.

Figure 2. Aerobic Cell of the FABCOM System in operation on a vegetable farm.

## 2. MATERIAL AND METHODS

Following the commissioning of the FABCOM® plant a series of trials of litter processing was performed. This involved processing chicken litter to determine effective operating mixtures and establish optimum process conditions. Poultry litter in Western Australia generally consist of 30% sawdust and 70% poultry manure on completion of the broiler growing cycle.

**Table 1. Overview of measured parameters during the project**

| | Parameter | Performed by |
|---|---|---|
| Fly breeding | Number of viable flies breeding in the product | WA Dept Agriculture |
| Odour | Odour concentration and intensity | The Odour Unit |
| Product analysis | Refer to table | WA Chemistry Centre (NATA) |

# 3. RESULTS AND DISCUSSION

Samples were taken from each batch on a weekly basis. Once collected, these samples were sent for testing to independent laboratories for a suite of biological, physical, and chemical tests. Sydney Environmental and Soils Laboratories determined the quality of the product in terms of its suitability for use in horticulture and against the criteria for the Australian Compost Standard. The Department of Agriculture assessed the fly breeding potential of the composted litter in accordance to the "Guidelines for registering a processed poultry manure product or a procedure for using raw poultry manure" (ISSN 1326-4168). The results of these tests are presented in the following sections.

## 3.1. Process Quality Monitoring

### *3.1.1. Process Control*

FABPro® is the process control software for the FABCOM® hybrid organic waste processing system. FABPro® enables user-friendly optimisation of the process. The FABPro® software uses temperature and oxygen concentration inputs to maximise the efficiency of the process via the novel aeration system. The parameters are individually measured and monitored and the FABPro® process control software program integrates the measured data and information and makes decisions to maintain optimum process conditions in each individual Aerobic Cell® .

The process temperature was maintained between 55 – 60°C for the 4 week duration of the process after which the temperature dropped to ambient levels. The oxygen levels were maintained at above 18% to ensure high rate of bio oxidation. The fluctuating turquoise line represents the ambient temperature which is subject to varying weather and day night conditions.

The graph illustrates the accurate constant process temperature control within the system compared to ambient temperature pattern.

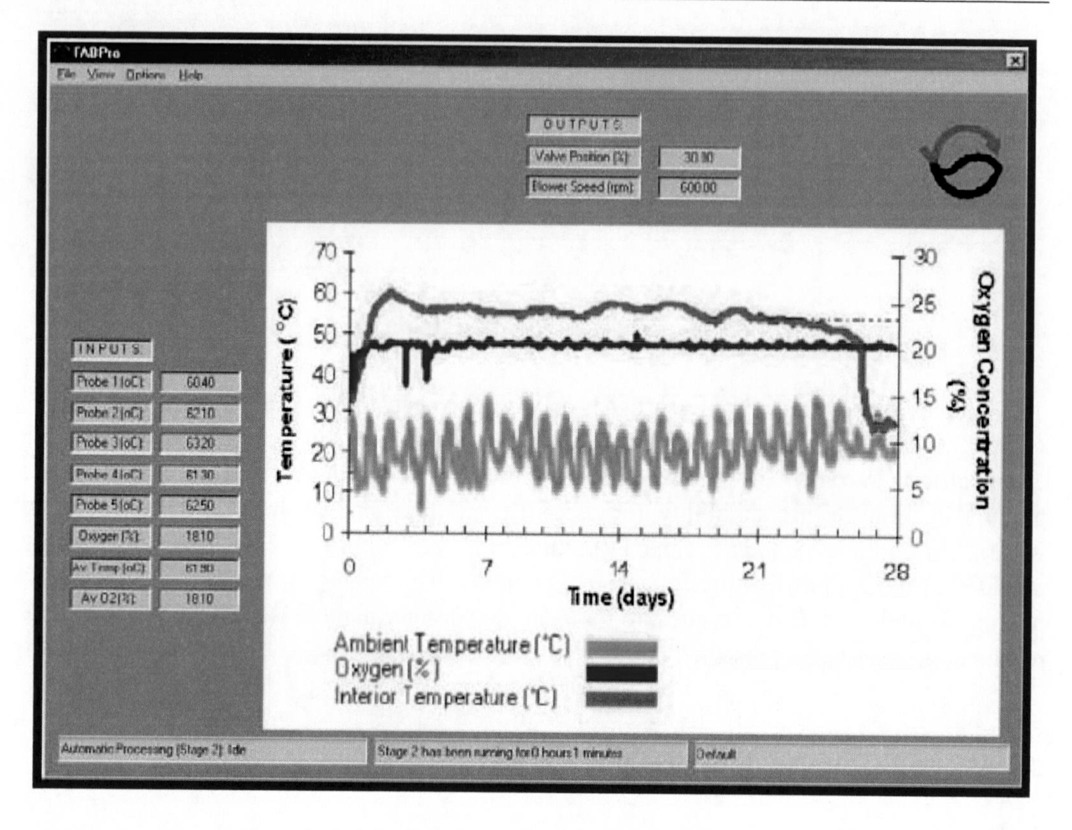


Figure 3. Process control monitoring showing oxygen and temperature profile of the complete process.

## 3.2. Fly Breeding Potential Reduction

The main obstacle preventing the application of chicken litter on farms is the ban associated with the litter's fly breeding potential. This section outlines the results the project has had in terms of reducing the fly breeding potential of the litter processed on the Farm site.

As illustrated below, the Agriculture Department of Western Australia tested the fly breeding potential of random samples taken from the processed litter. The Agriculture Department was supplied with composted litter samples from week one through to eight, as well as raw poultry litter samples which were used as the control. The samples were put into trays and set up on two farm sites, to emulate normal vegetable growing conditions.

After one week the trays were removed from the field, covered, and stored in a controlled environment fly emergence chamber at the Agriculture Department. Three weeks later, the samples were removed from the chamber and the number of emergent flies counted. This procedure was performed for samples taken from weeks three through to eight and was duplicated for each batch.

Results of fly breeding potential of composted samples are shown in table 1. The WA Health Department requires a 90 % reduction relative to raw chicken litter. After the initial calibration performed after Batch 1, fly breeding control in excess of the 90 % threshold was quickly achieved in each of the following batches. This is shown in table 2 where fly breeding reduction of up to 98 % has been achieved in three weeks.

Figure 4. Field incubation of testing trays containing poultry litter samples with varying processing times as well as raw litter.

**Table 2. Reduction in fly breeding (%) in the processed litter compared to raw poultry litter (control) based on fly numbers per kg of poultry litter**

| | Processing time (weeks) | | | | | |
|---|---|---|---|---|---|---|
| Sample | 3 | 4 | 5 | 6 | 7 | 8 |
| Test Batch 1 | 92 | 95 | 90 | 98 | 85 | 97 |
| Test Batch 2 | 99 | 94 | 97 | N/a | 77 | N/a |
| Test Batch 3 | 98 | 100 | 96 | 100 | 100 | 93 |
| Average fly Reduction % | 96 | 96 | 94 | 99 | 87 | 95 |

These results indicate that compliance with Health Department requirements has been achieved and signifies that the use of compost processed by the FABCOM® system can be approved for application to land in horticulture. The Health Department subsequently approved the process for converting poultry litter into compost without significant fly breeding potential.

## 3.3. Product Quality – Chemical Analysis

Random samples of the product were sent to NATA accredited laboratory for testing, the results of which have been summarised in table 3.

**Table 3. Analytical results**

| Parameter | Unit | FABFERT® Organic Soil Fertiliser® | Raw poultry litter | % change |
|---|---|---|---|---|
| Total Nitrogen (N) | % | 4.2 | 4.2 | - |
| Soluble Ammonia N-NH4 | mg/L | 1390 | 600 | + 232% |
| Soluble Nitrate N-NO3 | mg/L | 5 | 0 | |
| Total Soluble N | mg/L | 1400 | 600 | |
| Total Phosphorous | % | 2 | 2 | - |
| Phosphorous - soluble | mg/L | 15.4 | 135 | -90% |
| Total CaCO3 Equivalent | % | 5 | - | |
| Humic & Fulvic Acids | % | 25.5% | - | |
| Total Organic Carbon | % | 35 | - | |
| Boron | mg/kg | 20 | - | |
| Potassium | % | 2 | - | |
| Sodium | % | 0.6 | - | |
| Calcium | % | 3.65 | - | |
| Magnesium | % | 0.5 | - | |
| Copper | mg/kg | 69 | - | |
| Zn | mg/kg | 340 | - | |
| Selenium | mg/kg | 2.1 | - | |
| Electrical conductivity | mS/cm | 13.2 | 9 | +50% |
| Moisture Content | % | 25 | 25 | - |
| Organic Matter Content | % dm | 85 | 70 | +20%# |
| pH | pH units | 7.5 | 8.1 | -15% |

# It is acknowledged that this figure requires further research for verification.

### *3.3.1. Nitrogen*

High nitrogen retention is a key factor in measuring the value of an organic fertiliser produced from chicken litter as this is one of the most important nutrients for plant growth. Alternative composting methods tend to loose significant amounts of nitrogen due to ammonia volatilization during treatment making them less valuable to farmers. In this project, the nitrogen levels were maintained as high as that found in the raw chicken litter (~4%) this is because the integrated process control system associated parameters were able to prevent the loss of nitrogen during the stabilising phase.

#### *3.3.2. Phosphorous*

Phosphate solubility is a major barrier to litter use on productive land due to the high eutrofication risk potential of soluble phosphorus. The processed product showed a 90% reduction in soluble phosphate. Reduced soluble phosphorus levels signify reduced potential for groundwater pollution. This is a good result for horticulture as groundwater contamination with phosphates is a major environmental problem.

### 3.4. Pathogen Destruction

Effective control of pathogens in recycled organics applied to land is critical as a means of ensuring public health and safety issues and subsequent long-term sustainable land application (Wilkinson, 2003). Pathogens present in manures applied to land have been found to have contaminated plants intended for human consumption in several instances (Wilkinson, 2003). One particularly high-profile instance a recent series of food poisoning outbreaks in the UK were traced back to the use of manures on vegetables. To prevent the risk of contamination in primary production, compost produced from manures needs to be sterilised to eliminate the risk of infection to consumers.

An integral aspect of the FABCOM® system is its ability to sanitise the processed material to the point where the product complies with the most stringent Australian standards. This was confirmed by independent pathogen lab tests, the results of which are shown in table 3.

**Table 3. low detection levels indicated destruction of pathogens in treated chicken litter, compliant with AS 4454**

| | Sample | Batch 1 | Batch 2 | Batch 3 | Batch 4 |
|---|---|---|---|---|---|
| Thermotolerant Coliforms | MPN per gram | < 3 | < 3 | < 3 | < 3 |
| E.Coli | MPN per gram | < 3 | < 3 | < 3 | < 3 |
| Salmonella | in 25 grams | not detected | | | |
| Listeria | in 25 grams | not detected | | | |

These test results indicated that all pathogens in the litter were destroyed to a level that rendered them undetectable by the independent laboratory testing used to certify that the compost is pathogen free. Especially noteworthy is the destruction of thermo-tolerant coliforms. These bacteria are particularly resilient to traditional pasteurisation measures and their destruction demonstrates the effectiveness of the process.

These results have widespread ramifications for horticulture in that they demonstrate that the composted poultry litter has no risk of causing contamination of produce and is therefore safe for use on horticulture.

### 3.5. Effective Odour Reduction

The result of the odour control research component showed that the processed litter had odour emission reduction of 77 %. This is mainly relevant to public amenity impact of litter application in horticulture as well as during litter processing. Figure 14 shows the results of odour testing of poultry litter samples treated with the process control system in the FABCOM® cell; these results confirm a 77 % reduction in odour produced by poultry litter processed through the FABCOM® system.

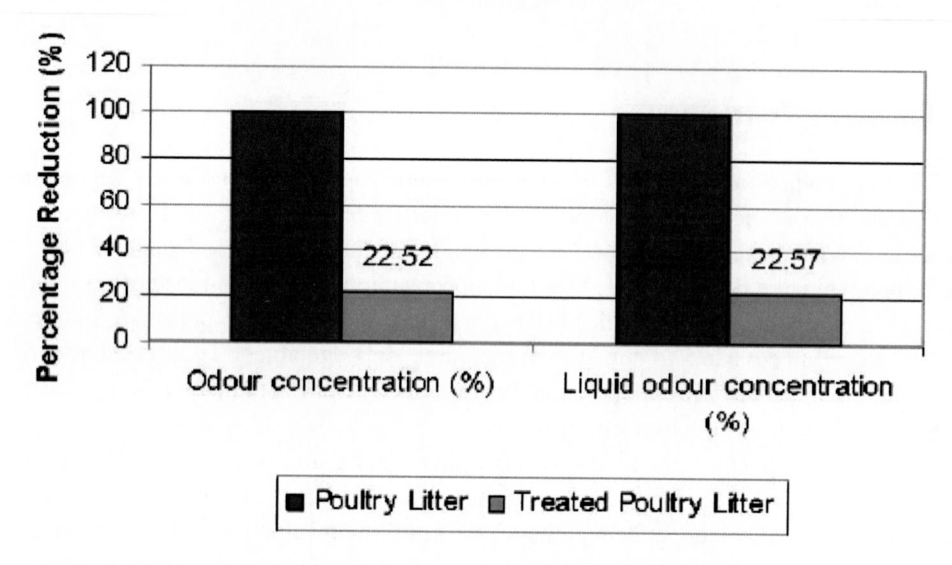


Figure 5. Odour reduction in processed poultry litter compared to raw poultry litter.

This result indicates that significant odour reduction potential exists in the current application of the FABCOM® system. As illustrated in figure 14 the tested materials were seen to change not only in the strength of the odour, but also in the nature of the odour. Raw, untreated litter was seen to have an odour of ammonia, while the final treated product had a smell of hay.

These results have significant implications for waste management in all industries, although especially for horticulture. The demonstrated ability to mitigate potential problems caused by odour generation on farms increases the life of farms in the rural urban fringe and has significant ramifications for the continued viability of the entire industry.

## 4. Conclusion

This project has successfully demonstrated an on-farm composting technology that provides a high performance and cost effective alternative to current methods of disposing of chicken litter and other farm wastes. In proving that a viable alternative exists for the processing of chicken litter, a considerable diversion away from less sustainable waste management processes is assured.

This project demonstrates that the on farm FABCOM® system produces a product that is an effective organic fertiliser that complies with the Health (Poultry Manure) Regulations 2001 and has reduced odour and pollution impact properties.

In confirming that a product can be produced cost effectively on site that exceeds the needs of the horticulture industry's organic fertiliser requirements, this project ensures that a large quantity of poultry litter can be recovered for nutrients and organic matter in a sustainable manner and improve environmental and public health conditions.

## ACKNOWLEDGEMENTS

I like to express appreciation and acknowledgement to the following organisations for their support: Individual vegetable growers, WA Vegetable Growers Association, WA Department of Environment, Australian Federal Department of Agriculture.

## 6. REFERENCES

Department of Health, Western Australia. *Health (Poultry Manure) Regulations*. 2001.

Hofstede, H.T. and Smith, J., 2000. Application Of FABCOM® Hybrid Composting System To Regional Organic Waste Management: Byron Shire Council Demonstration, In: *Proceedings of Compost 2000 Down Under*, Melbourne, AWMA.

Hofstede H T, 2006, Sustainable Poultry Litter Management, International Conference on Agricultural Waste, University Putrajaya Malaysia.

R. Paulin & D. Cook, Guideline for the Production of Conditioned Poultry Litter and for Testing other Poultry Manure Based Products, Misc Publication No 24/2001, WA Department of Agriculture, Perth WA.

Wilkinson K., et al. Strategies For The Safe Use Of Poultry Litter In Food Crop Production, Department of Primary Industry, State of Victoria Australia ISBN 1 74106677 8.

In: Technologies and Management for Sustainable Biosystems ISBN: 978-1-60876-104-3
Editors: J. Nair, C. Furedy, C. Hoysala et al. 

*Chapter 17*

# THE POTENTIAL HEALTH IMPACTS AND RISKS OF UTILISING BIOSOLIDS IN A TIMBER PLANTATION

***J. Levitan*[1], *J. Nair**[1], *G. Ho*[1], *N. Penney*[2] *and I. Dumbrell*[3]**
[1] Environmental Technology Centre, Murdoch University,
Perth, WA 6150, Australia
[2] Water Corporation, WA, Western Australia
[3] Forest Products Commission, WA, Western Australia

## ABSTRACT

During land application of biosolids in forestry, the management of the plantations following biosolids application and the harvesting of the timber, there is a potential for pathogen contamination. When using biosolids cake, the preferred method for land application is to use a 'manure spreader' which throws the biosolids out into the air and allowing it to scatter over the ground. Applications of liquid biosolids can become aerosolised when sprayed into the air, providing a near perfect medium for pathogens to be introduced into the air and be transported via the wind. However with biosolids cake the likelihood of the pathogens becoming aerosolised is minimised and therefore a lower pathogen risk is observed. Once the biosolids have been applied to the land, interactions with soil biota can result in limited growth of pathogens, which can reduce the risk. Studies have shown that residents living in the vicinity of land applied with biosolids and workers exposed to biosolids on a daily basis have reported coughing, sore throats, headaches and sinus effects. These findings indicate that there is a health risk associated with the use of biosolids and this review of the literature has shown that there are a number of significant points during the process of utilising biosolids that require some research into assessing the risks that accompany each point and action.

**Keywords:** pine plantation; municipal sludge; land-application.

* Corresponding author. Tel.: +61 8 9360 7322; fax: +61 8 9310 4997. *E-mail address:* j.nair@murdoch.edu.au

## INTRODUCTION

The first legislation that prohibited dumping of sewage sludge into the world's oceans came into effect in the 1980s (Renner 2000; Lewis and Gattie 2002). This left few options for biosolids; landfill, incineration, composting or land application. One of the most rapidly growing uses of biosolids is application to timber plantations. This process has value due to the forest ecosystem's ability to utilise the large organic input and to handle any contaminants (Magesan and Wang 2003). In developing a manual for the use of biosolids in short rotation forestry, Riddell-Black (1998) identified some significant advantages for the process. These included 1) the life-span of the plantation being over 20 years and therefore long-term operational planning for the use of biosolids can occur; 2) as plantations are a perennial crop, the application of biosolids is not limited by growth stages or ground conditions which can prevent safe application in regards to human health; 3) the use of biosolids for a non-food, renewable crop that has conservation and landscape benefits as well as sustainability goals. A number of studies have found that when biosolids were applied to timber plantations, the growth rates and the productivity of the trees increased significantly (Robinson *et al.* 2002; Kimberly *et al.* 2004; Wang *et al.* 2004; Wang *et al.* 2006). Wang *et al.* (2006) also found that *Pinus radiata* trees experienced a reduction in wood density from the increased growth rate, however the decrease was outweighed by the increase in wood volume. Furthermore, Kimberly *et al.* (2004) concluded that the land application of biosolids to a pine plantation can significantly increase the economic returns of the plantation through increased wood production.

Biosolids from any large urban area are highly likely to include a large number and variety of pathogens (Lewis and Gattie 2002; Magesan and Wang 2003) that will add to the soil microbial population if applied to the land. Whilst these pathogens are known to reduce in numbers and die-off over time (Rogers and Smith 2007), their potential threat to human health before they die is present and needs to be observed. As a result, many countries around the world regulate the use of biosolids (Renner 2000).

Minimisation of the health effects associated with biosolids application in timber plantations can be achieved very successfully through appropriate operational and management policies. Development of such policies should allow for the commercial value of producing and using biosolids to be optimised whilst ensuring that health concerns and issues are not overlooked and are known within the industry. This chapter highlights the main health issues of biosolids use and identifies the key areas that require monitoring and observation to provide for the development and improvement of management and operational policies for the use and application of biosolids.

## PATHOGENS IN BIOSOLIDS

Human pathogens that are present in biosolids and are considered a concern include viruses, bacteria, protozoa and helminths (National Water Quality Management Strategy (NWQMS) 2004). Table 1 shows the major pathogens potentially present in biosolids.

**Table 1. Major pathogens potentially present in biosolids**

| Bacteria | Viruses | Protozoa | Helminths |
|---|---|---|---|
| *Salmonella* spp. | Polioviruses | *Cryptosporidium* | *Ascaris lumbricoides* |
| *Shigella* spp. | Coxsackievirus | *Entamoeba histolytica* | *Ascaris suum* |
| *Yersina* spp. | Echovirus | *Giardia lamblia* | *Trichuris trichiura* |
| *Vibrio cholerae* | Hepatitis A virus | *Balantidium coli* | *Toxocara canis* |
| *Campylobacter jejuni* | Rotavirus | *Toxoplasma gondii* | *Taenia saginata* |
| *Escherichia coli* | Norwalk Agents | | *Taenia solium* |
| | Reovirus | | *Necator americanus* |
| | | | *Hymenolepis nana* |

adapted from United States Environmental Protection Authority (2000).

Around the world a number of national guidelines exist for the pathogenic content of biosolids. In the United States any biosolids that are intended to be used for land application must meet the pathogen criteria for either Class A or Class B according to legislation. Class B biosolids are treated using processes to significantly reduce pathogens and include aerobic digestion, anaerobic digestion, air drying and lime stabilization (United States Environmental Protection Authority 2000). Class A biosolids are treated using processes to further reduce pathogens and include composting, pasteurization, drying or heat treatment and advanced alkaline treatment (United States Environmental Protection Authority 2000). The guidelines also specify that standard pathogen density limits must be met (table 2).

**Table 2. Pathogen Density Limits**

| Pathogen or Indicator | Standard Density Limits (dry wt) |
|---|---|
| | Class A |
| Salmonella | < 3 MPN / 4g Total Solids or |
| Fecal Coliforms | < 1000 MPN / g and |
| Enteric Viruses | < 1 PFU / 4g Total Solids and |
| Viable Helminth Ova | < 1 / 4g Total Solids |
| | Class B |
| Fecal Coliform Density | < 2,000,000 MPN / g Total Solids |

adapted from United States Environmental Protection Authority (2000).

Class A biosolids are treated to a point where all pathogens, listed in table 2, are below detectable limits whilst Class B biosolids must be monitored as they may contain some pathogens. As a result, restrictions on land application sites in the form of personnel restrictions, prevention of planting specific crops, prevention of crop harvesting and animal grazing for specific time periods post-biosolids application are also enforced (United States Environmental Protection Authority 1994).

In Australia, national guidelines have been developed and the State/Territory Governments have the responsibility and authority to set specific guidelines for their individual needs based on these national guidelines (National Water Quality Management Strategy (NWQMS) 2004). The NWQMS identifies four pathogen grades that range from minimum pathogen reduction through to very high pathogen reduction with minimum potential for re-growth (table 3). The Australian State/Territory guidelines demonstrate combinations of both the United States guidelines and the Australian national guidelines. This has resulted in site-specific guidelines that address the issues and concerns of biosolids land application on a local scale (Environmental Protection Authority 1996; Environment 1999; Department of Environmental Protection 2002; EPA Victoria 2004).

**Table 3. Pathogen Grades for biosolids in Australia and allowable uses**

| Pathogen Grade | Description of Grade | Microbiological criteria (dry wt) | Allowable Biosolids Use |
|---|---|---|---|
| P1 | Very low pathogen levels with minimum regrowth potential | < 1 *Salmonella* per 50 gram final product<br>< 100 *E.coli* (or thermotolerant coliforms) per gram final product | Unrestricted, including residential<br>Institutional Landscaping (recreational) |
| P2 | Low pathogen levels but with some pathogen regrowth potential | < 10 *Salmonella* per 50 gram final product<br>< 1000 *E.coli* (or thermotolerant coliforms) per gram final product | Agriculture (Salad plants and root crops) |
| P3 | Established processes that achieve significant pathogen reduction | < 2,000,000 *E.coli* (or thermotolerant coliforms) per gram | Agriculture (Crops consumed cooked/processed, grazing animals, dairy cattle pasture and fodder)<br>Institutional Landscaping (non recreational)<br>Forestry and Land Rehabilitation |
| P4 | Minimum pathogen reduction | N/A | Landfill not including landfill final surface rehabilitation.<br>Secure landfill or other disposal options |

adapted from National Water Quality Management Strategy (2004).

Zaleski *et al.* (2005) performed studies on the potential re-growth and recolonisation of *Salmonella* spp. and faecal coliforms in biosolids and biosolids amended soils. They found that when Class B biosolids were solar-dried in field-scale drying beds, within 3-4 weeks Class A requirements were achieved. However, after rainfall events it was found that significant increases in both *Salmonella* spp. and faecal coliforms occurred and the increases returned the biosolids to Class B levels. The study concluded that the recolonisation of *Salmonella* spp. was most likely to be due to growth from faecal matter introduced by animals and birds rather than re-growth from the indigenous biosolids *Salmonella* spp. population. This conclusion points to possible explanations of re-growth and re-colonisation of *Salmonella* spp. at field sites. After land-spreading of municipal sewage sludge Pourcher *et al.* (2007) found that enteroviruses were undetectable after two weeks, whilst Enterococci and *Escherichia coli* did decrease gradually over a period of two months but they never reached their initial levels observed in the soil before application. *Clostridium perfringens* was also tested, and was observed as being present in the soil before the application took place. Over the testing period of two months, the levels of *Clostridium perfringens* were not observed to have decreased at all and at one point an increase occurred, however a reason for this increase was not observed. The longevity of pathogen survival seen in this study supports the necessity of utilizing several indicators and further treatment techniques that Godfree *et al.* (2005) concluded in their study to reduce the level of enteric micro-organisms in biosolids and sewage sludge before spreading on the land.

Godfree *et al.* (2005) identified three main barriers that are being used in the United States and Europe in an attempt to prevent transmission of pathogens from biosolids to humans. The barriers they identified are i) treatment to reduce pathogen content and vector attraction, ii) restrictions on crops grown on land to which biosolids have been applied, iii) minimum intervals following application and grazing or harvesting. Whilst these barriers seem common sense, it is a highly important conclusion that Godfree *et al.* (2005) draw in stating that when all three of these barriers are in place then the process of pathogen reduction in the sludge treatment is augmented.

## BIOSOLIDS ON THE LAND

Once applied to the land, there are a number of interactions that can occur between the biosolids and the soil. Clay soils usually provide a more protective environment for the microbial community than sandy soils. However, Lang *et al.* (2007) in a study in the United Kingdom found that the levels of *E.coli* were higher in the sandy loam soil type rather than the silty clay soil. They concluded that *E.coli* numbers were suppressed by the greater microbial numbers supported by a higher organic matter content of the silty clay soil. The same study also found that the mean *E.coli* population numbers were higher over a 91 day period in the dry silty clay and dry sandy loam than their respective wet samples. It was suggested that the large input of moist biosolids may have stimulated the growth of bacterial feeding protozoa that limited the *E.coli* numbers. This finding suggests that there is a threshold moisture level, above which *E.coli* is suppressed and is unlikely to survive successfully. Rogers *et al.* (2007) also state that the presence of increased protozoa levels is likely to reduce the population numbers of gram-negative bacteria, such as *E.coli* and *Salmonella* spp. Therefore a soil profile with a healthy microbial population is likely to cope with the large input of pathogenic organisms that may occur when biosolids are applied. This finding is important when decisions of whether to apply biosolids to a site are being considered, as it allows for risk assessments to be developed and undertaken before the application occurs.

However, biosolids application is not limited to healthy soils as shown by Meyer *et al.* (2004) who used biosolids to help in ecosystem restoration. Their study concluded that the presence of biosolids produced significant increases in plant biomass when compared to the untreated and unfertilised control sites. The study found that the nitrogen and carbon levels increased in the first two months following application of biosolids which led to a change in the plant community from predominantly shrubs to grasses. This increase in nitrogen was also observed in a pine plantation where increased tree growth was attributed to the improved nitrogen supply from the application of biosolids to the plantation (Wang *et al.* 2004).

In Western Australia, some of the timber plantations are found on sandy soils and are subject to a temperate climate. This would suggest that moisture and rainfall could be key factors for the survival, re-growth and die-off of pathogens.

# RISKS

## Land Applied Biosolids

There are two main methods of applying biosolids to the land. The first involves simple surface application of the biosolids, whilst the second involves injecting/incorporating the biosolids into the soil profile. Surface application and soil incorporation present separate issues especially in regards to human health. By incorporating the biosolids into the soil, the exposure of the biosolids to wind and personnel working in the application area is reduced and therefore the potential for the pathogens to be transferred to human populations is also reduced. However there is the potential for groundwater contamination from nitrate leaching, heavy metal leaching and the transferral of pathogens.

Applying biosolids to the soil surface exposes any personnel in the application site and any populations downwind of the application site to the biosolids. Exposure to human populations is increased through aerosolisation, direct contact with the biosolids and the attraction of vectors.

The preference for either application method is site dependant and there are benefits for each method, however in both situations the risk of pathogen infection is apparent and requires assessment to ensure minimal effects on the local personnel and populations. Whilst it is highly recommended that sites with land-applied biosolids are restricted to the public, there is a potential for the signage and warnings to be violated and direct exposure to the biosolids could occur

Studies have shown that it can take up to 3 months for *E.coli* to die-off in the soil following a biosolids application (Horswell *et al.* 2007; Lang *et al.* 2007) whilst a study by Eamens *et al.* (2006) found that *E.coli* and *Salmonella* spp. survived above baseline levels for up to 12 months post-application. It is recommended that a period of restricted access be applied to the application sites for at least 6 months which would allow the microbial contaminants to reduce to background levels (Horswell *et al.* 2007). This period however will depend on the local conditions such as rainfall, humidity, temperature and age of the tree stand.

## Aerosolized Biosolids

Biosolids are applied to the land either as a liquid or as a dewatered sludge. As a liquid the common method is to spray the biosolids over the land, whilst as a dewatered sludge the common method is use a spreader that 'throws' the biosolids out approximately 15 metres to the side of the vehicle. Both methods can potentially result in the biosolids becoming aerosolised. There have been many studies that have identified the health concerns caused by aerosolised pathogens (Epstein *et al.* 2001; Brooks *et al.* 2005a; Brooks *et al.* 2005b; Brooks *et al.* 2006; Robinson *et al.* 2006). Paez-Rubio *et al* (2007) identified that the aerosol emission rates of dewatered biosolids were higher than those of liquid biosolids on a dry weight basis with levels of approximately 10 milligram per second (mg $s^{-1}$) and 0.125 mg $s^{-1}$ respectively within wind speeds of above 0.8 metres per second (m $s^{-1}$) and below 2.5 m $s^{-1}$. The study importantly found that the majority of particles produced were inhalable (< 10 μm diameter) and may be respirable (< 4 μm diameter). However, the study significantly pointed out that these results were derived from measurements taken directly downwind of the emission source and noted that biosolids workers are unlikely to spend substantial amounts of time downwind from the emission source. Further investigations would be required to establish whether the workers exposure time and the emission rate is below the threshold health level.

These studies have found that whilst the risk of infection is minimal, there is a chance that infection could occur. Lewis *et al.* (2002) state that limiting the exposure to windblown dusts would reduce the risk to the general population, but also that properly treated Class B biosolids (table 2) should present no significant risk of infection. These results must be treated with caution however, as they relate to liquid biosolids land application and not dewatered biosolids land application.

### Vector Attraction

A significant issue in the ability to protect public health in regards to biosolids use is that of 'Vector Attraction Reduction' (United States Environmental Protection Authority 1994; National Water Quality Management Strategy (NWQMS) 2004). Vectors, such as rats and flies, can be attracted by inadequately stabilized biosolids and they can spread disease by carrying and transferring the pathogens to humans via direct and indirect means. Vector attraction reduction can be achieved through reducing the moisture content of the biosolids; reducing the organic content of the biosolids; adding alkalis; composting; and by incorporating the biosolids directly into the soil and thus providing a physical barrier to the potential vectors (National Water Quality Management Strategy (NWQMS) 2004). Physically removing the vectors is not recommended as this may effect local ecosystems, unless the vectors are not indigenous to the area and therefore not integral to the local environment. Measures to reduce vectors at an application site are likely to be site specific dependant on a range of factors that include proximity to residential areas and the effect of the vectors on the local ecosystem.

## HEALTH IMPACTS

Complaints associated with land-applied biosolids primarily involve the irritation of skin and mucous membranes and also infections of the respiratory tract (Gattie and Lewis 2004). Robinson *et al.* (2006) conducted a survey of workers exposed to biosolids on a daily basis and recorded the occurrence of coughing, sore throats, headaches and sinus effects. The workers who reported these symptoms were also found to be the least compliant with personal protective equipment supplied to them, such as respirators and masks to filter the air they breathe. Burton *et al.* (1999) suggested in a study on biosolids application that the symptoms observed by the workers were probably caused by inhalation of the biosolids.

Rylander (1999) attributed bioactive organic dust to the cause of various symptoms recorded by residents and more specifically endotoxins released by the pathogenic bacteria. This is significant as endotoxins can cause a weakened immune system therefore exposing individuals to possible health effects from pathogens. Endotoxins are found in the outer membrane wall of gram-negative bacteria commonly found in biosolids. When the outer membrane breaks down the endotoxins are released and are capable of causing large-scale immune reactions (Gattie and Lewis 2004; Brooks *et al.* 2007).

Lewis *et al.* (2002) also identified chemical irritants from the biosolids as a health concern. Lime stabilisation is a common method used to reduce the pathogen content of the biosolids, however this can result in the finished biosolids containing a number of contaminants such as ammonia, endotoxins and alkyl amines. Whilst the reduction of pathogens is essential before the biosolids can be applied to the land, the introduction of these chemicals has the potential to be just as hazardous. The study by Lewis *et al.* (2002) showed that residents living within one (1) kilometre of the application sites complained of irritation when the wind blew from the biosolids applied land. Studies (Lewis and Gattie 2002; Lewis *et al.* 2002) concluded that the combined effects of the chemical irritants and the pathogens infecting the irritated skin and membranes increased the infection rate within the local population. Rylander (1999) found that general symptoms (fatigue, diarrhoea) were more common amongst biosolids workers than municipal workers at sites with no wastewater or organic dusts. Whilst the effects at biosolids production sites might be expected to be higher than at application sites, the risk posed by the pathogens will be transferred with the biosolids.

## CONCLUSION

The benefits of applying biosolids to timber plantations are significant however without monitoring and adherence to guidelines on application and management, it is a process that could affect the health of personnel working with or living near the activity. Restricted access to a plantation application area is likely to reduce the risk of human contact with the biosolids. Whilst preventing direct contact between humans and the biosolids is possible, a concern lies with the aerosolisation of the biosolids and the pathogens. Inhalation of biosolids dust is one of the most likely pathways for the pathogens to enter the human body, and so the wearing of personal protective respiratory equipment is highly recommended. As the inhalation of endotoxins can cause ailments it is recommended that these substances also are monitored. Where biosolids are being applied, downwind monitoring must occur for all activities that are capable of producing airborne particles. This will ensure that activities can be stopped if increased levels of the indicators are observed. Vector attraction is a site-specific issue; however treatment of the biosolids should occur before the application to the land occurs to reduce the potential of attracting vectors. Soil type and climate conditions are significant in the survival of pathogens and as conditions vary from tropical zones to temperate climates, the risks of pathogen contamination will vary from location to location. Provided the appropriate guidelines for biosolids application are followed and the specific conditions of the application site are taken into account, then the risk to human health from biosolids application is significantly reduced if not removed. Revision of management and operational policy following the development of risk assessments on these key areas will allow the future of biosolids application to become a healthy, safe and commercially viable opportunity for successful waste management.

## References


Brooks, J. P., S. L. Maxwell, C. Rensing, C. P. Gerba and I. L. Pepper (2007). "Occurrence of antibiostic-resistant bacteria and endotoxin associated with the land application of biosolids." *Canadian Journal of Microbiology.* 53: 616-622.

Brooks, J. P., B. D. Tanner, C. P. Gerba, C. N. Haas and I. L. Pepper (2005a). "Estimation of bioaerosol risk of infection to residents adjacent to a land applied biosolids site using an empirically derived transport model." *Journal of Applied Microbiology.* 98: 397-405.

Brooks, J. P., B. D. Tanner, C. P. Gerba and I. L. Pepper (2006). "The measurement of aerosolized endotoxin from land application of Class B biosolids in southeast Arizona." *Can. J. Microbiol.* 52: 1-7.

Brooks, J. P., B. D. Tanner, K. L. Josephson, C. P. Gerba, C. N. Haas and I. L. Pepper (2005b). "A national study on the residential impact of biological aerosols from land application of biosolids." *Journal of Applied Microbiology.* 99: 310-322.

Burton, N. C. and D. Trout (1999). Biosolids land application process. NIOSH Health Hazard Evaluation Report HETA 98-0118-2748. LeSourdsville, Institute for Occupational Safety and Health Publications Office.

Department of Environmental Protection (2002). Western Australian guidelines for direct land application of biosolids and biosolids products. D. o. Health. Perth.

Eamens, G. J., A. M. Waldron and P. J. Nicholls (2006). "Survival of pathogenic and indicator bacteria in biosolids applied to agricultural land." *Australian Journal of Soil Research.* 44: 647-659.

Environment, Planning and Scientific Services Division (1999). Tasmanian biosolids reuse guidelines. Department of Primary Industry, *Water and Environment.* Hobart.

Environmental Protection Authority (1996). South Australian Biosolids Guidelines for the safe handling, reuse or disposal of biosolids. D. o. E. a. N. Resources. Adelaide.

EPA Victoria (2004). Guidelines for environmental management biosolids land application. E. Victoria. Victoria.

Epstein, E., N. Wu, C. Youngberg and G. Croteau (2001). "Dust and Bioaerosols at a biosolids composting facility." *Compost. Science and Utilization.* 9(3): 250-255.

Gattie, D. K. and D. L. Lewis (2004). "A high-level disinfection standard for land-applied sewage sludges (biosolids)." Environmental Health Perspectives 112(2): 126-131.

Godfree, A. and J. Farrell (2005). "Processes for managing pathogens." *Journal of Environmental Quality.* 34(1): 105-113.

Horswell, J., V. Ambrose, L. Clucas, A. Leckie, P. Clinton and T. W. Speir (2007). "Survival of *Escherichia coli* and *Salmonella* spp. after application of sewage sludge to a *Pinus radiata* forest." *Journal of Applied Microbiology.* 103: 1321-1331.

Kimberly, M. O., H. Wang, P. J. Wilks, C. R. Fisher and G. N. Magesan (2004). "Economic analysis of growth response from a pine plantation forest applied with biosolids." *Forest Ecology and Management.* 189: 345-351.

Lang, N. L., M. D. Bellet-Travers and S. R. Smith (2007). "Field investigations on the survival of *Escherichia coli* and presence of other enteric micro-organisms in biosolids-amended agricultural soil." *Journal of Applied Microbiology.* 103: 1868-1882.

Lang, N. L. and S. R. Smith (2007). "Influence of soil type, moisture content and biosolids application on the fate of *Escherichia coli* in agricultural soil under controlled laboratory conditions." *Journal of Applied Microbiology.* 103(2122-2131).

Lewis, D. L. and D. K. Gattie (2002). "Pathogen risks from applying sewage sludge to land." *Environmental Science and Technology.* 36(13): 287A-293A.

Lewis, D. L., D. K. Gattie, N. E. Novak, S. Sanchez and C. Pumphrey (2002). "Interactions of pathogens and irritant chemicals in land applied sewage sludges (biosolids)." BMC Public Health 2(11).

Magesan, G. N. and H. Wang (2003). "Application of municipal and industrial residuals in New Zealand forests: an overview." *Australian Journal of Soil Research.* 41: 557-569.

Meyer, V. F., E. F. Redente, K. A. Barbarick, R. B. Brobst, M. W. Paschke and A. L. Miller (2004). "Plant and soil responses to biosolids application following forest fire." *Journal of Environmental Quality.* 33: 873-881.

National Water Quality Management Strategy (NWQMS) (2004). Guidelines for Sewerage Systems: *Biosolids Management.* N. R. M. M. Council.

Paez-Rubio, T., A. Ramarui, J. Sommer, H. Xin, J. Anderson and J. Peccia (2007). "Emission rates and characterization of Aerosols produced during the spreading of dewatered Class B biosolids." *Environmental Science and Technology.* 41(10): 3537-3544.

Poucher, A.-M., P.-B. Francoise, F. Virginie, G. Agnieszka, S. Vasilica and M. Gerard (2007). "Survival of faecal indicators and enteroviruses in soil after land-spreading of municipal sewage sludge." *Applied Soil Ecology.* 35: 473-479.

Renner, R. (2000). "Sewage Sludge, Pros & Cons." *Environmental Science and Technology.* 34(19): 430A-435A.

Riddell-Black, D. (1998). "Development of a water industry manual for biosolids use in short rotation forestry." *Biomass and Bioenergy.* 15(1): 101-107.

Robinson, C., K. Robinson, C. Tatgenhorst, D. Campbell and C. Webb (2006). "Assessment of wastewater treatment plant workers exposed to biosolids." *AAOHN.* 54(7): 301-307.

Robinson, M. B., P. J. Polgase and C. J. Weston (2002). "Loss of mass and nitrogen from biosolids applied to a pine plantation." *Australian Journal of Soil Research.* 40: 1027-1039.

Rogers, M. and S. R. Smith (2007). "Ecological impact of application of wastewater biosolids to agricultural soil." *Water and Environment Journal.* 21: 34-40.

Rylander, R. (1999). "Health effects among workers in sewage treatment plants." *Occupational Environmental Medecine.* 56(5): 354-357.

United States Environmental Protection Authority (1994). A plain english guide to the Part 503 rule. USEPA, Office of Wastewater Management.

United States Environmental Protection Authority (2000). A Guide to Field Storage of Biosolids. *USEPA, Office of Wastewater Management.* EPA/832-B-00-007.

Wang, H., M. O. Kimberly, G. N. Magesan, R. B. McKinley, J. R. Lee, J. M. Lavery, P. D. F. Hodgkiss, T. W. Payn, P. J. Wilks, C. R. Fisher and D. L. McConchie (2006). "Midrotation effects of biosolids application on tree growth and wood properties in a *Pinus radiata* plantation." *Canadian Journal of Forest Research.* 36: 1921-1930.

Wang, H., G. N. Magesan, M. O. Kimberly, T. W. Payn, P. J. Wilks and C. R. Fisher (2004). "Environmental and nutritional responses of a *Pinus radiata* plantation to biosolids application." *Plant and Soil.* 267: 255-262.

Zaleski, K. J., K. L. Josephson, C. P. Gerba and I. L. Pepper (2005). "Potential regrowth and recolonization of Salmonellae and indicators in biosolids and biosolid-amended soil." *Applied and Environmental Microbiology*. 71(7): 3701-3708.

In: Technologies and Management for Sustainable Biosystems ISBN: 978-1-60876-104-3
Editors: J. Nair, C. Furedy, C. Hoysala et al. 

*Chapter 18*

# PERFORMANCE OF A PHYTOCAPPED LANDFILL IN A SEMI-ARID CLIMATE

***Kartik Venkatraman**[1], *Nanjappa Ashwath*[1] and *Ninghu Su*[2]**

[1] Centre for Plant and Water Science; Central Queensland University, Rockhampton, Queensland, Australia – 4702

[2] School of Earth and Environmental Science James Cook University, Cairns, Queensland, Australia – 4870

## ABSTRACT

Landfills have been the major repositories of urban wastes, and they will continue to be built, so long as the humans live in communities. The costs of construction, maintenance and remediation of landfills have escalated over the years and research is therefore required to identify alternative techniques that will not only minimise the costs, but also demonstrate increased environmental performance and community benefits. This chapter discusses the alternative landfill capping technique known as 'Phytocapping' (establishment of perennial plants on a layer of soil placed over the waste), which was trialed in Rockhampton, Australia. In this technique, trees were used as 'bio-pumps' and 'rainfall interceptors' and soil cover as 'storage' of water. Tree performance was measured based on their canopy rainfall interception and water uptake potential. The rate of percolation of water was modelled using HYDRUS 1D for two different scenarios (with and without vegetation) for the thick (1400 mm) and thin (700 mm) soil covers respectively. Evidence and simulations incorporating 15 years of meteorological data showed percolation rates of 16.7 mm $yr^{-1}$ in thick phytocap and 23.8 mm $yr^{-1}$ in thin phytocap, both of which are markedly lower than those expected from a clay cap.

Keywords: **HYDRUS 1D, landfill, percolation, phytocap, water balance, methane.**

* k.venkatraman@cqu.edu.au

## INTRODUCTION

Landfill capping is a mandatory post closure procedure to isolate the deposited wastes from external environment, mainly water (Vasudevan *et al.* 2003). Landfill capping involves placing a barrier, which acts as a raincoat over filled landfill to minimise percolation of rainfall or surface water into the waste (Scott *et al.* 2005). This is not only expensive but also not viable for small and medium sized landfills in Australia. In recent years, conventional capping systems made of compacted clay; Geosynthetic Clay Liners (GCL) and HDPE have been used extensively in many countries. Amongst these, the most popular practice in Australia has been the use of compacted clay caps. A typical compacted clay cap recommend in Queensland is given below (figure 1)

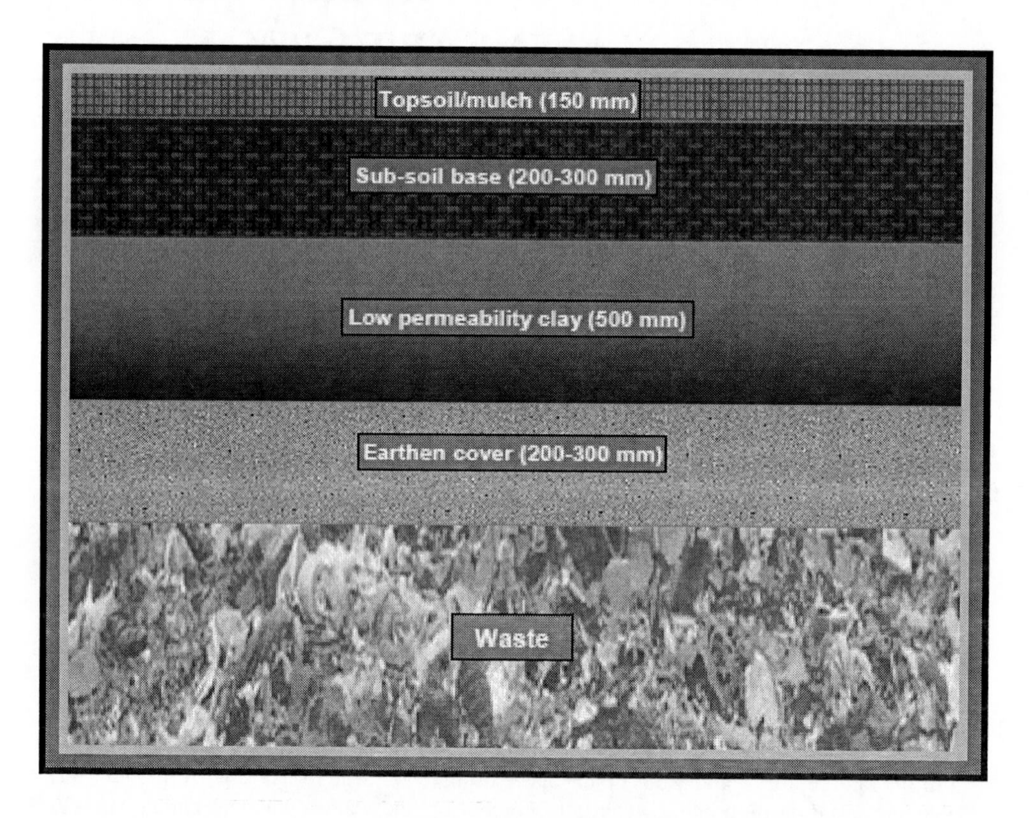


Figure 1. Various layers used in a typical clay capping system, permeability clay (Ks = $10^{-8}$ m/s) (EPA 2005).

In Australia, the caps constructed on landfills should be sustainable for at least 30 years. Recent studies however show that clay caps have shorter life span (Vasudevan *et al.* 2003) and fail to prevent percolation of water due to cracking (Khire *et al.* 1997, Melchoir 1997). Furthermore, clay caps do not allow optimal interaction of methane with oxygen, which is a must for methane oxidation (Abichou *et al.* 2004).

A new technology called 'Phytocapping' (Venkatraman and Ashwath 2007) was trialed at Lakes Creek Landfill, Rockhampton. In brief, phytocaps have two major components, viz. the trees that act as 'bio-pumps' and 'rain interceptors' and the soil that acts as 'storage' (figure 2). The research which was conducted over three years (2005 to 2007) examined various aspects of the performance of the phytocap, including soil water storage and tree performance along with the modelling for site water balance using HYDRUS 1D code.

Figure 2. The Phytocapping concept.

Water sources in a landfill include the waste itself, soil cover and precipitation (Bengtsson *et al.* 1994). It is essential to assess the effectiveness of the phytocaps with respect to the amount of water that percolates into the waste. For these reasons numerous water transport modelling softwares have been developed to predict percolation rates based on Richard's equation and Water Balance method (Albright *et al.* 2002, Williams 2005). The models using Richard's equation have shown higher accuracy (Albright *et al.* 2002) and proven to be better than those based on water balance method due to their ability to describe water flow in any direction (Jirka Simunek: pers. Comm.).

The performance of landfill caps have been evaluated either by qualitative (groundwater monitoring, leachate collection etc.), indirect quantitative methods (empherical methods, mass balance methods and unsaturated flow process methods using Richard's equation) or direct quantitative techniques that are based on measurements using lysimeters (Albright *et al.* 2002). However, lysimeters are very expensive to construct (Albright *et al.* 2002) and hence soil moisture measurements (qualitative) and HYDRUS 1D (indirect quantitative methods) were used to estimate percolation in this research.

A variety of models have been used to measure percolation rates in different scenarios. Several modelling softwares have been developed and compared for their accuracy and robustness (Ho *et al.* 2004). A few comparative studies have been undertaken on hydrologic performance evaluations of landfills cover (Chai and Miura 2002, Ho *et al.* 2004) with most measuring seepage production (Dho *et al.* 2002, Ham 2002). These models include UNSAT-H, HYDRUS 1D, HYDRUS 2D, Simulation of Heat and Water (SHAW), Vadose/W, Soil Water Balance and Infiltration Model (SWIM), and The Hydrologic Evaluation of Landfill Performance (HELP), TOUGH-2, MACRO and LEACHM (Fayer *et al.* 1992, Fayer and Gee 1997, Khire *et al.* 1999, Scanlon *et al.* 2002, Benson *et al.* 2004, Albright *et al.* 2002, Johnson *et al.* 2001, Ross 1998). Amongst these models, HYDRUS 1D, HYDRUS 2D, UNSAT-H and Vadose/W are used most frequently (Benson 2004). Erosion Productivity Impact Model (EPIC) (Williams 2005) uses the water balance method and has been extensively used in agriculture, but was found less robust than the models using Richards's equation and more accurate than HELP (Hauser and Gimon 2001). The performance of HELP was compared with that of Vadose/W (Chammas *et al.* 1999) and UNSAT-H (Khire *et al.* 1997) in Alternative Cover Assessment Program (ACAP) and the Alternative Landfill Cover Demonstration (ALCD) project (Khire *et al.* 1997). It was found that percolation rates were over-predicted by HELP in comparison with UNSAT-H (Khire *et al.* 1997). Hauser and Gimon (2001) compared HELP, HYDRUS and UNSAT-H and found that UNSAT-H and HYDRUS were more robust and accurate than HELP for phytocaps. Another study by Scanlon *et al.* (2002) compared HELP, HYDRUS, SHAW (Albright *et al.* 2002), Vadose/W, SWIM (Dwyer 2003) and UNSAT-H to model site water-balance from covers in semi-arid Texas, New Mexico, and Idaho, USA, over a period ranging from one to three years. Scanlon *et al.* (2002) concluded that models employing Richard's equation predicted water balance more accurately than the HELP model. Other models such as MACRO (Johnson *et al.* 2001) were not as robust as HYDRUS and TOUGH-2 (Albright *et al.* 2002), and Vadose/W did not effectively predict drainage (Albright *et al.* 2002, Benson *et al.* 2004). A few efficient models like LEACHM, Model for Effluent Disposal using Land Irrigation (MEDLI) (Tillman and Surapaneni 2002) and WATLOAD have not been used in landfill studies to date. Amongst all the above models, it appears that the UNSAT-H and HYDRUS predicted drainage effectively (Hauser and Gimon 2001, Benson *et al.* 2004, Albright *et al.* 2002, Scanlon *et al.* 2002). Comparison of various models in terms of their use, and various parameters used is shown in table 1.

With all the contradictions and inaccuracies in the models used so far, the Subsurface Transport over Multiple Phase (STOMP) (Oostrom *et al.* 2004) which takes into account gaseous, aqueous and solid phase in one single model was also trialled during the study. Due its complexity, however, HYDRUS 1D a model that can simulate water, heat and solute movement in the saturated zone (Simunek *et al.* 2005) was used. HYDRUS 1D has been extensively used in site water balance studies and is continuing to become popular amongst hydrologists and environmental engineers. This software is a finite element solution to Richard's equation for one dimensional flow in variably saturated media (Simunek *et al.* 2005).

**Table 1. Comparison of different models used in predicting site water balance**

| Model Acronym | Name | Developed by | Application | Plant Growth | Transpiration | Solute Transport | Water Retention Method | Reference |
|---|---|---|---|---|---|---|---|---|
| EPIC | Erosion-Productivity Impact Calculator | Texas A & M | Agriculture | Yes | Yes | No | Water Balance | Williams (2005) |
| HELP | The Hydrologic Evaluation of Landfill Performance | U.S. Army Engineer Waterways Experiment Station | Landfills | No | No | Yes | Water Balance | Scanlon *et al* (2002) |
| TOUGH-2 | Transport Of Unsaturated Groundwater and Heat | | Nuclear Waste | No | No | Yes | Richards Equation | Albright *et al* (2002) |
| MACRO | MACRO | Swedish University of Agr. Sc. | Soil water Balance | Yes | Yes | Yes | Richards Equation | Johnson *et al* (2001) |
| UNSAT-H | UNSAT-H | U. S. Dept of Energy | Landfills | Yes | Yes | No | Richards Equation | Albright *et al* (2002) |
| HYDRUS | HYDRUS 1D/2D | P.C Progress SRO and Simunek *et al* (2005) | Landfills | Yes | Yes | Yes | Richards Equation | Albright *et al* (2002) |
| LEACHM | The Leaching Estimation And Chemistry Model | Flinders University | Agriculture | Yes | Yes | Yes | Richards Equation | Albright *et al* (2002) |
| SWIM | Soil Water Balance and Infiltration Model | Ross, 1998 | Landfills | Yes | Yes | Yes | Richards Equation | Dwyer (2003) |
| MEDLI | Model for effluent disposal using land irrigation | DNR & DPI, Forestry | Piggeries, Sewage Treatment Plants | Yes | Yes | Yes | | Tillman and Surapaneni 2002 |
| WATLOAD | WATLOAD | CSIRO | Vegetation Management, Effluent Disposal | Yes | Yes | Yes | | Myers *et al.* 1999 |
| STOMP | Subsurface Transport Over Multiple Phase | U. S. Dept of Energy | Nuclear Waste | Yes | Yes | Yes | | Oostrom *et al.* 2004 |
| Vadose/W | Vadose/W | Geo-slope | Agriculture | Yes | Yes | No | | Benson *et al* (2004) |
| SHAW | Simulation of Heat and Water | USDA Agr. Rech Centre | Landfills | Yes | Yes | No | Richards Equation | Albright *et al* (2002) |

## MATERIALS AND METHODS

Details of establishing the Phytocapping trial are provided in Venkatraman and Ashwath (2007). Firstly an experimental site of 5000 m$^2$ was selected for the trial consisting of two soil thickness (thick cap: 1400 mm soil; thin cap: 700 mm soil), replicated twice with 21 tree species (18 seedlings of each) (figure 3). The experimental site was mulched with shredded green waste (100 mm deep), and the plants were drip irrigated. Various plant (plant growth, transpiration, canopy interception etc.) and soil parameters (soil compaction, hydraulic conductivity) were monitored over two and a half years, and the site water balance was predicted using HYDRUS 1D (Benson 2002, Albright *et al.* 2004). Climate data was acquired from both the Bureau of Meteorology (BOM) and the on-site weather station. The species were initially observed for their growth and survival in the landfill environment followed by a study on canopy interception, transpiration rate, biomass, root depth, mineral composition of plant and soil hydraulic properties, but only the results of modelling are presented in this chapter.

Figure 3. Establishment of the Phytocapping trial at the Lakes Creek landfill, Rockhampton. above: Placement of thin and thick soil caps over the waste, below: seedlings of 21 tree species were established on each of these soil treatments.

# MODELLING

HYDRUS 1D uses soil hydraulic parameters and various tree parameters such as transpiration rate and root depth, including climate data (rainfall and evaporation). Water balance was water predicted for two scenarios, without vegetation and with vegetation. Various plant and soil parameters required for the modelling were measured during the study. Canopy rainfall interception was measured for 50 rainfall events over two years. Transpiration was determined using Thermal Dissipation Probes (TDP) and Dynagauges. Root depth was measured during biomass estimation by excavation method. Soil hydraulic parameters were taken from the studies conducted by Dr Ian Phillips (Griffith University) and mulch hydraulic parameters were obtained from Findeling *et al.* (2007). Precipitation and evaporation data were obtained from the BOM and the weather station located at the landfill site. Final simulations were completed using the average values obtained for the selected ten tree species grown in the phytocapping system.

Before running the model, canopy interception (32%) was deducted from the actual rainfall data for the experimental the site. Irrigation values were added to the rainfall data, and the rate of soil evaporation was taken as 50% of that of un-vegetated site (worst case scenario), as the soil evaporation under agroforestry (Albright *et al.* 2002) systems will be much less than that under a tree canopy (reduced by 23% to 40%; Wallace *et al.* 2000, Jackson and Wallace 2000). Merta *et al.* (2006) found that the soil evaporation under agricultural crops was considerably low under high LAI. For example, the soil evaporation was 50% at a leaf area index (LAI) of 1.5 in comparison with 5% for denser crops (LAI>3). Based on these data, soil evaporation was taken as 50% of that reported by the BOM. LAI recorded in 19 different species during the study ranged 1.9 to 2.5. Based on these data, soil evaporation was taken as 50% of that reported by the BOM as the worst case scenario.

HYDRUS 1D simulated surface soil storage and drainage of the phytocaps. The site water balance was simulated for 15 years (1992 to 2006). Results from simulations using 15 years of data are presented in the current chapter. Since the established species grew at different rates, with some species growing slower than the others, transpiration data of only the 10 best performed tree species (*Dendrocalamus latiflorus, Casuarina cunninghamiana, Acacia mangium, Hibiscus tiliaceaus, Eucalyptus grandis, Syzigium australis, Acacia harpophylla var hillii, Ficus recemosa, Ficus mocrocarpa* and *Eucalyptus raveretiana*) in terms of canopy rainfall interception and transpiration were used in the simulation. These selected species can be used in the Central Queensland Region, as they perform well in a landfill environment and resilient to drought and fire both of which are very common in the landfills.

# RESULTS AND DISCUSSION

## Scenario 1

Percolation simulated for the thick cover (1400 mm) and the thin cover (700 mm) without vegetation was 133.3 mm $yr^{-1}$ and mm 153 mm $yr^{-1}$ respectively (figure 4). This difference in percolation rates between the two covers was expected, as the soil depth plays a vital role in retaining maximum amount of water (Warren *et al.* 1996). The thick cover could hold moisture up to 660 mm in comparison with 350 mm by the thin cover (figure 4). Surface runoff predicted for this site was infinitesimally small in both covers (ranging from 0.20 mm to 4 mm) (figure 4) due to flat surface and most importantly, due to the presence of 100 mm of mulch layer and a thick layer of litter fall under some tree species.

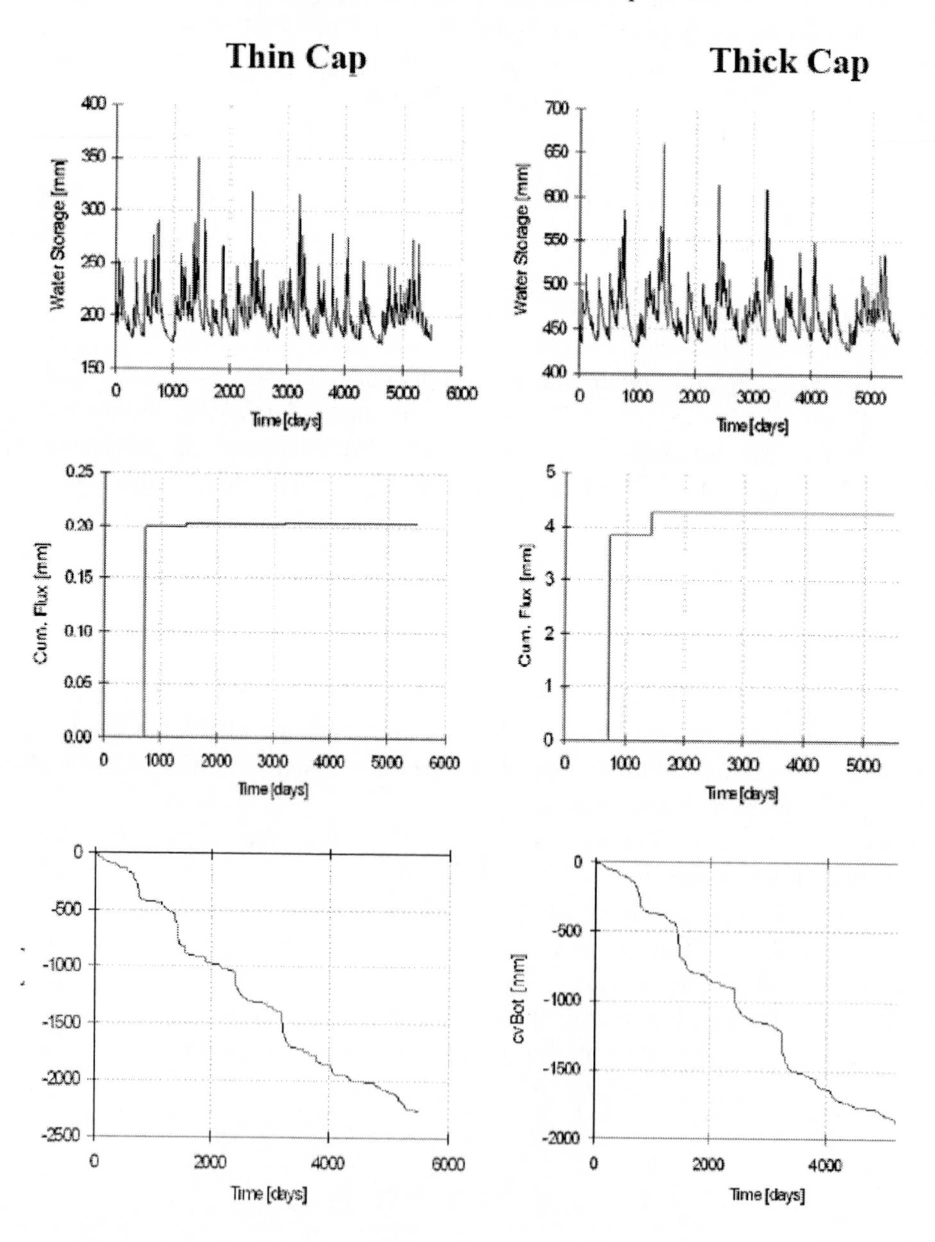


Figure 4. Simulated storage capacity of soil (top), cumulative runoff (middle) and percolation of water (bottom) in thin (left) and thick (right) covers, respectively in the absence of vegetation (cumulative of 15 years data; 1992 to 2006).

## Scenario 2

In this scenario, percolation was simulated using the same parameters as in scenario 1, but an additional component, vegetation was introduced. An average transpiration of 1.5 mm day$^{-1}$ was used, which was the average measured value from the top 10 selected tree species grown on the experimental site. The average rain intercepted by the 10 tree species was 32% and therefore, the incident rain was reduced by 32% and this corrected rainfall was used in the simulation. The HYDRUS 1D simulation for the vegetated site showed a percolation of 16.7 mm yr$^{-1}$ for the thick cover and 23.8 mm yr$^{-1}$ for the thin cover (figure 5). The 15 year rainfall data also included very dry periods and very wet periods (300 mm rain in three consecutive days in 2003). The percolation for vegetated covers was 8 to 10 times less than that simulated for non-vegetated sites. This clearly demonstrates the role played by the vegetation in phytocapping. Benson *et al.* (2004) has demonstrated the significance of vegetation in site water balance, particularly their role in soil moisture depletion and the relationships between the root depth and the soil moisture depletion. The soil storage capacity of the phytocaps was reduced from 350 mm (scenario 1, without vegetation) to 320 mm (scenario 2; with vegetation) in thin cover and from 660 mm to 570 mm in thick cover (figure 4). This is due to influence of tree roots on soil structure (Glinski and Lipeic 1990), bulk density (Kalman *et al.* 1996) and pore size (Johnson *et al.* 2003) which in turn affects the water retention property of the soil (Auge *et al.* 2001), as does the spatial variability (Shouse *et al.* 1995). Roots of trees can perforate tough soil layers thereby creating macropores (Lal *et al.* 1979, Glinski and Lipeic 1990) and allowing free water movement. The surface runoff decreased in the thick cover but slightly increased in thin cover (figures 4 and 5) in scenario 2 compared to scenario 1. The decrease in surface runoff can be attributed to the increased water uptake by tree (Freebairn *et al.* 1986) thus creating more space for water storage. The slight increase in surface runoff in thin cap may be due to soil saturation (Liang and Xie 2001) and when rainfall rate exceeds the infiltration capacity (Fiedler *et al.* 2002).

Results from the above simulation suggest that phytocaps are very effective in reducing percolation of water into waste. In this simulation, establishment of 10 selected tree species using 1400 mm layer of unconsolidated soil will allow a percolation of 251 mm in 15 years. This is equivalent to 16.7 mm yr$^{-1}$. This value is significantly lower than the percolation rate expected for a clay cover (c. 10%; Geoff Thompson; pers. Comm.). Comparing figures 4 and 5, it is clearly evident that the reduced percolation was due to the presence of deep rooted tree species.

A comparison of the above results with those reported for the ACAP sites (Benson *et al.* 2002) suggest that the percolation estimated for the phytocaps in this research are within the values reported for alternative covers (12 to 128 mm yr$^{-1}$) in the USA (Benson 2002). For example, the percolation rates at the ACAP site at Omaha, Nebraska, is comparable to Rockhampton site in terms of rainfall (760 mm yr$^{-1}$), and the measured percolation (using lysimeters) at Omaha was 60 mm yr$^{-1}$. A comparison of the features of this site and the Rockhampton site is shown in table 2.

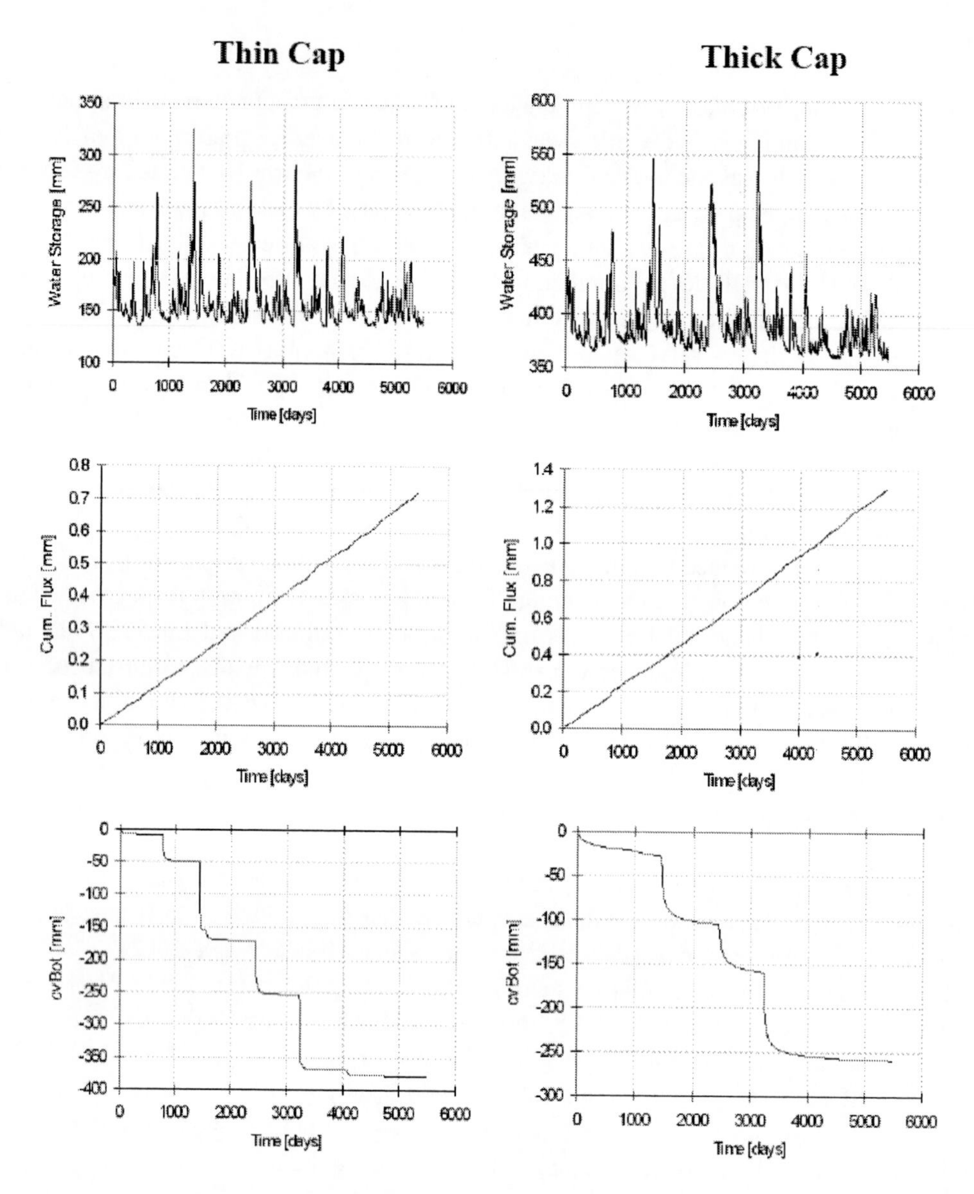


Figure 5. Simulated storage capacity, runoff and percolation in the thick and thin covers in the presence of vegetation (15 years data 1992 to 2006).

**Table 2. Comparison of Rockhampton and Omaha site with regards to percolation of water**

| Parameters | Rockhampton, Qld, Australia | Omaha, NE, USA |
|---|---|---|
| Rainfall | 780 mm $yr^{-1}$ | 760 mm $yr^{-1}$ |
| Soil thickness | 1400 or 700 mm | 1100 mm |
| Species grown | Trees | Grasses |
| Drainage | 16.7 – 23.8 mm $yr^{-1}$ | 60 mm $yr^{-1}$ |

The rainfall pattern at the Rockhampton site matches with the rainfall distribution at the Omaha site. The simulated percolation rate at Rockhampton ranges between 16.7 mm $yr^{-1}$ to 23.8 mm $yr^{-1}$ for thick and thin covers respectively. This rate therefore is much lower than the percolation rates reported for the Omaha site. The currently simulated percolation rates are also significantly lower than those expected for clay cover (78 mm at c. 10% of the rainfall; Geoff Thompson; pers. Comm.).

## CONCLUSION

Percolation rates estimated for the phytocaps using HYDRUS 1D were 16.7 to 23.8 mm $yr^{-1}$ for thick and thin covers respectively. The predicted percolation rate for the Rockhampton site is much lower than that expected from well constructed and maintained clay capped landfill (which is equivalent to 78 mm in Rockhampton; at 10% of the incident rain). This shows the better or equivalent ability of the phytocapping system to limit entry of water into the landfill (c. 50% of clay cap). The reduced cost of establishing phytocaps on landfills (compared to clay caps) show possibly the superiority of phytocaps over clay caps. Economical evaluation of phytocaps against clay caps is underway and will be reported shortly.

Evidences show that the thick and thin phytocaps perform well in maintaining a low percolation rate (less than 10% of the received rainfall). However, in events of heavy rainfall, thick cover has better moisture retaining capacity than thin cover. These trends and the consistency of the results amongst the thick and thin phytocaps clearly support the recommendation of the phytocapping technique for landfill remediation in many parts of Australia, especially in the drier regions of Queensland.

## ACKNOWLEDGEMENTS

This research was funded by the Rockhampton Regional Council (RCC) via Phytolink Australia Pty Ltd., and proudly supported by the Queensland Government's Growing the Smart State PhD Funding Program, and may be used to assist public policy development.

We are grateful to Mr. Craig Dunglison (RRC), Mr. Richard Yeates (Phytolink Australia Pty Ltd), Professor David Midmore, Central Queensland University (CQU), Dr Ram Dalal, Department of Natural Resources and Water, Dr. Jirka Simunek (University of California Riverside), Dr. Bill Albright (Desert Research Institute, Nevada, Mr. Roshan Subedi (CQU) and many others who assisted with this project in various ways.

## REFERENCES

Abichou, T., Palueson, D. and Chanton, J. (2004) Bio-reactive cover systems. Florida Centre for Solid and Hazardous Waste Management, Florida, pp. 1 - 37.

Albright, W. G., Benson, C. H., Gee, G. W., Roesler, A. C., Abichou, T., Apiwantrgoon, P., Lyles, B. F. and Rock, S. A. (2004) Field water balance of landfill final covers, *Journal of Environmental Quality*, 33: 2317 - 2332.

Albright, W. G., Gee, G. W., Wilson, G. V. and Fayer, M. J. (2002) Alternative cover assessment projects; Phase 1 Report. Desert Research Institute, Nevada, USA.

Auge, R. M., Stodola, A. J. W., Tims, J. E. and Saxton, A. M. (2001) Moisture retention properties of a mycorrhizal soil, *Plant and Soil*, 230: 87-97.

Bengtsson, L., Bendz, D., Hogland, W., Rosqvist, H. and Akesson, M. (1994) Water balance of landfills at different age, *Journal of Hydrology*, 158: 203-217.

Benson, C. H., Bohnhoff, G. L., Apiwantrgoon, P., Ogorzalek, A. S., Shackelford, C. and Albright, W. H. (2004) Comparison of model predictions and field data for an ET cover. In: *Tailing and Mine Waste*, pp. 137.

Benson, C. H., Albright, W. H., Roesler, A. C. and Abichou, T. (2002) Evaluation of final cover performance: field data from the Alternative Cover Assessment Program (ACAP). In: *WM 2002 Conference,* Feb 24-28, Tucson, pp. 1-17.

Chai, J. and Miura, N. (2002) Comparing the performance of landfill liner systems, *Journal of material cycles and waste management*, 4: 135-142.

Chammas, G. A., Geddis, M. and McCaulou, D. R. (1999) A comparison of two models for simulating the water balance of soil covers under semi arid conditions. In: *National Meeting of the American Society for Surface Mining and Reclamation,* Arizona.

Dho, N. Y., Koo, J. K. and Lee, S. R. (2002) Prediction of leachate level in Kimpo metropolitan landfill site by total water balance, *Journal of Environmental Monitoring and Assessment*, 73: 207-219.

Dwyer, S. F. (2003) Water balance measurements and computer simulations of landfill covers. PhD Thesis, University of New Mexico, Albuquerque.

EPA 2005 Guideline: landfill siting, design, operation and rehabilitation. Environmental Protection Agency, Brisbane, Australia.

Fayer, M. J., Rockhold, M. L. and Campbell, M. D. (1992) Hydrologic modelling of protective barriers: Comparison of field data and simulation results. *Journal of Soil Science Society*, 56: 690-700.

Fayer, M. J. and Gee, G. W. (1997) Hydrologic model tests for landfill covers using field data. *Landfill capping in the semi arid west: Problems, Perspectives and Solutions. Environmental Science and Research Foundation,* Idaho Falls, USA.

Fiedler, F. R., Reamirez, J. A. and Ahuja, L. R. (2002) Hydrologic response of grasslands: effect of grazing, interactive infiltration and scale, *Journal of Hydrological Engineering*, 7: 293-301.

Findeling, A., Garnier, P., Coppens, F., Lafole, F. and Recous, S. (2007) Modelling water, carbon and nitrogen dynamics in soil covered with decomposing mulch, *European Journal of Soil Science*, 58: 196-206.

Freebairn, D. M., Silburn, D. M. and Wockner, D. H. (1986) Effects of catchment management of runoff water quality and yield potential from vertisols, *Agricultural Water Management*, 12: 1-19.

Glinski, J. and Lipeic, J. (1990) *Soil physical conditions and plant roots,* CRC Press, Inc, Florida, pp 173 - 191.

Ham, J. M. (2002) Uncertainty analysis of the water balance technique for measuring seepage from animal waste lagoons, *Journal of Environmental Quality*, 31: 1370-1379.

Hauser, V. L. and Gimon, D. M. (2004) *Evaluating Evapotranspiration (ET) landfill cover performance using hydrologic models. Metretek Systems,* San Antonio, Texas.

Ho, C. K., Arnold, B. W., Cochron, J. R., Taira, R. Y. and Pelton, M. A. (2004) A probabilistic model and software tool for evaluating the long term performance of landfill covers, *Environmental Modelling and Software*, 19: 63-88.

Jackson, N. A. and Wallace, J. S. (2000) Soil evaporation measurements in an agroforestry system in Kenya, *Agricultural and Forest Meteorology*, 94: 203-215.

Johnson, C. A., Schaap, M. G. and Abbaspour, K. C. (2001) Model comparison of flow through a municipal solid waste incinerator ash landfill, *Journal of Hydrology*, 243: 55-72.

Johnson, A., Roy, I. M., Matthews, G. P. and Patel, D. (2003) An improved simulation of void structure, water retention and hydraulic conductivity in soil with the pore-core-three-dimensional network, *European Journal of Soil Science*, 54: 477-489.

Kalman, R., Sandor, K., Van Genuchten, M. and Per Erik, J. (1996) Estimation of water retention characteristics from the bulk density and particle size distribution of Swedish soils, *Soil Science*, 161: 832-845.

Khire, M. V., Benson, C. H. and Bosscher, P. J. (1997) Water balance modelling of earthen final covers, *Journal of Geotechnical and Geoenvironmental Engineering*, 123: 744-754.

Khire, M. V., Benson, C. H. and Bosscher, P. J. (1999) Field data from a capillary barrier and model predictions with UNSAT-H, *Journal of Geotechnical and Geoenvironmental Engineering*, 125: 518-527.

Lal, R., Wilson, G. F. and Okigbo, B. N. (1979) Changes in properties of an alfisol produced by various crop covers, *Soil Science*, 27: 377.

Liang, X. and Xie, Z. (2001) A new surface runoff parameterisation with sub grid-scale soil heterogeneity for land surface models, *Advances in Water Resources*, 24: 1173-1193.

Melchior, S. (1997) "*In situ* studies on the performance of landfill caps". In: *Proceedings of the International Containment technology Conference,* St. Petersburg, pp. 365-373.

Merta, M., Seidler, C. and Fjodorowa, T. (2006) Estimation of evaporation components in agricultural crops, *Biologia*, 61: 280-283.

Myers, B., Bond, W., Benyon, R., Falkiner, R., Polgase, P., Smith, C., Snow, V. and Theiviyenathan, S. (1999) *Sustainable effluent irrigated plantation: an Australian guide, CSIRO Division of Forestry*, Canberra.

Oostrom, M., Wietsma T.W. and Foster, N.S. (2004) The subsurface flow and transport experimental laboratory: a new department of energy users's facility for intermediate scale experimentation, *Hydrology Days,* pp 182-189.

Ross, P. J. (1998) SWIM a simulation model for soil water infiltration and movement. *CSIRO,* Townsville, Australia.

Scanlon, B. R., Marty, C., Reedy, R. C., Porro, I., Simunek, J. and Flerchinger, G. N. (2002) Intercode comparisons for simulating water balance of surfacial sediments in semi arid regions, *Water Resources Research*, 38: 1323 - 1338.

Scott, J., Beydoun, D., Amal, R., Low, G. and Cattle, J. (2005) Landfill management, leachate generation and leach testing of solid wastes in Australia and overseas, *Critical Reviews in Environmental Science and Technology*, 35: 239-332.

Shouse, P. J., Russell, W. B., Burden, D. S., Selim, H. H., Sisson, J. B. and Van Genuchten, M. (1995) Spatial variability of soil water retention functions in a silt loam soil, *Soil Science*, 159: 1-12.

Simunek, J., Van Genuchten, M. T. and Sejna, M. (2005) The Hydrus 1D software package for simulating the one dimensional movement of water, heat and multiple solutes in variably saturated media. University of California, Riverside, California, USA, pp. 240.

Tillman, R.W. and Surapaneni, A (2002) Some soil related issues in the disposal of land, *Australian Journal of Experimental Agriculture*, 42: 225-235.

Vasudevan, N. K., S. Vedachalam and D. Sridhar (2003). *Study on the various methods of landfill remediation*. Workshop on Sustainable Landfill Management, Chennai, India.

Venkatraman, K. and Ashwath, N. (2007) Phytocapping: an Alternative technique for reducing leachate and methane generation from municipal landfills. *The Environmentalist*, 27: 155 – 164.

Wallace, J. S., Jackson, N. E. and Ong, C. K. (2000) Modelling soil evaporation in an agroforestry system in Kenya, *Agricultural and Forest Meteorology*, 94: 198-202.

Warren, R. W., Hakonson, T. E. and Bostik, K. V. (1996) choosing the most effective hazardous waste landfill cover, *Remediation (online publication)*, Spring: 23-41.

Williams, L. L. (2005) Effect of plant intrusion on the water balance of landfill cover systems. In: *Graduate School of Vanderbilt University,* Vanderbilt University, Nashville, pp. 121.

In: Technologies and Management for Sustainable Biosystems ISBN: 978-1-60876-104-3
Editors: J. Nair, C. Furedy, C. Hoysala et al. 

*Chapter 19*

# TECHNICAL ADVICE ON WASTE MANAGEMENT LIVELIHOODS IN TSUNAMI AFFECTED AREAS OF NAD NIAS

***Le Ngoc Thu*, Jaya Nair and Martin Anda***
Environmental Technology Center, Murdoch University,
Western Australia 6150

## ABSTRACT

After almost three years after tsunami and earthquake in 2005 in Nias and Aceh, most rehabilitation efforts from international organizations are now addressing longer term economic development and livelihood improvements. As a part of UNDP project reference: RFP/UNDP/ERTP/013/2006 "Organisations/Firms to manage Tsunami Recovery Waste Management Program" (TRWMP), environmental experts from Murdoch University conducted a technical assessment for local NGOs to advise on waste management livelihood goals for the rehabilitation process. Pre-training visits in affected areas of Banda Aceh occurred in June 2007 to identify the potential project opportunities. The outcome of this mission was matched with local NGOs and community identification. These were 3 potential activities: biogas, coconut processing, and plastic recycling processing.

The two-day training workshops for each of three mentioned activities were conducted in February 2008 for NGOs and community representatives. These workshops have provided information to beneficiaries and NGOs staff on specific technical details for sustainable waste management livelihood opportunities.

**Keywords:** Biogas processing, coconut processing, plastic processing, sustainable livelihood, Nad Nias.

* { j.nair, m.anda}@murdoch.edu.au; Corresponding author. Tel: +61.8.93607322.

# 1. Introduction

Many donors and governments have provided support for the reconstruction process and livelihood improvement in Nias and Aceh after the tsunami and earthquake in 2005. Austcare, an Australian non-government international development organization, was one of the earliest international organizations to contribute the finance and human resources for the rehabilitation efforts in devastated areas. Austcare is working with local NGOs to develop many activities to improve the income of local people as well as protect environment. With Austcare supporting, technical advice has been conducted in June 2007 and February 2008 by environmental consultants from Murdoch University for local NGOs: PINBIS, Palapa Plastic Recycle (PPR), Yayasan Paramadiana Semesta (YPS), Ikatan Pemuda Keudee Aron (IPKA) to implement UNDP project reference: RFP/UNDP/ERTP/013/2006 "Organisations/Firms to manage Tsunami Recovery Waste Management Program" (TRWMP). The TRWMP is a joint initiative of UNDP and BRR (*Bandan Rehabilitasi dan Rekonstruksi NAD-Nias* which is an Agency for rehabilitation and reconstruction of Aceh-Nias) funded by the Multi Donor Fund (MDF). This is a part of a broader 5 year commitment to a livelihoods/waste management recovery programme.

The first mission in June 2007 mainly included the discussion, meetings and site visits across 3 regions (Kota Banda Aceh and Aceh Besar; Aceh Jaya / Aceh Barat / Nagan Raya / Simeulue; Nias / Nias Selatan) including a number of districts around the affected towns of Banda Aceh, Muelaboh, Calang and Gunung Sitoli and Teluk Delam(Nias) to identify:

i. Existing and potential waste related livelihood opportunities for target grantees that could be funded by Austcare
ii. Existing and potential waste related project opportunities across the 3 regions; and
iii. Gender considerations.

The second mission was conducted in February 2008 with training workshops for three potential subjects: Biogas, coconut processing, plastic recycling processing were introduced to local NGOs staff, other stakeholders as well as representatives of beneficiaries. These livelihood activities have been identified by the community and local NGOs and are matched with outcome from the first mission. The main beneficiaries will be the villagers in the Nias project area in which biogas digesters are planned to be installed; the villagers in Aceh Barat District who have many coconut trees and coconut waste; the Naga Saewe industry who are developing poly bags; as well as, scavengers in Aceh Barat District. Integrated with technical information, the sustainable livelihoods assessment strategy has applied to help implementers understanding the overall aspects of the project with the target being the local people in project areas. With the success from demonstration of these activities, the Austcare programme will be extended to support ongoing implementation and expansion at a local level.

This paper discusses the technical advice and training programme undertaken to establish livelihoods from waste management projects.

## 2. METHODOLOGY

The training programmes were developed from the outputs of the previous and first visit by the consultants and the identifications of local NGOs and community; they were identified for the second mission which is the training programme.

As the first step, the consultant under the arrangement and participation of Austcare local staffs, NGOs staffs, conducted site visits in the intended project areas to increase the understanding of the existing situation of the local people and activities relate to desired projects. The next step was organizing two day workshops in each identified activity with relevant NGOs staff and beneficiary representatives. The workshop was organised as a session to ask questions, to understand the knowledge and understanding of the participants and the desired outcomes expected by the participants through implementing the projects. This was an important stage to understand the expectations as well as potential livelihood opportunities identified by different stakeholders.

The third step was the technical presentations which covered detailed suggestions and potential analysing activities, the required inputs, the operation and maintenance, the possible outputs and the applications of desired project in the simplest way that was suitable to audience knowledge level. Also, many closely related case studies, demonstration videos on the livelihood activities and relevant projects in other places in the world were introduced to provide better graphical effects of the presentation to audiences. Three potential projects which were assessed and demonstrated to increase the sustainable livelihood in Banda Aceh were the Biogas plants for treating organic waste, coconut waste processing and plastic waste recycling.

### 2.1. Biogas

Biogas is a sustainable energy source which has been widely applied at the global level, especially in rural areas of developing countries. In rural areas, daily domestic energy needs for lighting, cooking, heating are sourced from wood, agriculture residues, charcoal and animal dung which are collected by women and children.. Traditional burning of biomass causes many direct adverse impacts to human health, mainly women and their children due to inhalation of smoke and soot followed by several long term environmental issues such as deforestation, soil corrosion, climate change, flooding. It has been proved that instead of using traditional energy sources, farmers can use animal dung, agriculture residues and even human excreta as the input for an anaerobic process which produces biogas (60% $CH_4$) (*Chongrak, 2007*), enough flammable gas for all their daily energy needs. The slurry, the secondary output, is a stable fertilizer which is suitable for agriculture crops or other purposes like food for fish, nutrients for aquatic plants, mushrooms and earthworm. The overall picture of biogas processing is shown in figure 1.

With simple operation conditions, sufficient gas for cooking or lighting for rural villagers can be obtained. Constructing biogas plants can reduce the cost for energy and increase the income from their more productive crops or selling fertilizers from slurry. Businesses can be developed in association with the projects in monitoring the performance of the plants and constructing plants. These plants are also efficient systems to treat organic waste from households and farms which will prevent the environmental pollution. When observing the benefits from biogas, farmers will transfer knowledge to other neighbours, not only concerning operation and maintenance but also how to develop the new biogas system by themselves. Whenever the demonstration project is successfully demonstrated, it can be diffused to other places, which will create the local jobs for livelihood.

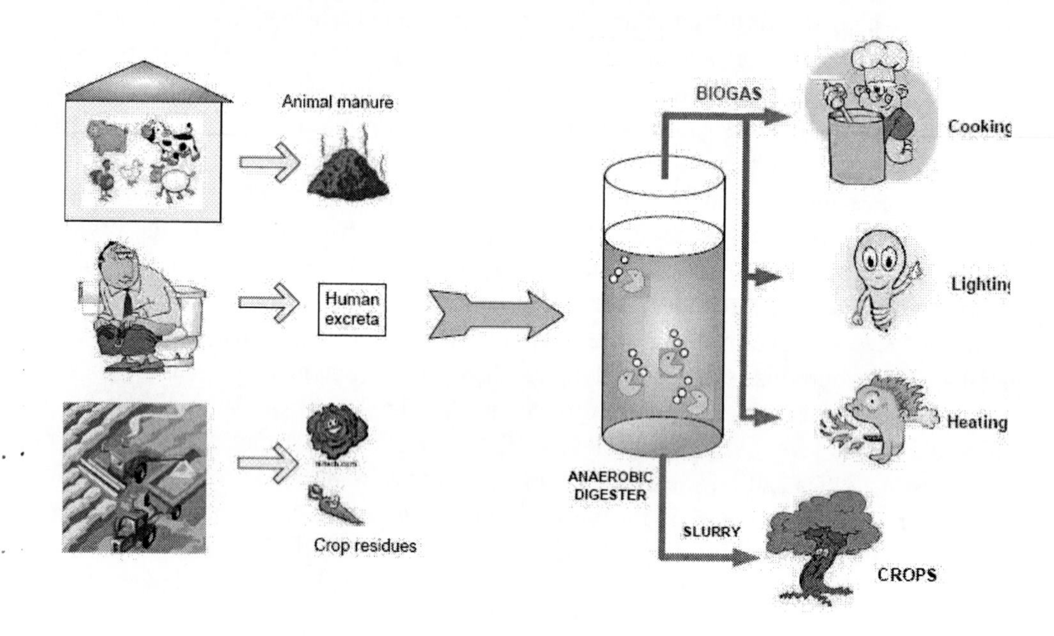


Figure 1. overview of application of biogas plants to rural livelihood.

## 2.2. Coconut Waste Processing

Banda Aceh being a tropical place with plenty of coconut trees, livelihood projects around this tree was considered an ideal project utilising local expertise. The coconut palm is a versatile plant with a variety of uses. Every part of it can be utilized in one form or the other, supplying food, drink, shelter and also raw materials for a number of industries. Coconut meat can been used to make coconut oil, coconut juice, coconut milk or other food products that are widely consumed in the local market or exported.

These can be set up as small scale business to provide livelihood for local people. The coconut waste such as the husk and shell are normally dried and used for cooking, as they have low economic value and are difficult to transport for selling in the market. Most cases it can be left without being properly utilised. Many value added products can be manufactured from coconut waste, such as activated carbon from the shell and fibre from the husk. These potential activities can increase the income for local people and create job opportunities from coconut waste, as shown in figure 2.

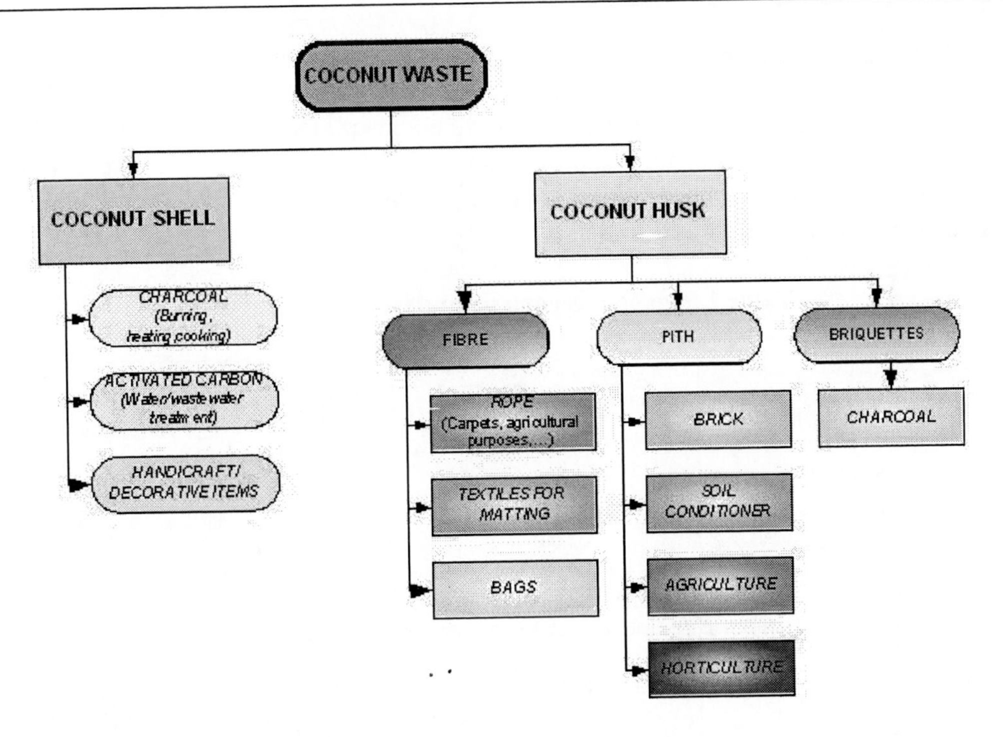


Figure 2. Potential products from coconut waste.

## 2.3. Plastic Recycling

In Banda Aceh after the Tsunami, there has been a high influx of plastic brought in by the aid workers as well as bottles and other packagings. This has created severe pollution problem in the province which was not so prominent in the pre- Tsunami period. Being light, durable, malleable, hygienic and economic, plastics are suitable for a wide variety of applications including food and product packaging, car manufacturing, agriculture and housing products. Most plastics are not biodegradable so it can cause many problems to environment. Handling plastic waste, both collection and recycling them was therefore considered as another potential livelihood project for Banda Aceh. Recycling plastic not only protects the environment but also creates the income for many people in the society. Currently, plastic recycling in Aceh, Nias are run by individuals unaware of how to access larger markets, how to be economical sustainable or how to reduce the volume of recyclable materials for more efficient transportation (UNEP Indonesia, Press Releases, 11/7/2007). The training provided the knowledge to change such issues and return the value of recycling plastic back to local people. Different plastic codes are separated and used to produce different products as shown in table 1. Normally, a discarded end product – such as plastic water bottle – is chopped, washed, melted and formed into plastic pellets. These pellets are then sold to plastic reprocessors to create a "new" or "recycled" product. Plastic is also recycled in making roads (Central pollution Control Board, Government of India) or concrete or substrate in constructed wetlands for wastewater treatment (Stewart et. al., 2004).

## 2.4. Workshop Sessions

The technical aspects of the three projects were introduced through 3 different workshops to provide background information and awareness of new technologies. The products that have been developed and applied in other countries and the potential to use those technologies in sustainable livelihoods projects in Banda Aceh.

**Table 1. Plastic categories and potential products**

| Code | Plastic categories | Recyclability | Potential products |
|---|---|---|---|
| (1) | PETE- Polyethylene Terephthalate | Commonly recycled | Textiles, polyesters carpets; fiber filling for pillows, quilts, jackets; clear sheets or ribbon for VCR and audio cassettes; bottles |
| (2) | HDPE –High density Polyethylene | Commonly recycled | Plastic pipes, lumbers, flower pits, trash cans, non food application bottles |
| (3) | PVC- Vinyl/Polyvinyl Chloride | Generally not recycled, except perhaps in small test programmes | |
| (4) | LDPE –Low density Polyethylene | Less commonly recycled | Plastic trash bags; grocery sacks, plastic tubing, agricultural film, and plastic lumber. |
| (5) | PP- Polypropylene | Generally not recycled, except perhaps in small test programmes | |
| (6) | PS- Polystyrene | Generally not recycled, except perhaps in small test programmes | |
| (7) | Other- Usually layered or mixed plastic | Generally not recycled, except perhaps in small test programmes | |

The sustainable livelihood framework (figure 3) was the underlying tool guiding workshop task, providing the systematic knowledge for NGO staff and participants who will be responsible for undertaking proposed activities. The instructor briefly provided background on this tool and gave handouts which have practical examples or information directly related to the tasks. The output of these tasks was the completed livelihoods framework which included the strategies, institutions needed to get the expected targets based on the existing resources. The core of all these activities was the people in project areas. The workshop also analysed the requirements for further sustainability of all activities after funding has ceased.

Another important part of workshop was the presentation from selected local participants with relevant expertise and already well understanding with their Austcare funded project. They contributed to the final outcomes which are more suitable to the local conditions and other less experienced grantees. That was a main factor to ensure the success of the demonstration projects.

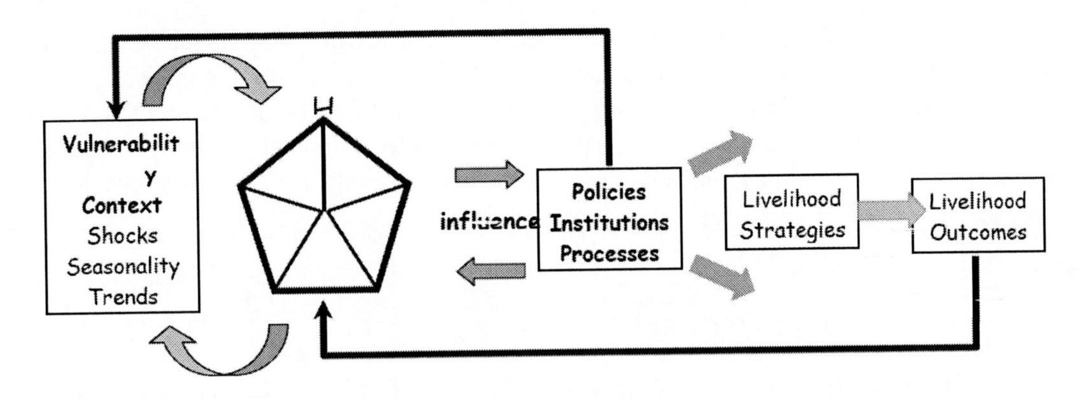


Figure 3. Sustainable livelihood framework (Sustainable Livelihoods Guidance Sheets).

## 3. Results and Discussion

The training programme has laid out the technical foundation for representatives of NGOs staff, main beneficiaries who participant in the planning and implementation this project. The conducted workshops have drawn out the clearer picture of the actual situation and the potential opportunity as well as benefits in terms of job creating for both men and women in rural areas. The equipped technical information is essential for them to select and develop the suitable livelihood activities to their potential conditions. Project implementers have approached to the advantages and disadvantages of their intended activities as well as the new technologies and equipment applied in other places. Through the site visit and discussion, the international environmental experts have provided their precious ideas of livelihood activities to the specific local conditions and the located fund accordingly.

The Austcare investments have led to the establishment of significant projects that represent major steps forward in innovative waste management livelihoods. Ongoing monitoring by Austcare project officers will hopefully determine whether they are sustainable at the respective sites and if it seems they can be independently transferred to other sites. Also, the training workshops were independently evaluated by Austcare and some of the participants' feedbacks are presented.

The conclusions and recommendations for each workshop derived from running the workshops and interacting with the participants and the sites are as follows:

### 3.1. Biogas Workshop

The biogas project on Nias if sustainable and transferable has the potential to revolutionise rural village-based energy supplies as well as reduce the current environmental impact of manure disposal in waterways from piggeries. The achievements of local initiative in these early stages of implementation (2 out of 5 digestors installed) were impressive. Successful commissioning and sustained operation of the digestors will be critical. To ensure successful commissioning, five Nias pilot biogas digestors must be close monitored. Once these are running successfully promotional activities should be considered to enable the diffusion of this technology to new sites. Consideration should be given to the use of the plastic bag digestor model if the pilot is a success as these will be significantly cheaper. Research into this model can commence now.

For participants, the training workshop created awareness of biogas and its benefits. They expected further similar training particularly with the wider community and for the local government officers and other supports not only for biogas development but also for animal breeding.

### 3.2. Plastic Processing Workshop

The plastics collection and sorting centre by PPR at Kampong Jawa was a great achievement and would contribute to increased outputs of processed plastics from the Lhokseumawe plant as well as improved livelihoods to 'scavengers' (waste pickers).

The mission of PPR to grow collection capacities is clearly achievable and the goal to secure new investment to add new processing machines at Lhokseumawe should be supported. The latter will lead to important industrial and economic development for Aceh. Additional support should be provided for the establishment for plastics wastes reprocessing facilities and machines to make products such as plastic bags in Banda Aceh and Meulaboh and plastic plant pots in Nias. Research into these opportunities can commence now.

The feedback from nine participants showed that they wanted assistance to develop the cleaning tool and more practical application case studies like recycling in onstruction and agriculture.

### 3.3. Coconut Processing Workshop

The successful commissioning of the new coconut wastes processing facility by IPKA at Meulaboh with machines from Bogor/Jakarta (Pt Hinoka Asindo Technik) is a great step forward in industrial development in this region. It appears this may give a good livelihoods boost to the 30 communities from which the facility draws its wastes. IPKA was pleased that the visit to the Department of Industry during the workshop has led to tentative arrangements for further training and support from government officials in Meulaboh which could be funded by Austcare as per the existing budget.

Once the coconut processing machines and livelihoods are well established in Meulaboh serious consideration can be given to the acquisition of additional machines for charcoal briquette making, steam activated carbon, coconut shell fibre. Feedback from four participants showed that research into these machines and markets can commence now.

The supporting sources from international organizations and governments are essential for redevelop the poor areas such as Banda Aceh especially after terrible damages from natural disaster. The livelihood can be improved through well-establish and suitably running activities which have been identified as mentioned mainly in this paper that are: biogas, coconut processing, recycling paper. These activities will not only bring the income for local people but also help to improve the sanitation and environmental conditions of the project area. Other aspects of the living standard such as water supply, energy supply also need to be concentrated to ensure the good living conditions for local people.

## REFERENCES

Chongrak Polprasert, 2007. Organic Waste Recycling: *Technology and Management.* IWA Publishing, 3rd edition.

Central pollution Control Board, Ministry of Environment and Forest, Government of India (http://www.cpcb.nic.in/Plastic%20Waste/default_Plastic_Waste.html).

Stewart Dallas, Brian Scheffe, Goen Ho, 2004. Reedbeds for greywater treatment—case study in Santa Elena-Monteverde, Costa Rica, Central America. *Ecological Engineering,* Vol. 23, 55–61.

A.S.F. Santos, B.A.N. Teixeira, J.A.M. Agnelli, S. Manrich, 2005. *Characterization of effluents through a typical plastic recycling process: An evaluation of cleaning performance and environmental pollution.* Resources conservation & Recycling, Vol. 45, 159–171.

Sustainable Livelihoods Guidance Sheets, Department of international Development, http://www.livelihoods.org/info/guidance_sheets_pdfs/section2.pdf

# SECTION FIVE: COMMUNITY GOVERNANCE

In: Technologies and Management for Sustainable Biosystems ISBN: 978-1-60876-104-3
Editors: J. Nair, C. Furedy, C. Hoysala et al. 

*Chapter 20*

# BIOENERGY THROUGH SUSTAINABLE MANAGEMENT OF BIOTECHNOLOGY AND BIODIVERSITY

***Charles K. Twesigye****
Department of Biological Sciences, Kyambogo University,
P.O. Box 1 Kyambogo, Kampala, Uganda

## ABSTRACT

The aim of this chapter is to provide a conceptual framework to help mitigate the risks of interventions to promote bioenergy development in developing countries which still have substantial amount of biomass resources. Biotechnology is increasingly recognized as a critical aspect of the development process, especially with the growing awareness of the role of science, technology and innovation in economic renewal. There are technological and economic challenges that have to be overcome in developing sustainable bioenergy production systems. The need to build capacity to harness biomass for bio-fuels and other forms of bio-energy through biotechnology and bioengineering, but without harming the natural resource base upon which life depends, is highlighted. In the context of globalization, this chapter calls on developing countries, especially from the tropics, and their development partners, to take advantage of abundant biomass resources and develop appropriate technologies for bioenergy production that will improve their incomes and alleviate poverty levels in the region. Conservation and rational utilization of natural resources in Africa and other communities throughout the world require the involvement of all stakeholders, regional and international cooperation and private-sector investment. Although there seems to be potential in developing bionenergy in the continent of Africa, it is not possible at this time to assess the balance of benefits and costs. However, there are undoubtedly risks entailed in the application of the technologies now being developed, and African countries need a way to assess these risks and try to reduce them, since it seems inevitable that international agencies, companies and governments are going to push ahead with applying biotechnologies in the region. The need for a conceptual framework as a starting point to identifying and reducing risks associated with bio-energy development is highlighted.

---

* Phone: 078-2353775,Fax:+256-414 220464, E-mail:twesigyeck@yahoo.com, cktwesigye@mulib.mak.ac.ug

# Introduction

Bioenergy is a term used to describe energy derived from organic materials such as living plants and plant components. Crops grown for bio-energy include traditional crops such as corn, sugar cane, wheat and oilseed rape, but also dedicated energy crops, like short-rotation willow-coppice, unusual grasses and forestry products. The production of bio-energy is gathering more and more attention as a feasible way of reducing dependence on imported oil and gas and is even being hailed as one of the potential key weapons in the battle against global warming. As bio-energy comes into the spotlight with a view to promoting its production in developed and developing countries, especially from Africa and Asia, it is important for the developing countries to explore the issue in depth, to discover what growth in the 'new industry' could really mean for the environment and biodiversity. If managed sustainably, the use of biomass, biogas and biofuels could help to cut greenhouse gas emissions. However, without adequate regulation and a well developed policy on the industry, widespread and rapid uptake of growing crops for bio-energy production could have devastating impacts on the environment.

Many countries recognize biomass as a domestic energy resource, and some see opportunities for exports of liquid biofuels. With political goals of many countries to increase the use of biofuels in the transport sector and domestic biofuel quota systems being introduced in many other countries as well, there is little doubt that biomass use for liquid transport fuels, as well as for electricity and heat production, will continue to rise in the future. The global trade with bioenergy will also rise in parallel. This will pose both opportunities and risks for sustainable development for regions, countries, and the world as a whole.

Economic and geopolitical factors which include high oil prices, environmental concerns, and supply instability, have been prompting policy-makers to put added emphasis on renewable energy sources. For the scientific community, recent advances in biotechnology and bioengineering which can be applied to metabolic engineering, are generating considerable excitement. There is justified optimism that the full potential of biofuel production from cellulosic biomass will be obtainable in the next 10 to 15 years(Stephanopoulos 2007).

Biodiesel has become more attractive recently because of its environmental benefits and the fact that it is made from renewable resources. The cost of biodiesel, however, is the main hurdle to commercialization of the product. Research is needed to improve our understanding of the mechanism and kinetics of bio-energy production processes. The great appeal of bioenergy is that it is theoretically a renewable source of energy. Crops can be converted to energy either by being processed into liquid fuel for the transport sector (biofuels) or by being burnt in power plants (biomass). With combustion, carbon dioxide is released. Because they are derived from biological material, the carbon released on combustion comes from carbon dioxide absorbed by plant material through the process of photosynthesis. Effectively, producing energy from biofuels or biomass could be seen as recycling carbon dioxide.

However, this rosy picture is over-simplistic. Bioenergy production is never a neutral process when it comes to “greenhouse gases”. During production, processing and transportation of the crops there are many other inputs to consider. Life-cycle analyses of bioenergy production have shown that with poor management methods, production can actually result in a net increase in the emission of greenhouse gases.

## Global Implications of Increasing Use of Bio-Energy

The rush to produce bio-energy crops is alarming for biodiversity on a global scale. Some of the most promising crops for producing biofuels are oil palm and Soya, two rapidly expanding tropical monocultures which are amongst the chief causes of tropical deforestation. A huge surge in demand for biodiesel could drive even further the large scale clearing of forests in key biodiversity hotspots such as Indonesia, Brazil and Uganda. A number of countries have made plans to clear millions of hectares of forest to create palm oil plantations dedicated to biodiesel production. Losing such large areas of precious habitat will serve a death sentence to threatened bird species like the Sumatran Ground-cuckoo *Carpococcyx viridis* and also threatened mammal species such as the Orang-utan *Pongo pygmaeus*.

Increasing energy use, climate change, and carbon dioxide ($CO_2$) emissions from fossil fuels make switching to low-carbon fuels a high priority. Biofuels are a potential low-carbon energy source, but whether biofuels offer carbon savings depends on how they are produced. Converting rainforests, peat lands, savannas, or grasslands to produce food crop–based biofuels in Brazil, Southeast Asia, and the United States creates a biofuel carbon debt by releasing 17 to 420 times more $CO_2$ than the annual greenhouse gas (GHG) reductions that these biofuels would provide by displacing fossil fuels. In contrast, biofuels made from waste biomass or from biomass grown on degraded and abandoned agricultural lands planted with perennials incur little or no carbon debt and can offer immediate and sustained GHG advantages (Fargione *et al.*2008).

Recent studies have revealed that current corn ethanol technologies are much less petroleum-intensive than gasoline but have greenhouse gas emissions similar to those of gasoline(Farrell *et al.*2006). However, many important environmental effects of biofuel production are poorly understood. New metrics that measure specific resource inputs are developed, but further research into environmental metrics is needed. Nonetheless, it is already clear that large-scale use of ethanol for fuel will almost certainly require cellulosic technology. Negative environmental consequences of fossil fuels and concerns about petroleum supplies have spurred the search for renewable transportation biofuels. To be a viable alternative, a biofuel should provide a net energy gain, have environmental benefits, be economically competitive, and be producible in large quantities without reducing food supplies. These criteria have been used to evaluate, through life-cycle accounting, ethanol from corn grain and biodiesel from soybeans. Ethanol yields 25% more energy than the energy invested in its production, whereas biodiesel yields 93% more (Hill *et al.*2006). Compared with ethanol, biodiesel releases just 1.0%, 8.3%, and 13% of the agricultural nitrogen, phosphorus, and pesticide pollutants, respectively, per net energy gain. Biodiesel provides sufficient environmental advantages to merit subsidy. Transportation biofuels such as synfuel hydrocarbons or cellulosic ethanol, if produced from low-input biomass grown on agriculturally marginal land or from waste biomass, could provide much greater supplies and environmental benefits than food-based biofuels.

## Bioenergy and Global Energy Use

The role of biomass in future global energy provision has been analyzed quite well, while the respective impacts in changing food, feed and fiber markets are yet under discussion. Bioenergy is seen by some to be a panacea for a range of energy, environment and poverty problems. However, the sustainability performance of bioenergy depends on where and how it is produced, processed, and used.

The foundation of a bio-based industry depends on an abundant supply of biomass. The most widely used liquid biofuels are bioethanol and biodiesel. Biofuels are at present classified in two categories: (i)first-generation and (ii) second generation fuels. While there are no strict technical definitions for these terms, the main distinction between them is the feedstock used. A first-generation fuel is generally a fuel made from sugars, grains, or seeds; that uses only a specific (often edible) portion of the aboveground biomass produced by a plant; and that is the result of a rather simple manufacturing process. First-generation fuels are already being produced in significant commercial quantities in a number of countries. Second generation fuels are generally those made from non-edible lignocellulosic biomass, either non-edible residues of food crop production (e.g., corn stalks or rice husks) or non-edible whole-plant biomass (e.g., grasses or trees grown specifically for energy).

There are two routes for producing second generation biofuels: enzymatic hydrolysis, which leads to biological second generation biofuels; and gasification which leads to thermochemical second generation biofuels. Cellulosic ethanol is an example of a biofuel obtained through a biological process. While cellulosic ethanol can be produced today, producing it competitively (i.e., without subsidies) from lignocellulosic biomass still requires significant successful research, development, and demonstration efforts. Second-generation fuels are not yet being produced commercially in any country. First generation biofuels have several limitations. They compete with food uses and plants have been optimized for food, not energy use. Only part of the plant is converted into biofuel. They bring only modest greenhouse gas (GHG) emissions mitigation benefits, except for sugarcane ethanol. While first generation biofuels can be blended with existing fossil fuels and so require minimal infrastructure change, there is limited large-scale experience outside Brazil and the United States. Second generation biofuels have some clear advantages. Plants can be bred for energy characteristics, and not for food, and a larger fraction of the plant can be converted to fuel. Lignocellulosic crops can be grown on poor quality land, requiring fewer fertilizers. There are substantial energy and environment benefits compared with most first generation biofuels due primarily to greater biomass usability per unit of land area. Second generation biofuels have greater capital-intensity than first generation biofuels, but lower feedstock costs.

## Sustainable Bioenergy Potentials

Given the substantial global potentials for sustainable provision of bioenergy, it could significantly contribute to transport fuel needs, and overall energy supply. The sustainable potential of bioenergy depends on the developments in agriculture and forestry, as well as the overall dynamics of the food, feed and fiber markets. Its potential further depends on the impact of global climate change, and the regionally differentiated adaptation measures to adjust to that change. There is a complex interaction of various driving forces, and massive feedback loops which make projections a matter of large uncertainty. However, the development of the bioenergy potential could increase income of exporting countries, and revenues for farmers and the forestry sector, favor rural job creation, and reduction of import bills for fossil fuels (UN Energy 2007). The other side of those opportunities is severe risk. Since current biofuels stem from agricultural crops, arable land use competition, rising food prices and food insecurity (FAO 2008a), water resource depletion, and deforestation could arise (CBD 2008). Similarly, increased bioenergy from agricultural and forest residues, industrial wastes and even from marginal and degraded lands could impact on local communities, and negatively affect poor people as well as soils and biodiversity.

Effects of bioenergy on the environment and natural resources can be either positive or negative, strongly depending on location, agricultural and forestry practices, previous land-use, and downstream conversion systems, including distribution and consumption. As current bioenergy production is closely related to agricultural crops, environmental impacts tend to be similar. Bioenergy offers significant opportunities to reduce greenhouse-gas emissions when replacing fossils fuels, provide low-sulfur biofuels which are biodegradable, and diversify plant varieties and cultivation practices, thus increasing agrobiodiversity. Furthermore, bioenergy production could make use of perennial plants requiring less agrochemical and water inputs than traditional crops, and could help reduce soil erosion. In addition, bioenergy feedstocks could come from cultivating degraded land, and from agricultural, forestry, and other organic residues and wastes, thus relieving pressure on arable land, and respective price and land-use change impacts. Depending on the developing path of bioenergy, the overall environmental impacts could be similar to agriculture, could add pressure, or could be positive. Furthermore, they depend on the production levels, scales, and conversion routes. Large-scale bioenergy production bears the risk of trade-offs between e.g., GHG savings, and the protection of natural resources such as biodiversity, soil and water.

Small-scale, distributed provision of bioenergy could be less effective in terms of land productivity and GHG reduction per hectare, but might adjust better into ecosystems and landscapes, and might offer more (agro)biodiversity. The promise of non-food ligno-cellulosic feedstocks and advanced biofuel production methods using forest, crop, and urban residues is yet a claim.

## Bioenergy and Land-Use Impacts on Biodiversity

Biodiversity is directly linked to properties and quality of habitats, and habitat loss is the most important threat to biodiversity. Given the globally rising demand for biofuels, increasing amounts of land will be used for respective feedstock production which could both directly and indirectly result in further habitat loss if forest, grass, peat or wetlands are affected. With the target to hold global biodiversity loss until 2010, there is an urgent need to protect land with high biodiversity value and ecosystem services from further deterioration. In that regard, agricultural cultivation for food, feed, fiber or fuel needs to avoid such areas unless biomass extraction conforms to protecting or enhancing biodiversity. In other areas, cultivation practices should respect biodiversity and agrobiodiversity in using many varieties of crops, adequate rotation schemes, minimum agrochemicals, and include specific landscape elements.

Key elements to be considered should be opportunities for sustainable biomass feedstock provision which have no negative or even positive environmental, biodiversity, climate, and social trade-offs. Efforts in seeking alternative sources of renewable energy should include developing technologies to harness biomass, solar, hydro, and wind resources. Solar power plants in the Sahara desert can supply the world with vast quantities of energy. Each square kilometre of African desert every year receives solar energy equivalent to 1.5 million barrels of oil. Solar energy received by deserts worldwide is nearly 1,000 times the world's entire annual energy consumption.

## Conceptual Framework for Bioenergy Development

There is need for a conceptual framework to guide sustainable biomass production. The aim of the framework on sustainable biomass production should be to mitigate risks. Land has to be categorized in areas where no bioenergy should come from, and those were biodiversity-friendly bioenergy production or residual extraction is possible. In a first step, relevant data need to be collected on global, national and local scales. A focus should be set on the characterization of areas relevant for the protection of biodiversity and critical watersheds as well as on environmentally "compatible" practices for biomass production. This information needs to be stored in a comprehensive GIS-database. In a second step, Protected Areas (PA) and biodiversity-relevant areas as well as prior bioenergy cultivation systems (including landscape structure) and residual extraction with low negative or positive impacts on biodiversity need to be identified. This screening, however, must be based on internationally accepted criteria and indicators for the principle of sustainability. The most common definition of Sustainability or Sustainable Development was given by the World Commission on Environment and Development (the Brundtland Commission) in 1987. It means to satisfy our present needs without compromising the future generations' ability to meet their own needs. The definition implies the balance of three components, stated in the Declaration of Rio on Environment and Development in 1992: 1) Environmental protection, 2) Economic growth and, 3) Social development.

Sustainable development demands joint reliance on the three criteria: preserving the environment, satisfying the human needs in a social fairness way and stimulating progress. In this context, the sustainable production of bioenergy is defined as "the production of biomass-based fuels for transportation, heat and electricity generation that allows an economic growth preserving the natural environmental and promoting a well balanced social development". In this chapter, biofuels stand for liquid fuels derived from biomass and used for transport purposes.

A principle is a premise of reasoning or action that is formulated based on social values, tradition and scientific knowledge (FAO, 2002). A set of principles define the contextual framework for developing a sustainability assessment system. It provides the basis of indicators, criteria, compliance checkers and verifiers. In order to assess the fulfillment of sustainability principles it is necessary to determine what the conditions that verify them are. A criterion is a set of conditions by which an object is assessed for given dimensions. It defines the rules to be satisfied in order to accomplish the sustainability principle and operationally translates the meaning of the principle. To satisfy one principle a set of criteria has to be verified. The measurements for verifying the criteria are called indicators. An indicator provides a consistent and clear measure of an attribute of the system under study that when satisfying the sustainability rules (criteria) it contributes to the accomplishment of a sustainability principle. An indicator is a quantitative or qualitative variable that can be measured. A verifier is a set of data that provides meaning, precision and site-specificity to the indicator.

For example recent field studies based on Remote Sensing Techniques (RSTs) and GIS data bases have been used in mapping and assessing degradation levels in the Lake Victoria Basin and the Nile Basin so as to inform decisions on land use. Global GIS data is needed on the following aspects:

- Degraded land, abandoned farmland
- Priority farming systems (conservation, multi-cropping, low-input etc.)
- Landscape elements within agricultural land
- Compatible residue extraction rates
- Country and ecoregion boundaries
- Protected areas (PA)
- Forests, wetlands-important bird areas
- Global land-cover maps
- Biodiversity-rich areas
- Location of PA and surrounding buffer zones
- Crop cultivation & residue extraction conforming with protection or enhancing of Biodiversity -friendly cultivation of crops or residue extraction
- Agro-Ecological Zones

Soil degradation, and especially soil erosion, can increase from annual bioenergy crops due to, e.g., tilling, excess irrigation, agrochemicals, and heavy farm equipment. In contrast, perennial bioenergy crops could improve soils and help to reduce erosion by creating year-round soil cover. For all cultivation systems, extraction of agricultural and forestry residues needs to reflect soil carbon and nutrient flows. Organic wastes and residues from agriculture, forestry, industry and households are prime options, as they offer very low GHG profiles, and do not induce risks for indirect land-use through displacement. Other possibilities are: (i) to develop low-input bioenergy cropping systems and the use of conservation agriculture practices;(ii) to sequester carbon in forests, grasslands and agroforestry systems using perennials, including short-rotation coppice plantations, and (iii) degraded lands for cultivation of bioenergy feedstocks without displacement risks.

## SUSTAINABLE BIOMASS PRODUCTION AND BIODIVERSITY CONSERVATION

It is well known that biomass production for biofuels can have both positive and negative impacts on biodiversity (CBD 2008). The challenge is to mitigate negative effects and to promote the positive ones, especially those that arise from direct and indirect land-use change.

Sustainable strategies should include protection of natural habitats which serve as cornerstones of regional conservation strategies. They are dedicated to the protection of biodiversity, agrobiodiversity, and natural and associated cultural resources. These areas should represent the biodiversity of each region, and they should separate this biodiversity from processes like habitat loss, habitat fragmentation and isolation, land- use intensification and overexploitation as well as species invasions threatening its persistence, e.g., by enforcement of land-use restrictions.

Existing Protected Areas (PA) throughout the world contain only a biased sample of biodiversity, usually that of remote places and other areas unsuitable for commercial activities. Thus, they do not come near to fulfilling global biodiversity commitments, nor the needs of species and ecosystems, given that a large number of these species, ecosystems and ecological processes are not adequately protected by the current PA network (Dudley&Phillips 2006).

To mitigate risks from bioenergy on biodiversity, areas need to be evaluated that are of importance for the protection of biodiversity, but that are currently not protected. Both, PA and currently unprotected biodiversity-relevant areas need the same strict protection status in order to withstand additional land-use pressure occurring from biomass production. In addition, forests are not allowed to be converted to agricultural land or plantations.

## CULTIVATION PRACTICE FOR BIOMASS PRODUCTION

Today, it is widely accepted that the implementation of conservation goals for the protection of biodiversity requires systematic planning strategies for managing landscapes, including areas allocated to both production and protection (Benedict & McMahon 2006). The CBD recognizes the limitations of PA as the sole tools for conservation, and promotes an Ecosystem Approach which seeks to mainstream biodiversity conservation into broader land- and seascape management (Smith &Maltby 2003, Dudley & Phillips 2006).

For successfully meeting development and sustainability goals, a fundamental shift in agriculture is needed that protect the natural resource base and the ecological implications of agricultural systems. Cultivation practices which respect biodiversity and agrobiodiversity require broad varieties of plants, adequate rotation schemes, low-erosion land-use methods (e.g. no-till systems), and minimal agrochemical application. Furthermore, the inclusion of specific landscape elements (e.g., stepping stones, corridors, buffer zones etc.) in the cultivation area must be considered. Approaches for environmentally “compatible” biomass production systems which include biodiversity concerns have been suggested (EEA 2006, 2007), but are still far from implementation. The cultivation of biomass on degraded land or abandoned farmland can safeguard against negative indirect land-use change effects from bioenergy development (Searchinger 2008).

Several global and local data exist that can directly be used in the location of natural habitats that are suitable for production of bioenergy crops and data on land-cover, degradation and environmental suitability of land for agriculture. Though a great deal of data are available, much of the data lack the correct resolution to be informative enough or they do not directly cover the correct scope. But even if agreed criteria and indicators and all spatial data would be available for a "screening" of land with regard to its biodiversity relevance, the second limitation needs to be considered. Local "hot spots" of biodiversity might easily be overlooked, and the social situation regarding land-use is of importance for sustainability as well. Therefore, stakeholder involvement and "bottom-up" knowledge from the ground are required to make the conceptual framework a sound tool for sustainability. One of the largest projects in Africa that aims at assessing sustainable management and utilization of renewable natural resources is the Lake Victoria Research Initiative(VIRES) based in the Lake Victoria Basin. The project uses Remote Sensing Techniques and GIS based analysis in mapping land use patterns and predicting the rate and dynamics of land cover changes and their impact on biodiversity. The generated data is used to inform decision making for future land use based on ecological principles and environmental restoration objectives (Twesigye *et al.* 2007). Lake Victoria, the second largest fresh water lake, has a very narrow watershed that is rich in biodiversity which has become under pressure due to increased urbanization and deforestation as a result of using solid fuel in form of charcoal and firewood. The development of more efficient forms of biofuel would help to alleviation the situation.

## THE BIOENERGY POTENTIAL IN AFRICA

Africa's bioenergy potential is almost unlimited. Harnessing it requires, at least, a two-pronged strategy: (i) promoting smallholder production and processing schemes; and (ii) encouraging socially and environmentally-sustainable large scale investment (Ejigu 2007). Of the total energy Africa consumes, traditional biomass (solid wood, twigs and cow dung) accounts for 59 per cent providing fuel to about 320 million people and, according to forecasts by the International Energy Agency (IEA), this figure is set to increase. By 2015, the IEA predicts a further 54 million Africans will be dependent on traditional biomass. This surge in consumption comes primarily from two sources: (i) a rapid population growth estimated at 2.5 per cent (the world's highest); and (ii) slow industrial and service sector growth and subsequent failure to create off-farm employment opportunities.

Widely available crops and plants are the basis for bioenergy production; crops that can be grown in many areas under rainfed conditions, with low input costs (fertiliser). Because bioenergy can be developed in many areas with low input costs, it can benefit both small and large scale farming operations. It can open up new livelihood opportunities and domestic and export markets, enhancing rural economic transformation. More than 80 per cent of the population of many African countries is engaged in agriculture, which is primarily smallholder and largely subsistence. Modern bioenergy can be produced at this smallholder level, creating possibilities for generating cash income. It also encourages the formation of agricultural marketing and processing cooperatives, which, in turn, facilitate adoption of modern agricultural inputs and better farming practices. Africa's total cultivable land is estimated at 840 million hectares. Of this land, only 27 per cent is used (cultivated), compared to 87 per cent for Asia and 97 per cent for Southwest Asia. The development of bioenergy offers opportunities to convert the uncultivated area into energy wealth, given appropriate socially and environmentally responsible policy measures.

## BIOENERGY INITIATIVES IN AFRICAN COUNTRIES

The Government of Uganda has planned for launching a large-scale biogas initiative countrywide in order to reduce the use of firewood and charcoal for cooking and lighting. The initiative aims at improving the health and living conditions, soil fertility and agricultural production, reducing green house gas emissions and creating new jobs through the development of a robust biogas sector. The programme aims at installing 20,000 quality biogas plants by 2010. The Renewable Energy Policy published in November 2007 by the Ministry of Energy is centered around biofuels and the improvement of biomass utilisation efficiency. The Ministry aims to blend 20 per cent biofuels into all gasoline fuel and diesel while for petrol will be blended with ethanol.

If all the available molasses were converted to ethanol in 2009, Uganda could produce 35 million litres of ethanol. The Ugandan government has passed the Renewable Energy Policy to encourage the development of the country's abundant renewable energy resources, especially biofuels. An important objective of the policy is to acquire advice on how to mitigate the environmental and social impacts that may arise from the unregulated development of an indigenous biofuels sector. The two most serious risks are rainforest destruction and possible negative impacts on food supply and food prices. Grants have been obtained to support Uganda's Renewable Energy Policy, and particularly to enhance the government's efforts to design a regulatory framework that will encourage the development of a biofuels industry and increase energy security without jeopardising the country's food supply. Among the non-food crops that have been selected for biofuel production in Uganda include Jatropha, a drought- resistant perennial crop with a 50-year life span. It is a member of the Europhobiaceace family. Its seeds contain over 35 per cent non-edible oil. A local company in Mukono district (Uganda) produced 3,000 litres of oil from 9,000kgs of seeds collected from vanilla farmers in the districts of Mukono, Kayunga, Jinja, Iganga, Kamuli and Bugiri. It has also planted over 40 hectares (about 60,000 trees) of jatropha at their farm in Mukono. The company has a tractor that has been running on the biodiesel for three months. *Jatropha carcus* is preferred by most companies for biodiesel projects because it can be inter-cropped with other food crops.

## REGIONAL SPECIFICS FOR ASSESSING SUSTAINABILITY

Even though the general objective and the principles are universal and common to all locations, the criteria, the indicators and the verifiers are specific for each region. Principles may have different degree of importance depending on the local context and social values. The best example in regional specifics is the case of the North-South differences, where industrialized countries are more concerned about the social and environmental constraints of biofuels production while developing and emerging economies, on the other hand, focus on the economic growth opportunity. For example, the European Union promotes the development of biofuels in order to reduce the vulnerability of the energy supply and to be able to achieve its GHG emissions reduction targets. This policy focuses more on environmental and social issues than on economic growth. On the other hand, the main driving force for biofuels production in developing and emerging countries (i.e.Malaysia, Brazil, India and Indonesia) is the promotion of the economic development.

The general consensus from past analysis and future projections is that biofuels production will raise food prices and consequently will threaten food security, especially in biofuels producing countries and in poor countries where food imbalances prevail (Tokgoz *et al.*, 2007; Kojima *et al.*, 2007; OECD, 2007; OECD/FAO 2007; FAPRI, 2007). Agricultural commodities used for biofuels production will increase their price and, due to the correlation between agricultural commodities market prices, the price of other agricultural commodities will also increase. While biofuels production contributes to reduce in short term the agricultural products available for food, other factors may also contribute similarly such as the weather conditions and the short term yield of agricultural production. The analysis must be deepened in order to identify the role of biofuels among all other factors. Presently, the share of arable lands used in the world for biofuels production is about 1% (IEA, 2006).

## CONCLUSION

Bioenergy has been in the spotlight for the last few years. A number of countries are keen to promote its production and use. The greener energy production has been generally welcomed by many communities around the world as a step forward in the fight against climate change, but there are serious concerns about the delivery on the ground. Threats posed by the development of the bioenergy industry to biodiversity are widely acknowledged but fail to outline a clear strategy to address them. There is need to set clear environmental standards to ensure that bio-energy production really does bring a significant decrease in greenhouse gas emissions while preventing harmful impacts on biodiversity. An accreditation system that traces how bioenergy and biofuels have been produced and screened against environmental standards is required.

There is need for developing countries to ensure that biofuels are sustainably produced so that environmental protection is not lost in the global drive to promote biofuels. Under sound strategies and appropriate regulatory frameworks, increased use of bioenergy may represent a way to alleviate the serious problems that most countries face at present with high prices and possible supply disruptions in international petroleum markets, and may facilitate access to energy, especially for poor people in developing countries. Africa has huge bioenergy potential and technological advances make it highly promising. The possibilities for developing non-food crops and perennial plants as biofuel feedstocks are broad and within reach, given the extensive research work underway at the global level. What is needed is not only political goodwill but also an enabling policy and institutional environment (on the part of key players) towards meeting Africa's sustainable energy and development objectives through promoting environmentally and socially sustainable large scale investments, while at the same time creating opportunities for smallholders to be energy producers. Largely in response to high oil prices, extensive research is underway on first and second generation technologies. To fully benefit from this, African countries need to have the human resources and the institutional infrastructure to facilitate the transfer, adoption, application, and development of bio-energy technologies.

## References

Benedict, M.A./McMahon, E.T. 2006: Green Infrastructure: *Linking landscape and communities;* Washington DC

CBD (Secretariat of the Convention on Biological Diversity) 2008: The potential impacts of biofuels on biodiversity Dudley, N &Phillips, A. 2006: *Forests and Protected Areas: Guidance on the use of the IUCN protected area management categories;* IUCN; Gland and Cambridge

EEA (European Environmental Agency) 2006: How much bioenergy can Europe produce without harming the environment? *EEA Report* No 7/2006; Copenhagen

EEA (European Environmental Agency) 2007: Estimating the environmentally compatible bio-energy potential from agriculture; EEA Technical Report, Copenhagen

FAO (Food and Agriculture Organization of the United Nations) 2008a: The State of Food and Agriculture 2008 - *Biofuels: prospects, risks and opportunities*; Rome

FAO (Food and Agriculture Organization of the United Nations) 2008b: *Bioenergy Environmental Impact Analysis* (BIAS); draft report prepared by Oeko-Institut

Farrell, AE, Plevin, RJ, Turner, BT, Jones, AD, O'Hare, M Kammen, DM, (2006) Ethanol can contribute to energy and environmental goals. *Science*, Vol. 311, No. 5760. pp. 506-508.

Fargione,J . Hill,J, Tilman, D,Polasky,S, Hawthorne, P (2008) Land Clearing and the Biofuel Carbon Debt, *Science*, Vol. 319. no. 5867, pp. 1235 – 1238

FAO, 2007. Crop prospects and Food Situation. No5 October 2007.

FAO, 2002. Criteria and Indicators for Sustainable Forest Management: Assessment and Monitoring of Genetic Variation. *Forestry Department, Working Paper* FGR/37E, November 2002.

FAPRI, 2007. U.S. and World Agricultural Outlook. FAPRI Staff Report 07-FSR 1. Food and Agricultural Policy Research Institute, Iowa State University, University of Missouri-Columbia. January 2007.

Hill,J ,Nelson,E Tilman, D ,Polasky S, Tiffany, D (2006) Environmental, economic, and energetic costs and benefits of biodiesel and ethanol biofuels. *Proc. Natl. Acad. Sci. U. S. A.* Vol. 103, No. 30

IEA, 2006. *World Energy Outlook 2006*, Chapter 14, The Outlook for Biofuels, OECD Publications, Paris.

Kojima M., D. Mitchell, W. Ward, 2007. Considering Trade Policies for Liquid Biofuels. Renewable Energy Special Report 004/07. *World Bank, Energy Sector Management Assistance Program (ESMAP).* Pp. 128.

OECD/FAO (2007), *Agricultural Outlook 2007-2016*, OECD/FAO, Paris, Rome

OECD, 2007. Round Table on Sustainable Development. Biofuels: is the cure worse than the disease? R. Doornbosch and R. Steenblik. Report SG/SD/RT(2007)3. Paris, 11-12 September 2007. pp. 57.

Searchinger, T, Heimlich, R,. Houghton, R. A., Dong, F. Elobeid, A., Fabiosa, J., Smith, RD& Maltby E. 2003: Using the Ecosystem Approach to Implement the Convention on Biological Diversity: Key Issues and Case Studies; IUCN; Gland, Cambridge

Stephanopoulos, G (2007) Challenges in Engineering Microbes for Biofuels Production. *Science,* Vol. 315. no. 5813, pp. 801 – 804

Tokgoz S., A. Elobeid, J. Fabiosa, D.J. Hayes, B. A. Babcock, T-H. Yu, F. Dong, C. E. Hart, J.C. Beghin, 2007. Emerging Biofuels: Outlook of Effects on U.S. Grain, Oilseed, and Livestock Markets, Staff Report 07-SR 101, Center for Agricultural and Rural Development, Iowa State University, Ames, Iowa.

Twesigye C.K, S. Onywere,Z. Getenga, S. Mwakalila ,J.Nakiranda (2007) The Potential of Satellite Imagery and Policy Framework in Addressing Sustainable Management of Watershed Resources of the Lake Victoria Basin. *The Uganda Journal.* 51, 74- 85

UN-Energy 2007: Sustainable Bioenergy: A Framework for Decision Makers; United Nations; New York

In: Technologies and Management for Sustainable Biosystems ISBN: 978-1-60876-104-3
Editors: J. Nair, C. Furedy, C. Hoysala et al. 

*Chapter 21*

# DEVELOPING INDONESIAN CAPACITY IN SANITATION: CONSTRUCTED WETLANDS FOR THE TREATMENT OF WASTEWATER

***D. Boyd*[*1], *S. Dallas*[1], *G. Ho*[1], *R. Phillips*[1], *G. Wibisono*[2] *and E. Wahyuniati*[*2]**

[1] Environmental Technology Centre, Murdoch University, Murdoch WA 6150, Western Australia
[2] Department of Civil Engineering, Technical Faculty, Merdeka University, Jl. Terusan Raya Dieng 62-64 Malang, Indonesia 65146

## ABSTRACT

Constructed wetlands have long been internationally recognized as a simple biological wastewater treatment system, but their recognition is limited in Indonesia. To encourage the use of this technology a full-scale constructed wetland using local reeds and treating effluent from a septic tank was designed and constructed with local participants. This initiative was the result of the Sustainable Sanitation and Wetland Technology (SSWT) Project which was a capacity development initiative aimed at improving Indonesian capacity in sanitation. The program combined theory and practice; the participants were provided with information about sustainable sanitation, constructed wetlands and awareness raising methods. At the same time, participants gained practical experience in constructing a wetland and raising awareness of sanitation.

The design and delivery of this SSWT Project was based on the application of an emerging Framework for Capacity Development. This Framework was developed as part of an ongoing research project into improving capacity development practice. In this chapter, both the technological and sociological aspects of constructed wetlands and their effectiveness in terms of providing successful sanitation outcomes in Indonesia are presented.

* **Email: d.boyd@murdoch.edu.au; t: 61-8 9360 6399; f: 61-8 9360 7311**
* **Phone and Fax: +62 341 564 994, email: gunawan269@gmail.com**

**Keywords** Capacity development; Indonesia; technology demonstration; training; wastewater treatment; constructed wetland, reedbed.

## INTRODUCTION

It is widely recognised that constructed wetlands have enormous potential for wastewater treatment and reuse in developing countries (Denny 1997; Haberl, 1999; Kivaisi, 2001). They are particularly well suited to tropical regions and represent ideal candidates for sustainable, ecologically engineered systems (Dallas, 2004), and yet their uptake "has been depressingly slow" (Denny, 1997). This has indeed been the situation in Indonesia, the site of this research, despite the successful use of constructed wetlands in neighbouring tropical Asia for example (Koottatep and Polprasert, 1997). In this chapter the potential to stimulate the uptake of constructed wetlands for wastewater treatment through an innovative capacity building approach is described.

It is widely agreed, if the Millennium Development Goals (MDGs) and related targets such as those for sanitation are to be achieved, that capacity development is one of the highest priority areas (UNDP, 2005). As reported by the Organisation for Economic Cooperation and Development (OECD, 2006) "adequate country capacity is one of the critical missing factors in current efforts to meet the Millennium Development Goals". The seventh MDG goal aims to ensure environmental sustainability; associated with this goal is the target to halve by 2015 the billions of people that lack access to basic sanitation (UN, 2000). At the midpoint between the adoption of the MDGs and the target date, estimates indicate that if trends since 1990 continue the target for sanitation is likely to be missed by almost 600 million people (UN, 2007).

In response to this need for capacity in sanitation, a capacity development intervention, the Sustainable Sanitation and Wetland Technology (SSWT) Project, was carried out. The project was implemented in Indonesia by staff from Murdoch University's Environmental Technology Centre (ETC), Perth, Western Australia and Merdeka University's Institute for Environmental Management and Technology (IEMT), Malang, East Java. This project aimed to develop Indonesian capacity in water and sanitation in addition to serving as a case study in research projects aimed at improving the practice of developing capacity and stimulating the uptake of constructed wetlands as a sustainable wastewater treatment technology.

Java is the most densely populated island and one of the most densely populated regions in the world with a population of over 125 million on a land area of 127,000 $km^2$. Understandably its environment is significantly affected and water and sanitation are critical issues with severely polluted rivers and groundwater. Wastewater treatment systems of any kind are rare (excepting septic tanks) and as has been found in many developing countries conventional 'hi-tech' wastewater treatment schemes are likely to be abandoned within several years of commissioning.

Two constructed wetlands have been installed recently in and around Malang, East Java in Indonesia: the one described in this research and a second as part of a United States Agency for International Development (USAID) sponsored pilot system built in the village of Temas to treat a blend of wastewater and some polluted river water (Dallas, 2007). The latter project is part of a larger USAID Environmental Services Program (ESP) in the larger catchment to tackle both diffuse and point sources of pollution. It has been a concerted effort to engage with the local community and establish and trial an appropriate low-cost wastewater treatment system. The construction of the wetland has not been without its share of problems including a difficult steep site and burrowing animals damaging the original clay lining system which resulted in it being replaced by concrete. The treatment system consists of a large septic tank followed by nine constructed wetlands all of similar dimensions in series (varying free water surface and subsurface flow) until discharging into a final containment tank and finally a nearby stream. Both projects have endeavoured to increase community awareness concerning the importance of effective sanitation and increase capacity building through constructed wetland technology.

## Materials and Methods

The SSWT Project consisted of a single activity, a workshop. The target audience for the workshop included individuals from local government, non-governmental organisations, academia and community-based organisations. Fourteen people participated in the workshop. In this chapter an overview of the workshop is presented and factors that contributed to its success are highlighted. The Framework for Capacity Development that underpins the workshop design and rationale is described in Boyd (2009).

## Workshop Program

The workshop program was developed collaboratively by the project members based on the local needs identified as part of a prior project. The participants in this project identified a need for simple low-cost sanitation technologies and awareness raising to address the problems associated with sanitation in Indonesia. The choice of constructed wetlands as a suitable technology was also identified by these participants and was key to informing the SSWT Project.

The resulting four day workshop program aimed to:

- Develop the skills and abilities of the participants to implement constructed wetlands
- Develop the skills and abilities of the participants to raise awareness in the area of sanitation
- Support participants in the development of learning resources that could be used to raise awareness of sanitation
- Provide an opportunity for participants to practice and apply what they learnt

The workshop program combined presentations, discussion sessions and practical activities.

## THE SSWT WORKSHOP

*Day 1:* Aside from formalities, such as the workshop registration and opening ceremony, the first day of the workshop consisted primarily of presentations. These were intended to provide participants with an overview of sustainable sanitation and wetland technology and to activate their prior experience of sanitation. Day one also included opportunities to get to know the participants.

*Day 2*: The program for day two was designed to demonstrate wetland technology and provide participants with an active role in the construction of the wetland. This activity also provided people with an opportunity to interact informally and provided a basis for fostering relationships.

In preparation for the wetland construction component a suitable site for the technology demonstration was suggested by one of the workshop participants. The site, a seminary, "Passionis" was located on 16 hectares of land approximately four kilometres from Merdeka University. The seminary itself housed approximately 50 individuals. Aside from religious training, the seminary operated as a small farm or 'ecovillage' with livestock, an orchard, a coffee plantation, and cheese making business. The site also included a centre for retreats. The site was considered appropriate for the following reasons:

- The wetland would be in keeping with the environmental ethos of the site which already practised greywater recycling, where water from the laundry was used to water 'salak' (spiny palm) plants.
- The site met the requirements for wetland construction e.g., the waste stream could be easily accessed and the available areas were sunny and downhill of the waste source eliminating the need for a pump.
- The workshop participant from Passionis would be in a position to maintain the wetland.

Though originally it was thought that the wetland would treat greywater, the greywater was already being reused. After consideration of alternative treatment options project members determined that as there was a leaking soakage pit onsite where septic effluent was discharging into the open drain a system to treat this wastewater (i.e. post septic tank effluent) would have relevant environmental and health benefits. The wetland design (i.e. size) was finalised based on estimated wastewater volumes which due to the fluctuating nature of the seminary's accommodation introduced some risk in terms of undersizing the system. The project leader and wetland expert decided that two wetlands of approximately two by three metres and sixty centimetres deep would be required to treat the wastewater. Subsequently, the materials required were purchased and the site was prepared. Given that the wetland was to be constructed in a single day it was decided to dig the holes prior to the workshop. Not only would this reduce the amount of labour, but also eliminate the risk of hitting hard rock, roots etc. In addition, some of the plumbing would also be pre-prepared. The materials selection and purchasing of materials and site preparation was carried out by the wetland expert and a contractor employed to provide assistance. At this time it was also decided that given that the wetland requires time to 'settle', the wastewater would only be allowed to enter the wetland two weeks after the workshop and the contractor would assume responsibility for this.

On the day of the workshop, the 14 participants, along with the three core project members and nine support staff actively participated in the tasks required to complete the construction. The process of constructing the wetland involved the following seven steps, steps two to six of which were completed by the participants:

1 Digging the hole
2 Lining the hole with plastic and sand
3 Determining the level of the outlet
4 Installing the inlet pipe
5 Filling the wetland with gravel media
6 Planting the reeds in the wetland
7 Allowing wastewater to enter the wetland

The two wetlands were built sequentially over a period of five hours with breaks for morning tea and lunch. All participants and project members helped line the wetlands. For each wetland a group of two or three participants worked with the wetland expert on steps three and four. The main task of filling the wetland with media was completed by everyone forming a chain to transport the six cubic metres of gravel the 25 metres from the gravel pile to the wetland by bucket. The participants took turns planting the plants in the wetland. The end result was two wetlands for treating blackwater (septic effluent) from the seminary. In this fashion all participants were exposed to the details and logistics necessary for constructing a successful reedbed.

*Day 3:* The program for day three included two sessions. In the first session, the participants were introduced to a range of methods that could be used for awareness raising. Drawing on the experience of local practitioners and experts including the workshop facilitator and researcher, the session included presentations on different approaches including the use of mass media, multimedia, edutainment, traditional song and dance, public speaking and posters.

In the second session participants worked together in groups to develop a resource or activity that could be used to raise awareness of sanitation. This activity or resource would be delivered to a group of community members by the participants on the fourth day of the workshop.

*Day 4:* On the final day of the workshop, the participants delivered the awareness raising session they had created to the community. The session provided the participants with an opportunity to practice and apply the skills they had learnt about raising awareness. Following the awareness raising session the SSWT participants completed a workshop evaluation questionnaire.

## RESULTS AND DISCUSSION

From the perspective of the participants and project members the overall impression was that the workshop was a success. Three criteria provided the basis for judging success the success of the intervention.

A successful invention:

1 Met the needs of the participants
2 Developed the capacity of the participants to facilitate learning and change
3 Facilitated ongoing action and change (i.e., encouraged the use of constructed wetlands and promoted awareness raising)

In the workshop evaluation questionnaire participants identified the workshop of value for the knowledge they obtained; the practical experience; and, because it raised their awareness. The valuable aspects of the workshop corresponded with the identified needs (criterion one).

The participants engaged fully in all the activities of the workshop. In doing so they demonstrated through their actions that they had the knowledge and skills to implement constructed wetlands and raise awareness of sanitation (criterion two).

A fully functional two stage reedbed was completed in one full day (excluding some prior site preparation as described) with all workshop members participating in its installation. The participants also confidently delivered an awareness raising session to 46 individuals from the local community.

While it is not expected that participants go away fully trained in the building of constructed wetlands it is a grassroots means by which to expose and educate people who are able to promote awareness and facilitate change in their communities to alternatives to high-tech wastewater treatment systems.

The workshop evaluation provided some indication that the participants were motivated to take action beyond the timeframe of the workshop (criterion three). Seven participants said they would like to use what they had learnt to raise awareness and five participants indicated that they would like to apply the technology. This included the following:

- Building a community wetland to treat wastewater from eight houses;
- Constructing a wetland in a central Malang park in order to research appropriate plants for wetlands in Indonesia;
- Building a wetland in another seminary (South East Asia Bible Seminary);
- Installing wetlands at home;
- Seeking land on the Brantas River in order to trial a wetland for wastewater treatment as part of a government funded project.

Eight months after the SSWT Project an outcomes assessment was conducted. An initial analysis of the results of this assessment indicates that the participants were motivated and empowered by the workshop. Many participants had shared what they had learnt with colleagues, family, neighbourhood administrators and friends. Several participants had funding, plans and support to implement wetlands as part of a government project, and in the seminary, and to carry out research on appropriate plant species. However, some of the participants identified barriers to implementing constructed wetlands. These included the availability of land in urban areas, a lack of funding for projects and a lack of data regarding the effectiveness of constructed wetlands.

Several factors were identified as contributing to the success of the SSWT Project, including:

- *Involvement of motivated participants with the ability to take action after the workshop:* The selection of appropriate participants was observed by all project members to be central to the success of the project. As expressed by the project leader "this is like an input of the system, the input should be good". The participants were carefully selected by the project leader. The participants were all known to the project leader and were considered "trustworthy, honest, genuine and responsive". Further, the participants all expressed an interest in the project and indicated that they were motivated to take action.
- *Delivering an intervention based on participants' need:* the SSWT Project responded to participants' need for affordable sanitation technology, awareness raising and practical experience in both these areas.
- *Designing activities based on adult learning principles:* the workshop was designed according to adult learning principles. There are many adult learning theories, according to the model used to develop the SSWT Project effective learning environments engage learners in four distinct phases of learning (Merrill, 2002). The workshop included three of these four phases. The workshop: 1) activated the participants' prior experience of sanitation and awareness raising; 2) demonstrated the skills necessary to implement constructed wetlands and raise awareness; 3) provided participants with an opportunity to apply those skills. The workshop program did not integrate these skills into real world activities (phase four) because of the timeframe and available resources.

The factors presented here are not the only factors that contributed to the successful implementation of the SSWT Project. Other factors including teamwork, local ownership, effective project planning and management are included in the Framework for Capacity Development (Boyd, 2009).

## CONCLUSION

The SSWT Project was successfully implemented based on application of the Framework for Capacity Development. In this chapter, several aspects of the Framework that contributed to the success of the project have been highlighted. This included: the careful selection of appropriate participants who were willing to engage fully in the activities of the workshop and were motivated to take action beyond the timeframe of the workshop; designing a workshop to respond to local needs; and, the delivery of a training workshop based on adult learning theory that innovatively combined training in wetland technology and awareness raising.

The Indonesian wetland described here is primarily a proof of concept system. That is, it is endeavouring to receive validation from the local communities and authorities that wetlands are viable and effective low-cost wastewater treatment systems. It can be argued that the system as installed is too expensive for local communities which is true as at this stage of the technology uptake cycle economies of scale are not achieved. A recent analysis of the construction costs of various sized subsurface flow wetlands in Nicaragua and Portugal by WERF (2006) highlighted this fact. Not until larger systems designed to treat all the wastewater from a village or villages for example are installed will the true cost of construction and operation and maintenance be accurately known. In rural Indonesia for example where villages are commonly surrounded by rice paddies belonging to members of that village, it would seem eminently feasible to transform a section of nearby rice paddy into the village's own wastewater treatment system.

Many of the main issues that confront constructed wetlands in Indonesia are common ones such as availability of land, suitable lining materials, affordable gravel media, and the need for a satisfactory level of wastewater pre-treatment. If wastewater recycling can be achieved in conjunction with useful wetland crops that can be harvested then the wetland systems can start to head towards a level of economic self-sufficiency.

Conventional wastewater treatment systems have been shown to be unsustainable in developing countries due to a combination of high operating and maintenance costs, lack of suitably skilled personnel and spare parts (Matsui et al, 2001). Constructed wetlands can provide an alternative to this paradigm. They are proven low-cost, low-maintenance systems that hopefully will go a long way towards achieving sustainable sanitation outcomes in developing countries.

## REFERENCES

Boyd, D. (2009). A framework for capacity development: closing the gap between theory and practice. Doctoral Dissertation, Perth: Murdoch University.

Dallas, S. (2007). A brief look at some new constructed wetlands in tropical Asia. *EcoEng. Newsletter* No.13, September. Switzerland: International Ecological Engineering Society.

Dallas, S., Scheffe B. and Ho G. (2004). Reedbeds for greywater treatment—case study in Santa Elena-Monteverde, Costa Rica, Central America. *Ecological Engineering.* 23 (1), 55-61.

Denny, P. (1997). Implementation of constructed wetlands in developing countries. Water Science and Technology, 35 (5), 27–34.

Haberl, P. (1999). Constructed wetlands: a chance to solve wastewater problems in developing countries. *Water Science and Technology.* 40 (3), 11–17.

Kivaisi, A. (2001). The potential for constructed wetlands for wastewater treatment and reuse in developing countries: a review. *Ecological Engineering.* 16 (4), 545–560.

Koottatep, T. and Polprasert, C. (1997). Role of plant uptake on nitrogen removal in constructed wetlands located in the tropics. *Water Science and Technology.* 36 (12), 1–8.

OECD (2006). The challenge of capacity development: working towards good practice. Paris: OECD.

Matsui, S., Henze, M., Ho, G. and Otterpohl, R. (2001). Emerging paradigms in water supply and sanitation. In: C. Maksimovic, and Tejada-Guibert, J.A. (Eds). *Frontiers in Urban Water Management: Deadlock or Hope.* United Kingdom: IWA Publishing.

Merrill, M. D. (2002). First principles of instruction. *ETR&D,* 50 (3), 43-59.

UN (2000). 55/2. United Nations Millennium Summit United Nations - Millennium Declaration. http://www.un.org/millennium/

UN (2007). The Millennium Development Goals Report. United Nations, New York. http://www.un.org/millenniumgoals/pdf/mdg2007.pdf

UNDP (2005). United Nations Development Program Millennium Development Goals, About the MDGs: Basics. http://www.undp.org/mdg/basics.shtml

WERF (2006). Small Scale Constructed Wetland Treatment Systems: Feasibility, Design Criteria and O&M Requirements. In Wallace, S.D. and Knight, R.L. (Eds). Report 01-CTS-5. USA:Water Environment Research Foundation.

In: Technologies and Management for Sustainable Biosystems ISBN: 978-1-60876-104-3
Editors: J. Nair, C. Furedy, C. Hoysala et al. 

*Chapter 22*

# CASE STUDY OF TECHNOLOGY TRANSFER TO A FIJI RURAL VILLAGE USING AN IMPROVED 'SUSTAINABLE TURNKEY APPROACH'

***Leslie Craig Westerlund*[*1], Goen Ho[*2], Martin Anda[*2], M. Daniel Wood[*3] and Kanayatha Koshy[*4]***

[1] Technologies and Strategic Management of Sustainable Biosystems, First International Conference, Murdoch University, W.Australia
[2] Environmental Technology Centre, Murdoch University, Murdoch, W.Australia
[3] School of Engineering & Physics. University of the South Pacific, Suva, Fiji.
[4] University of the South Pacific, Suva, Fiji

## ABSTRACT

A modified appropriate 'sustainable turnkey approach' (STA) was developed, trialled and used to introduce new cottage papermaking technology to an existing papermaking village in a remote highland part of Fiji, an example of a less developed country. The new research explored making a high technology piece of equipment in Suva, the nearest city to the Wainimakutu Village. The STA has empowered a local engineering business and the engineering department of the local university to be able to understand and transfer the 'hardware' (equipment) and 'software' (skills) of these technologies and actually make the machines. The principle Author helped them to adapt and improve the design, to trial it, to test it, and then network with a village to complete the ideals of using the 'best available technology' that is also sustainable, eco-friendly, appropriate and financially viable. The STA has proved a far superior process. When something goes wrong with the machine it can be adjusted, fixed or modified to work: first in the village; second, with local industry help; third with local university help; and fourth with contacting the inventor for advice and networking with all stakeholders.

* L.Westerlund@murdoch.edu.au
* G.Ho@murdoch.edu.au
* M.Anda@murdoch.edu.au
* michael.wood@usp.ac.fj
* koshy_k@usp.ac.fj

**Keywords:** case study; Fiji; papermaking; sustainable turnkey approach; technology transfer.

## INTRODUCTION

In technology transfer (TT) there are several approaches to transfer the hardware (equipment) and software (skills) of new innovative technologies to less developed countries (LDCs). The established industry approach is to transfer small increments of innovation in 'software' or 'hardware'. The next level is argued to be larger levels of innovative technology transfer with more risks and benefits. The modern perspective includes using the 'best available technology' from the start of a project and also pioneering sustainable and eco-friendly principles in the 'life cycle analysis'. Within this range of technology options is importing new high technology advanced equipment from developed countries and with a bit of local training it might work in the village by 'turning the key' and miraculously working.

## FIVE PHASES OF SUSTAINABLE TECHNOLOGY TRANSFER

The 'turnkey approach' is part of the third phase of transfer technology to developing countries. This is based on a introduction to a trial of 'five phases of sustainable technology transfer' to LDCs [Westerlund 2008], summarised for this project as:

- The 'first phase' is the selection of the project site. This is first based on the existing papermaking technology hardware and software; then the village infrastructure, culture, skills, needs, wants, management and marketing.
- The 'second phase' is to further examine methods of improving existing hardware and software technology levels of use and understanding; then to actually select and improve examples in TT in small and bigger steps using the research on specialised handmade papermaking technology from Westerlund [2005];
- The 'third phase' is to introduce new or more radical higher levels of hardware and software in TT. This can include a "turnkey approach" for hardware; and "copy it exactly" approach for software.
- The 'fourth phase' is a 'vertical transfer' of technology. This is to empower a village with new skills and tools to value add a raw product into a range of more useful and valuable merchandise. This project aimed to make a value added product range from the base A3 size paper product: to be guillotined into quality standard A4 paper, then printed into useful books, poems and publications;
- The 'fifth phase' is a 'horizontal transfer' of technology. This is to then help empower the village to own, modify and transfer the holistic range of new technologies to the next village.

## TURNKEY APPROACH

The 'turnkey approach' becomes very important as it could transform a LDC cottage industry into a more competitive production advantage. This is levered further by: using local cheaper labour ; and new free raw material (post industry waste paper) ; and using locally grown village flowers and feature fibres in the paper. This could gain a competitive edge in local, regional and international markets.

The 'turnkey approach' may not be initially considered relevant to LDCs because a village does not have the money or expertise to buy sophisticated equipment to be just delivered to them – and – expect it to seriously work in the harsh environment of LDCs. Likewise the village has no technology skills to re-design, and engineer the range of technologies – so there is a level of co-dependence to make the new technology relevant to them, fixable by them, with training to use it safely, and phone support for maintenance advice or breakdown services. Stewart [1987] argues that the "turnkey plant are typically built by engineering firms with no proprietary technology rather than by the innovative firm" that compromises the quality and integrity of the holistic system. He goes on to be disillusioned with the whole incentive system for technology diffusion or transfer and suggests "efforts to promote technology transfer are either useless or counterproductive." A useful case study of the 'turnkey approach' is the Cypriot 'advanced manufacturing technology' researched by Efstathiades et al.,[2000]. They argue that it is preferable to develop the technology and expertise in one city, make units, then export the finished product to either developed countries, intermediate DCs or LDCs.

There are also many challenges to the 'turnkey approach' so a compromise would be to get the inventor on-side, to re-design the plans of new technology for a LDC, and make it with industry in the nearest city to the project. This would empower TT at the grass roots levels and allow the machines to be made and serviced in a developing country. However, who is going to give away good new up-to-date or 'best available technology' to a LDC? Who will pay for it? Who will teach them? Who will profit? Who will answer their phone calls for help when it breaks down? Who will translate complicated science and engineering to a village unskilled person using the equipment with limited training? Which village would appreciate it? Which traditional hand-made papermaking village will truly appreciate the opportunity:

## BACKGROUND ON A (MODIFIED) TRADITIONAL PAPERMAKING CRAFT IN FIJI

The traditional papermaking involves the following steps: (1) collecting plant fibres and glue from the village gardens; (2)soaking, pounding and boiling into softer, small fibres; (3) further pre-treatment in a 'Hollander beater'/ grinder (500g/20litres); (4) diluting the fibres into 99% water solution in a vat; (5) scooping with a sieve-like mould and deckle to collect a thin layer of pulp; (6) flat couching onto a cotton couching cloth; (7) pressing a batch of fifty in a screw press; (8) peeling each off and pasting onto a hot metal surface for drying; and (9) when dry, the rougher textured paper is sorted and graded for wrapping paper.

## BACKGROUND ON NEW TECHNOLOGY

The new technology involves a range of new equipment to make the process quicker, easier, and cheaper. It derives a superior quality paper smooth surface texture. The new technology would have the following features locally: (1) 95% free post industry waste paper could be used with local village grown fibres and pretty flowers as 5% features within the paper; (2) new 200 litre hydropulper that chops 3kg of paper into 150 litres of pulp; (3) new aluminium moulds and deckles that do not rust or warp; (4) new transfer roll-over couching curve; (5) new guided stacking system of couching clothes; (6) new pressing system with a 20 tonne hydraulic press; and (7) traditional loft drying avoids fires and resultant OHS smoke issues.

The new hydropulping machine allowed the new source of free waste paper to be quickly, cheaply, easily used and compliments the existing Hollander technology that slowly grinds fibres for use. This process is further explained in Westerlund [2005,2007,2008] and available on the www home page.

From research of the greater Pacific region as an example for regional LDCs, Wainimakutu Village, Fiji was chosen as the best place to launch the project, Kassim [1992], Liebregts [1998], Liebregts and Townsend [1998]. They were already making a decorative hand made A3 wrapping paper and had the interest and incentive to diversify their cottage (business) enterprise. The traditional protocols were followed and the project had the Headman and Chief's blessing. This part of the greater project, which was to investigate better ways of transfering low and high technology equipment – hardware and software – to remote villages in a less developing country.

## METHOD

The project involved designing a hydropulper machine (LHP) to chop up paper into pulp, engineering and building, financing and then giving away the new technology hardware (equipment) and software (training) to an existing village-based papermaking project in Fiji. The propriety 'intellectual property' to the new smooth papermaking process and equipment is with the Author* [Westerlund 2007]. This 'creative commons' approach to technology transfer was freely offered to help the village acquire the best available technology and training [Bunt 2007]. This can give credit to the person(s) technology and transfer it with their blessing. This acknowledgement is empowering for all concerned. If the next group wish to further consult with the original inventor/designer than that can be done in a fair and equitable way.

The idealised: Plan A was to make it in the nearest city, Suva with some village help. Plan B: the business engineers can redesign and build a hydropulping papermaking machine near the city of Suva, Fiji in Jan 2007. Plan C: to network with the engineering section of the University of the South Pacific to make one as a student project in 2007. Plan D: where the original hydropulper (B) was remade as a university student project in 2008 and given to the national training provider. Plan B: was chosen as best and we sought the help of a small manufacturing company. The Plan was discussed with the manager of 'On Time Engineering' and set in place – with a hand shake–and a traditional bowl of 'yogona-grog' to make the

LHP at their small engineering company. This also involved redesign it using locally available material and tradesmen [Sharma, 2006].

## EVALUATION

The evaluation included field trials in the village and another trial of a questioner. It was based on five points per question per level of competency: Questions like...Can you fix the motor/electrics / weld the frame / sharpen the blade/ engineer the drum/ plumbing/ oil seals/ drive shaft/ bearings/ pulleys...? Each person could get up to 25 points per equipment. Then up to 100 points for making four pieces of low technology equipment; 4 pieces of high technology equipment; then making the four stages of smooth papermaking skills. An individual could get 300 points if they participated in learning all facets of the new technology. This was compared against the inventor at 25 points x4 =100/ (equipment); then x3 phases=300/(equipment and skills) total transfer for three phases. This system can keep going on to include a 100 points for each new significant phase of technology transfer. The system has allocated the same points to each phase as all phases are important to any holistic project.

## RESULTS

Plan B was successfully implemented and resulted in a book being published on "How to Make a Hydropulper" [Westerlund 2007]. The small engineering company had a suitable workshop, tools and tradespersons to help design and make a hydropulper using locally available parts.

Figure 1. Hydropulper (200 litre) being made by the Author* with industry help in one week in Suva.

There were many challenges from this first hand experience of TT. At an engineering level there were several design issues to be understood and overcome and the Authors can be contacted to get more details on the case study. One design challenge is if there is a problem in the village then they can at least courier the unit back to Suva and get it serviced and re-engineered locally. A photo of the group was taken with it working in the engineering business [Westerlund 2007]. There was only so much the Author* could do with the engineering company before the Author* completed the rest at his Suva home base.

The final tests were in the village. The important moment came to start the generator and 'christen' the new motor-hydropulper in the remote village... and... it 'did not work'! After all the pre-planning and testing, it did not work! The local village portable generator was not powerful enough to start the motor. We had overlooked that issue. The Author was not an electrician and did not fully appreciate using non standard portable 240 volt systems. The pulley had to be literally 'kicked' to make it move and help start the motor.

The motor was then making some noises after the rough road trip up the mountain. The Author* asked if there was an electrician in the village. There were two! The hydropulper motor was soon stripped down and realigned to work better. Several batches of paper were then made to trial the machine under operating conditions. The machine was used for a year until a better machine was redesigned and made. In part summary, a local business was empowered to help make a suitable and functional hydropulper and with local village electrical support it could just work in the harsh village environment.

## DISCUSSION

The idealised Plan A, was too difficult to be implemented in the village. The villages did not have the equipment, experience or available skilled tradespersons to start or complete the project.

The next level, Plan B, was successfully implemented with the help of a local engineering business, co-designed with the inventor / Author, using locally available parts. This involved some input from the villages to help ensure it could be fixed and serviced in the village or the nearest city (Suva). This project of making a hydropulper was quickly done within two weeks against many challenges. Executive decisions could be discussed and made with the industry Manager and Authors and staff several times during the week. It also became a first hand experience for the other staff who looked on and helped out in different engineering, electrical, welding, lathing, and mechanical skills [Radesh 2007].

The next level of help, Plan C, from the University of the South Pacific was very important to introduce the engineering section to the project and make a better hydropulper. This took a year to make and network as a student project. It was delivered to the village in January 2008 and was happily accepted in the village and now an integral part of their new improved processes. Two students produced a document as part of their research on making a hydropulper at USP [Dayal and Khan 2007]. This will be further discussed in another paper.

The next level, Plan D, is for all Fiji stakeholders to be more in control of the technology while networking with Community Education Training Centre and Murdoch University and for ongoing support of new projects in Fiji and the Pacific Region [Bola 2006, Maka 2007-8]. The hydropulper could be re-engineered and donated to this group for ongoing research and development. This will be further discussed in another paper.

## EVALUATION

The hydropulper was successfully made and trialled in the remote papermaking village in 2007. The USP hydropulper was then trialled from January 2008. This shows the importance of getting the inventor to help in the re-design of the equipment and linking with local industry to make it work against many odds.

The results show that for the hydropulper stage of the project; the village contribution from 13 men, seven helped and were allocated 5,5,5,5,10,10,15 points= only 55 points (15%); the industry and university generated 364 points from over 20 tradesmen, staff and students (85%),of which 6 were at 25; 5 at 20 points and the majority below 10 points, (or seven times more transfer potential than the village contribution). This highlights the invaluable contribution of industry partners and the university in making the higher technology equipment in the nearest city to the village.

The combined total for the hydropulper of 419 (55+364) points divided by 25 equals (16.76) over sixteen times effective transfer of the inventors/Authors ideas and skills to various stakeholders in the LDC of Fiji. This should empower the ongoing success of technology transfer over many years. The analysis of the overall results suggest that a factor of 5:1 for low technology equipment and 10:1 for higher technology equipment or skills is sufficient buffer to effectively transfer technology to LDCs.

## CONCLUSION

- In conclusion, the modified appropriate 'sustainable turnkey approach' worked to introduce a new sophisticated piece of technology to a village craft papermaking cottage industry in Fiji, a developing country.
- The remote village does not have the expertise, skills and equipment to design and make the more advanced hardware. The best way forward is for the inventor of the technology to empower a local engineering business and local university in the project. The inventor is acknowledged under the 'creative commons' and this process was successfully trialled to work best for a LDC. The process involves redesigning the equipment, using local parts, engineering local parts, learning new skills, learning the limitations of local machinery to fabricate parts and finally new traditional protocols of doing business.
- A combined collaborative approach involving the scientist /inventor, the papermakers of the village, headman, chiefs, then local engineering business, then local universities, then local government or NGOs training agents, are important to combine all stakeholders to link in with a long term village project.
- The project was evaluated with a survey that reinforced the results that the technology transfer process could be evaluated and objectively judged to be successful. The values of 5:1 and 10:1 could be a new base line for the evaluation of success in transfer of low and high levels of technology transfer to developing countries.
- There were many champions in this 'sustainable turnkey approach' project and at least one is needed at every phase to push-pull the project along.

## REFERENCES


Bola,J. (2006) (Personal conversation) Director: Centre for Appropriate Technology and Development. *Nausuri. Suva.* Fiji.

Bunt, S. (2007) Creative commons and commercialisations. Intellectual Property Forum. Murdoch University, Perth W.Australia. Law School. 15th November 2007

Dayal,L.K. and Khan,A.A..(2007) Design and manufacture of a hydropulper. The University of the South Pacific. School of Engineering and Physics. Fiji. (Student Project under Dr D.Wood)70p.

Efstathiades, A; Tassou,S; Oxinos,G; Antoniou,A (2000) Advanced manufacturing technology transfer and implementation in developing countries (Cypriot) Technovation V20/2.

Kassim, H. (1992) Transfer and utilization of technology: A country study on Fiji. UNCTAD/ITP/TEC/30: UN. New York.

Maka, L. (2007-8) (Personal communication and emails). Head of Community Education Training Centre. Human Development Programme. Secretariat of the Pacific Community. Suva, Fiji.

Radesh, L.(2007) (Personal communication and help modifying equipment) Technician, University of South Pacific Engineering. Suva. Fiji.

Stewart. (1987) Technology transfer and diffusion: A conceptual clarification. *Journal of Technology Transfer.* V12-1.

Sharma. (2006) (Personal communication): Manager of 'On Time Engineering', Suva Fiji.

Liebregts, W and Townsend, P. (1998) Handmade paper production in Fiji. A sustainable resource project. Stage1.Eco-Consultants. Suva. Fiji.

Liebregts,W.(1998) Training and support for handmade paper production to Sandollars Ltd, Fiji Phase 1. Report to Development of Industries, Brussels, Belgium. Suva. Fiji.

Westerlund, L.C.(2005) Science and practice of smooth handmade paper. Westerlund Eco Services. Rockingham. Perth. W.Australia.

Westerlund, L.C.(2007) How to make a hydropulper. Westerlund Eco Services. Rockingham. Perth. W.Australia. (Available online at http://leslie_westerlund.tripod.com)

Westerlund,L.C.(2008) How to make smooth papermaking technology. Westerlund Eco Services. Rockingham. Perth. W.Australia. (Available online at http://leslie_westerlund. tripod.com)

In: Technologies and Management for Sustainable Biosystems ISBN: 978-1-60876-104-3
Editors: J. Nair, C. Furedy, C. Hoysala et al. 

*Chapter 23*

# INTEGRATED BIOSYSTEMS - MONTFORT BOYS TOWN, SUVA, FIJI ISLANDS

***Kamal Khatri****

## ABSTRACT

The Montfort Boys Town located near Lami approximately fifteen kilometres away from Suva, Fiji is a vocational school that offers free technical courses for disadvantaged teenagers. The integrated biosystem at Montfort Boys Town is an installation where a range of technologies have been utilised so there is no (minimal) waste or zero emissions. Systems used include anaerobic digester, algae settling ponds and refeeding. Agricultural activities at the school provide food for the students as well as feedstock for the school's farm animals. The residual material from the food and feedstock is utilised for the production of biogas that provides an energy source where the balance is used through the process of assimilation and lost to the environment contributing to the production of biomass. An integral component of the biosystem is management of wastewater. Water quality in different treatment ponds was assessed through physical water testing and microbiological analysis for faecal coliform bacteria. Evidence has shown significant improvement in water quality upon digestion and passing through treatment ponds with higher coliform levels closer to the digester. The students need to be trained on operation and maintenance components of the biosystem to enable them to manage and monitor levels of the effluent released.

**Keywords**: biosystem, digesters, water quality, samples, dissolved oxygen, faecal coliform.

## 1. INTRODUCTION

There are three digesters in the Montfort Boys Town, two made out of concrete and one made out of fiberglass. This chapter, looks at the effectiveness of one of the operational

* World Health Organization, PO Box 113, Suva, Fiji. Email: khatrik@wpro.who.int, Phone: (679) 3234100, Fax: (679) 3234166.

concrete digesters being used from July 1998 (has a size of 15 cubic meters), and the fibreglass digester (size of 20 cubic meters). A schematic showing the location of the biodigesters, the treatment ponds and sampling sites chosen at the Montfort Boys Town can be referred in Figure 1. A rapid analysis was done on a field visit (8 January 2004) to the farm which included sampling (for microbial analysis) and *in-situ* water testing for selected physical water quality parameters. The concrete digester was not operational at the time of visit therefore has not been considered in this chapter.

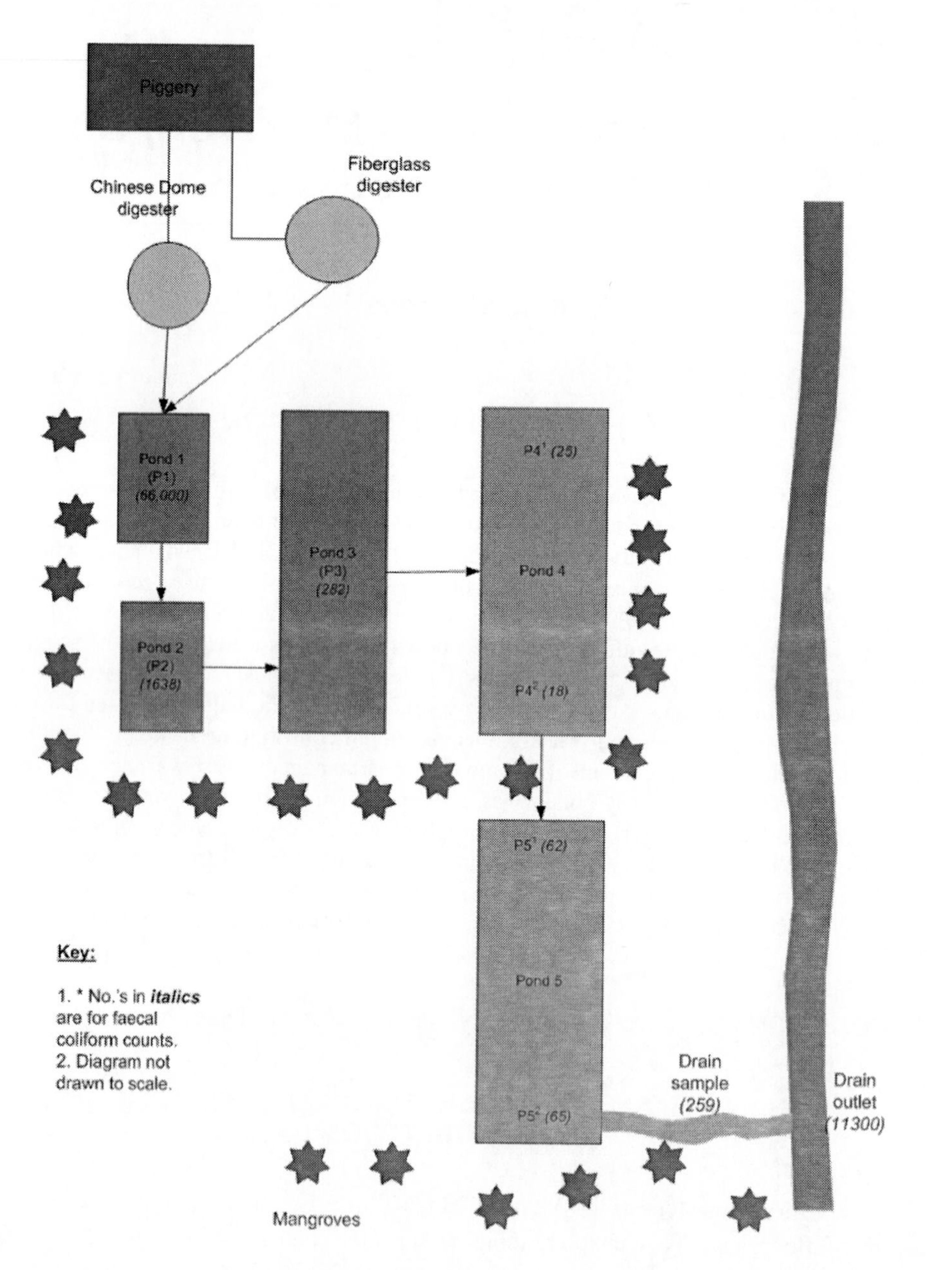


Figure 1. Schematic showing the location of the biodigesters, treatment ponds and sampling sites at Montfort Boys Town.

The digesters at Montfort were inspired from traditional integrated systems from China, and had been constructed by Professor George Chan. It is built so that no air can get in providing anaerobic conditions. The working principle is to digest organic matter into nutrients, nitrogen, phosphorous and carbon, with the help of anaerobic bacteria. The bacteria help in converting the 60% of the organic matter in the waste into nutrients in 5 to 6 days.

The aims for this paper are 1. to find out whether the water quality improves after it passes from one treatment pond to the other, and 2. to see the changes that occur in the physical water quality parameters and the faecal coliform concentration.

The faecal coliform bacteria are those members of the coliform group, which are capable of fermenting lactose at the elevated temperature of 44.5 ± 0.2 °C. Colonies produced by faecal coliform are blue in colour. The microbiological analysis for faecal coliform counts was used in this research as it acts as a very good indicator of water pollution from faecal origin. If high levels are found in the water it could have detrimental effects on marine life and also render the water unsafe for recreational activities.

## 2. METHODOLOGY

The physical and chemical parameters were measured using the YEO-KAL (YK-611) water quality analyzer. The instrument was calibrated using standard solutions provided by manufacturers to ensure integrity of data collected. The multi-meter can be calibrated on a regular basis or when deemed necessary. The measurements included variables such as temperature, conductivity, salinity, dissolved oxygen (DO), potential hydrogen (pH), oxidation-reduction potential (ORP) and turbidity. The data obtained will be discussed later in this chapter.

The sampling was designed to ensure critical points are covered with consideration to getting representative sample from the ponds. Strategies for monitoring of microbiological quality of water should include hazard identification and risk assessment (Mosley *et al.* 2005). Equally important to monitoring and assessments is the need to educate the students on water treatment methods/benefits for integrated biosystems and encouraging maintenance of records for periodic reviewing (e.g. once each year).

The water samples were collected in *bacto* bottles obtained from the National Water Quality Laboratory located in Kinoya, Suva for faecal coliform analysis. A detailed methodology on the faecal coliform filter method that is used by the laboratory in Kinoya could be referred in (Clescari *et al.* 1998; Singh 2002). Other materials used included an ice container and ice packs to maintain low temperature, as the samples were taken from Montfort Boys Town approximately fifteen kilometres from the Kinoya laboratory.

### 2.1. Field Testing

Analyses for many of the physical and chemical variables can be carried out in the field using equipment that is specifically made for field use. A significant advantage of field analysis is that tests are carried out on fresh samples whose characteristics have not been contaminated or otherwise changed as a result of storage in a container (UNEP/WHO 1996).

The physical properties of the water were tested *in-situ*, on field. The water samples for the faecal coliform analysis was collected in 250 ml borosilicate glass bottles called *bacto* bottles provided by the National Water Quality Laboratory in Kinoya who carried out the coliform counts.

## 2.2. Types of Water Samples

Two different types of samples can be taken from surface waters. The simplest, a "grab" sample is taken at a selected location, depth and time (Arizona Water Resources Research Center 1994; Clescari *et al.* 1998; State of Art 1992; UNEP/WHO 1996). This type of water sample collection is quite convenient and the quantity of water is sufficient for all physical, chemical and microbial analyses. The other type of sample is called composite or integrated, which are made of several parts, and are often needed to fulfil specific monitoring objectives (UNEP/WHO 1996). Composite samples can be of 4 types, depth integrated, area integrated, time integrated and discharge integrated. Composite samples are more complexed and very specific to certain variables.

For this research, "grab" samples had been collected from the selected algae ponds. Grab samples are also known as "spot" or "snap" samples. There are several reasons to why "grab samples" had been collected in this research. This sampling method was most efficient, as the ponds were relatively shallow and the main purpose of collecting the water sample was for microbiological analysis of faecal coliform concentration. Whereas, in the composite samples there are certain criteria's that needs to be fulfilled before a sample is taken, for example, if depth integrated samples are to be collected, it needs to be collected from predetermined depth intervals between surface and the bottom.

Figure 2. Near algae pond 3, the concrete digester and the piggery can be seen in the background.

One of the aims of this paper is to test the improvement in water quality and effectiveness of the biodigester to treat biological waste. For this reason, water samples were only taken from those ponds releasing effluent directly from the biodgester near the piggery for water quality analysis. Figure 2, is taken at the Montfort Boys Town showing the location of pond 3 relative to the biogas digester and the piggery in the background. Figure 3, shows the researcher placing the probe of the YK-611 water quality analyzer to measure physical water quality at algae pond 4. Once the values get stabilized the water quality readings are saved and written manually as well. The YK-Multigraph software allows one to download the data on computer but writing it down ensures that there is a back up in case the data gets deleted accidentally or if the battery runs out.

Figure 3. Measuring the physical and chemical water quality parameters using the YK-611 analyzer.

## 2.3. Transportation and Storage Of Samples

Proper sampling protocols were strictly followed to ensure integrity of the data collected. Sample bottles were placed in an ice container for transportation, to protect them from sunlight as well as prevent any possible breakages. For bacteriological analysis a temperature between 0°C and 4°C is necessary (ANZECC and ARMCANZ 2000b; State of Art 1992; UNEP/WHO 1996). This temperature was maintained in the ice container using ice packs. In practice, it is difficult to ensure the transport of samples under conditions that do not affect the bacteriological quality. Optimum quality control was kept to prevent errors in results.

# 3. Results and Interpretation

The treated waste from the digester flows into a circuit of three algae basins (refer Figure 1). This setup enhances the efficiency of the aerobic digestion of the waste through photosynthesis. The algae assimilate energy from the sun, fix nitrogen, oxidize and treat the waste. The treated waste is used for composting and acts as high-grade fertilizer for vegetables and fruit growing on the dykes around the ponds, and as feed for livestock.

**Table 1. Physical and chemical water quality parameter measurements from treatment ponds at the Montfort Boys Town, Suva, Fiji**

| STORE KEY DATA | | | | |
|---|---|---|---|---|
| SERIAL NUMBER: 287 | | | | |
| CAL DATE/TIME | | SENSOR | OFFSET | SLOPE |
| 27/05/03 | 08:12 | TEMPERATURE | 1535.666 | 77.206 |
| 17/09/03 | 09:05 | SAL/COND MSCM | 0.000 | 23574.564 |
| 17/09/03 | 10:03 | COND USCM | 1.000 | 0.396 |
| 08/01/04 | 06:23 | DISSOLVED OXYGEN | 2358.000 | -5.566 |
| 16/09/03 | 11:53 | PH | 4137.032 | -602.370 |
| 28/05/03 | 07:40 | ORP | -287.732 | 1.339 |
| 16/09/03 | 11:41 | TURBIDITY | -87.000 | 6.720 |

| | SAM NUM | DATE | TIME | TEMP C | COND ms/cm | COND us/cm | SAL ppt | DO %sat | DO mg/l | PH pH | ORP mv | TURB ntu |
|---|---|---|---|---|---|---|---|---|---|---|---|---|
| 1 | P3 | 08/01/04 | 10:12:07 | 30.80 | 0.1 | 54 | 0.04 | 63.6 | 4.7 | 8.76 | -8 | 8.9 |
| 2 | P41 | 08/01/04 10: | 10:19:19 | 31.24 | 0.1 | 31 | 0.03 | 69.4 | 5.1 | 8.41 | 43 | 49.3 |
| 3 | P42 | 08/01/04 10: | 10:23:51 | 31.20 | 0.1 | 29 | 0.03 | 71.7 | 5.3 | 8.18 | 74 | 59.8 |
| 4 | P51 | 08/01/04 10: | 10:27:39 | 30.47 | 0.1 | 29 | 0.03 | 79.5 | 6.0 | 7.90 | 77 | 63.8 |
| 5 | P52 | 08/01/04 10: | 10:37:07 | 30.98 | 0.1 | 22 | 0.03 | 55.9 | 4.1 | 9.28 | 55 | 39.6 |
| 6 | D1 | 08/01/04 | 10:52:18 | 26.17 | 0.4 | 217 | 0.19 | 138.4 | 11.2 | 7.21 | -121 | 37.6 |
| 7 | D2 | 08/01/04 | 10:56:12 | 27.60 | 0.5 | 293 | 0.24 | 121.6 | 9.6 | 6.85 | -65 | 8.2 |
| 8 | D3 | 08/01/04 | 11:02:06 | 29.42 | 1.1 | 648 | 0.54 | 92.3 | 7.0 | 6.80 | -144 | 5.8 |
| 9 | D4 | 08/01/04 | 11:05:56 | 31.13 | 2.1 | 1251 | 1.06 | 76.6 | 5.6 | 6.54 | -196 | 239.7 |
| 10 | DP | 08/01/04 | 11:10:45 | 33.90 | 13.9 | 8000 | 7.99 | 67.8 | 4.6 | 6.49 | -87 | 51.6 |

Table 1, shows the results that were obtained when the algae ponds were tested for various physical water parameters. A lot of fluctuation was seen in each of the parameters that were measured. The YK-611 meter was used from algae pond three onwards, as the first 2 ponds (P1 and P2 in Figure 1) had a lot of algae present, the water colour was green and seemed quite polluted. This was done as a precautionary measure to prevent the probe from getting damaged. However, water samples were collected from all the ponds, the drain and drainage outlet for microbiological analysis of faecal coliform bacteria.

The water temperature has a direct influence on other water quality factors such as dissolved oxygen (DO) and biological oxygen demand (BOD) as well as on aquatic life (Stevens Institute for Technology 2002). All the physical properties were measured *in-situ* as the water properties will vary over time. Sample 1 was tested at algae pond 3 (P3), samples 2, 3 were tested at pond 4 ($P4^1$ and $P4^2$), likewise samples 4, 5 were tested at pond 5 ($P5^1$ and $P5^2$), the other samples 6-9 were tested at the drain and lastly sample 10 was measured near the drain pipe outlet, closer to the mangroves and the sea (see Figure 1 for corresponding pond locations and sample points).

Dissolved salt concentration in water can be measured by conductivity. For this research, the conductivity has been reported as millisiemens per centimeter (ms/cm) as well as in microsiemens per centimeter (μs/cm). Generally, the conductivity was seen to be relatively stable when measured in the algae ponds, though when water was tested from the drain it increased as one moved to its outlet. Salinity is a measure of total concentration of dissolved ions (salts) in the water (ANZECC and ARMCANZ 2000a; Johnson 1985) and in this chapter it has been measured in parts per thousand (ppt). The salinity results obtained were similar to what had been obtained for the conductivity. The values for salinity are as expected since closer to the sea, the salinity levels increase (refer Table 1 for values).

The DO concentration measured in a water body reflects the equilibrium between oxygen consuming processes (e.g. respiration) and oxygen releasing processes (ANZECC and ARMCANZ 2000a). The DO level can be an indication of how polluted the water is, and its ability to support marine life. The DO levels, generally increased from pond 3 onwards, except for one anomaly seen in pond 5 ($P5^2$) and the drain water samples tended to have lower DO values. pH is a measure of the acidity or alkalinity of water and has a scale from 0 (extremely acidic to 7, neutral), through to 14 (extremely alkaline) (RIC 1998). The pH changed from being slightly basic (near algae pond 1) to fairly neutral when the water from the drain was tested.

Lastly, TURB in Table 1, refers to turbidity which is a measurement of the suspended particulate matter in a water body, that interferes with the passage of a beam through the water (RIC 1998). In this chapter, it has been represented in nephelometric turbidity units (ntu). The acceptable turbidity limit is <5 ntu for aesthetic reasons (Robillard *et al.* 1991; Suslow 2004). Very high turbidity values were found, and this could indicate presence of algal blooms as well as contaminated water. High levels of turbidity increase the total available surface area of solids in suspension upon which bacteria can grow. High turbidity reduces light penetration; therefore it impairs photosynthesis of submerged vegetation and algae (ANZECC and ARMCANZ 2000a; RIC 1998). However, with the background turbidity unknown it is uncertain whether values obtained would be detrimental for fish production (*tilapia*).

The temperature levels had many fluctuations. Sample 6, onwards the temperature level increased dramatically. Generally at higher temperatures the dissolved oxygen levels are expected to be lower. This correlation was seen from sample 6 onwards, with higher temperatures the DO levels decreased. Increased temperatures elevate the metabolic oxygen demand, which in conjunction with reduced oxygen solubility, impacts many species (RIC 1998).

To see how the temperature and dissolved oxygen reacted with each other a graph was plotted to see whether any relationship could be established. Graphs can also be plotted using the YK Multigraph software, using the YK-611 water quality analyzer as it allows one to download data directly to the computer. Figure 4, shows on the x-axis the sample numbers and on the y-axis the temperature and dissolved oxygen. The graph revealed that generally when the temperature is high the dissolved oxygen levels drop and vice versa. Upon data analysis, correlation value came to (r = -0.91167), indicating a negative correlation, meaning both the variables behave in the opposite manner.

A logarithmic graph was plotted to see the trend of faecal coliform concentration against the different algae ponds tested (refer figure 5). The results showed that as the wastewater moved from one pond to the other there was a dramatic reduction seen initially and later the concentration stabilized. One anomaly was seen with the drainage outlet sample and this could have been due to wastewater accumulating from other sources (that is, other ponds that do not have a biodigester for treatment). For most part it can be concluded that the treated wastewater can be safely used for crop irrigation if desired as the recommended microbiological quality guideline value is ≤ 1000 (faecal coliforms/100ml) (Ayres and Mara 1996).

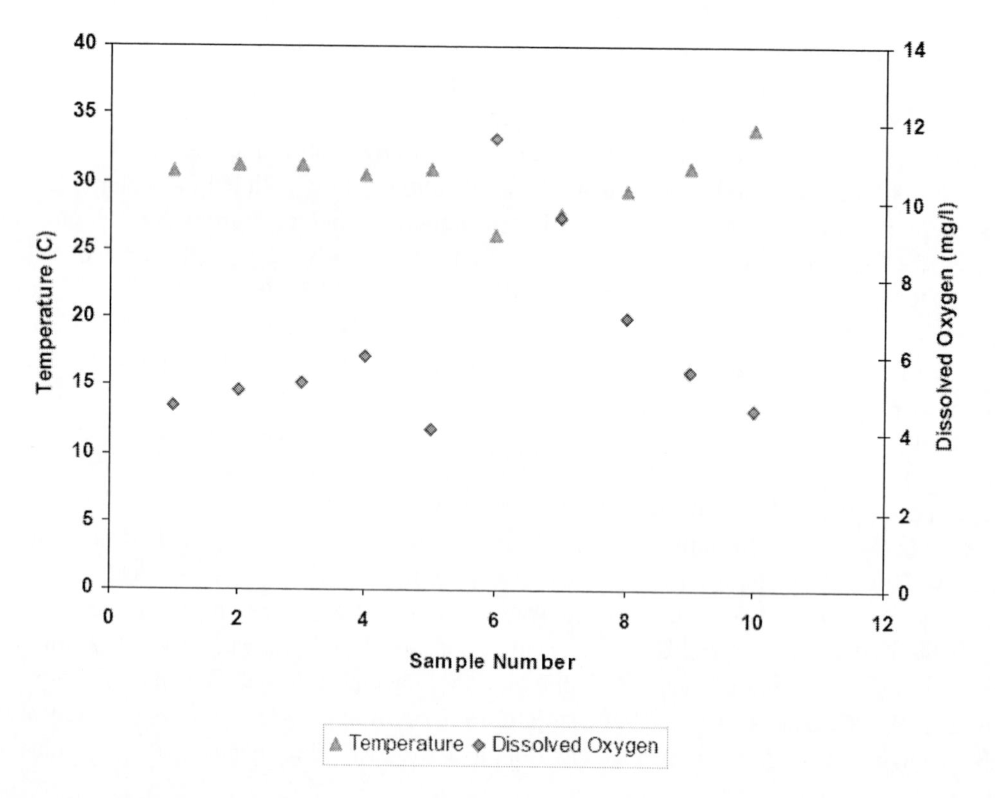


Figure 4. Graph showing the correlation between Temperature (°C) and Dissolved Oxygen (mg/l) at the algae ponds and sites tested at the Montfort Boys Town.

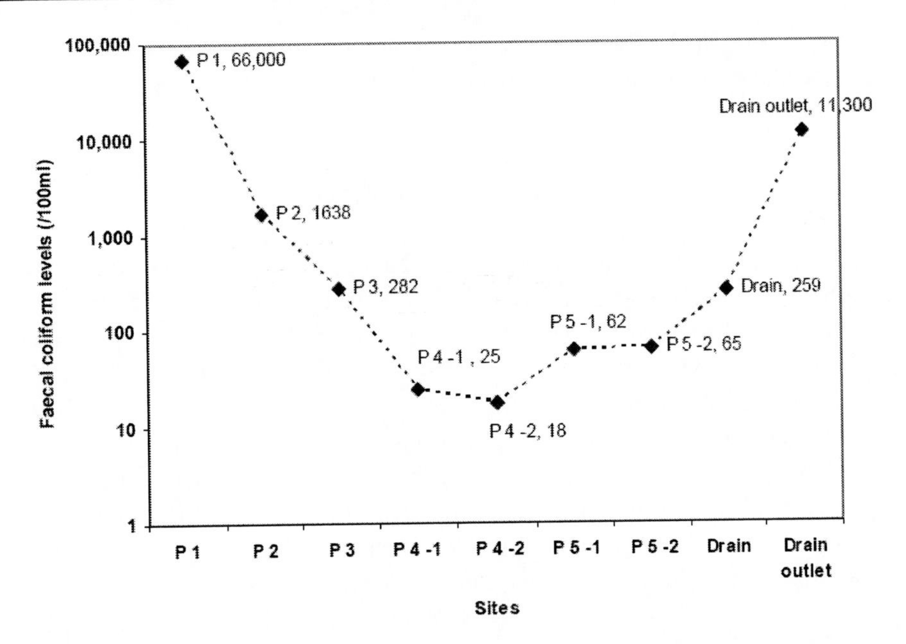


Figure 5. Graph showing the sampling sites vs. the faecal coliform concentration (/100ml) at the Montfort Boys Town in Suva.

Water quality studies within Fiji, along the Coral Coast in Sigatoka have revealed high levels of nutrients above guidelines for coral reefs with the potential to damage the reef ecosystems and biodiversity (Coral Cay 2001; Mosley and Aalbersbery 2003; Thaman and Sykes 2004). These surveys also found high faecal coliform levels in some places. Case studies from several Pacific islands like Tuvalu, Kiribati and Cook Islands have showed the adverse implications to human/environmental health and economic costs associated with pig husbandry practices especially if wastewater is not managed properly (Khatri 2007). This problem is further amplified through uncontrolled sewage pollution.

## 4. Conclusion

The general trend that was observed in the physical water parameters and faecal coliform levels, after getting processed from the algae ponds was positive in nature. There is insufficient data to comment on the effectiveness of the biogas digester to treat the wastewater, as influent concentration would need to be known for comparison purposes. However, results indicated the ability of ponds to improve the water quality through reduction of faecal coliform counts. The proposed faecal coliform Concentration Standards for Fiji is 400 col/100ml for general and 200 col/100ml for significant ecological zone conditions (Kirkwood 2006). This is similar to the WHO guideline for faecal coliform levels in recreational waters being $\leq$ 200 coliforms/100ml. The discharge point at the Montfort Boys Town has minimal natural flow thus very susceptible to stress from pollutants at current coliform concentration and the Fiji Ministry of Environment could declare it a 'Sensitive Ecological Zone'. This would mean that the current system would need to be upgraded or alternative systems put in place to either treat or manage wastewater to achieve lower concentrations in the final effluent. This can be confirmed through a similar analysis of nitrates and phosphates (nutrients) over a period of time.

## ACKNOWLEDGEMENT

This research would not have been possible without the support of the Australian Centre of International Agricultural Research (ACIAR) and the Secretariat of the Pacific Community (SPC). Thank you to the National Water Quality Laboratory, Kinoya. Deep appreciation goes out to Andrew Tukana who was jointly involved in the field sampling. Tasleem Hasan for useful discussions and assistance offered. Atish Sewak, who assisted with drawing the schematic for Montfort pond sites. Finally, I would like to thank Paul Fairbairn and Marc Overmars, of SOPAC for their continuous support and encouragement. Special thanks to AusAID, who provided support to participate and present at the International Organisation for Biotechnology and Bioengineering (IOBB), 1st International Conference on Technologies and Strategic Management of Sustainable Biosystems (6-9 July, 2008).

## REFERENCES

ANZECC and ARMCANZ. 2000a. *Australian and New Zealand guidelines for fresh and marine water quality*. National Water Quality Management Strategy Paper No. 4, Australian and New Zealand Environment and Conservation Council and Agriculture and Resource Management Council of Australia and New Zealand, Canberra.

ANZECC and ARMCANZ. 2000b. *Australian guidelines for water quality monitoring and reporting*. National Water Quality Management Strategy Paper No. 7, Australian and New Zealand Environment and Conservation Council and Agriculture and Resource Management Council of Australia and New Zealand, Canberra.

Arizona Water Resources Research Center.1994. *Field manual for water quality sampling*. Arizona Board of Regents, The University of Arizona, Water Resources Research Center.

Ayres, R.M. and D. Duncan Mara. 1996. *Analysis of wastewater for use in agriculture – A laboratory manual of parasitological and bacteriological techniques*. World Health Organization.

Clescari, L.S., Greenberg, A.E., Eaton, A.D. and Franson, M.A. 1998. *Standard methods for the examination of water and wastewater*. 20th Edition. Prepared and published jointly by American Public Health Association, American Water Works Association, Water Environment Federation.

Coral Cay. 2001. Mamanuca Coral Reef Conservation Project – Fiji 2001: Pilot project final report. Coral Cay Conservation Limited, U.K.

Johnson, S.K. 1985. *Understanding water analysis reports: Water from freshwater fish ponds and their water supplies*. Reprinted from the Proceedings of the 1985 Texas Fish Farming Conference, at College Station, Texas.

Khatri, K. 2007. *People, Pigs and Pollution – A participatory approach to pig-waste management at the village level in Fiji*. Master of Arts thesis. School of Geography, Faculty of Islands and Oceans, the University of the South Pacific.

Kirkwood, A. 2006. *Review of Proposed Liquid Effluent Wastewater Standards and Regulations for the Fiji Government.* Ministry of Environment, Suva, Fiji.

Mosley, L., S. Singh and B. Aalbersberg. 2005. *Water quality monitoring in Pacific Island Countries: Handbook for water quality managers and laboratories, public health officers, water engineers and supplier, environmental protection agencies and all those organisations involved in water quality monitoring.* SOPAC and the University of the South Pacific.

Mosley, L.M. and W.G.L. Aalbersberg. 2003. Nutrient levels in sea and river water along the 'Coral Coast' of Viti Levu, Fiji. *South Pacific Journal of Natural Science* 21:35-40.

Resources Inventory Committee (RIC). 1998. *Guidelines for interpreting water quality data.* Ministry of Environment, Lands and Parks Land Data BC, Geographic Data BC for the Land Use Task Force.

Robillard, P.D., Sharpe, W.E. and Martin, K.S. 1991. *How to interpret a water analysis report.* The Pennsylvania State University, College of Agricultural Sciences, U.S. Department of Agriculture, and Pennsylania Counties Cooperating.

Singh, S. 2002. *Laboratory procedures for wastewater analysis.* National Water Quality Laboratory, Kinoya, Suva. Fiji.

State of Art. 1992. Sampling Methods: ISO International Standard # 5667/10. *Water quality–sampling–Part.10: Guidance of sampling of wastewater.* Available from Internet: http:www.stateofheart.it/samp-met10.htm.

Stevens Institute of Technology. 2002. *The global water sampling project: An investigation of water quality.* Center for Improved Engineering and Science Education (CIESE). Available from Internet: http://riee.stevens.edu/index.php?id=mini_grants

Suslow, T.V. 2004. *Introduction to ORP as the Standard of Postharvest Water Disinfection Monitoring.* UC Davis, Vegetable Research and Information Center. Available from Internet: http://vric.ucdavis.edu/veginfo/foodsafety/orp.pdf.

Tanner, C.C. and A.J. Gold. 2004. *Review and recommendations for reduction of nitrogen export to the Coral Coast of Fiji.* Prepared for Coastal Resource Center, University of Rhode Island, U.S.A. National Institute of Water and Atmospheric Research (NIWA).

UNEP/WHO. 1996. Water Quality Monitoring – *A Practical Guide to the Design and Implementation of Freshwater Quality Studies and Monitoring Programmes.*

In: Technologies and Management for Sustainable Biosystems ISBN: 978-1-60876-104-3
Editors: J. Nair, C. Furedy, C. Hoysala et al. 

*Chapter 24*

# BIOTECHNOLOGY AND RURAL SUSTAINABILITY IN BANGLADESH

***Dora Marinova*[*1] *and Amzad Hossain*[*1,2]**
[1] Curtin University Sustainability Policy (CUSP) Institute,
Curtin University of Technology; GPO BoxU1987, WA6845
[2] Institute for Sustainability and Technology Policy (ISTP),
Murdoch University, Murdoch, WA6150

## ABSTRACT

The analysis of trends in recent biotechnology patenting in the USA reveals that many patents issued between 1976 and 2005 are based on biological material sourced from the Indian subcontinent. The chapter also warns about the probability of deploying these newly emerged technologies in developing countries and causing unsustainable practices. It refers to the basic ecological services provided by the natural environment and stresses the need for understanding their economic dimension, particularly in relation to rural people. Building self-reliance and using local knowledge and skills are suggested as a way to counteract the pressure from the West. The case of rural Bangladesh where the Baul culture is highly influential is used as a case study.

**Keywords:** Baul, biosystems, indigenous, patents, self-reliance, technology.

## INTRODUCTION

The term biotechnology, understood as the use of biosystems for products and processes for human benefits (Convention of Biological Diversity, 1993), has evolved immensely throughout the centuries with many new and emerging fields (Uber and Sell, 2007). While most of the new biotechnologies are developed in the West, developing countries continue to

* Australia; phone: +61 8 9266 9033; fax: +61 8 9266 9031
* Australia phone: +61 8 9360 6266, fax: +61 8 9360 6421; email: A.Hossain@murdoch.edu.au

provide a rich environment for these innovations. Firstly, local knowledge and traditional use of certain plants, substances and processes in these societies often generate the ideas and directions for further research and inventiveness. Moreover many seeds, plants and animal varieties consequently need to be sourced from developing countries because of their climatic characteristics and/or better biodiversity conditions. Secondly, many of the developed new biotechnologies are ultimately aimed for application in developing countries as was the case with the Green Revolution.

This situation reduces farmers and growers in countries such as India, Pakistan or Bangladesh to suppliers of biomaterial to western companies and later on establishes dependence on the products created by the West. Questions such as whether the new biotechnology really provides human benefits for all and whether there is an alternative way of using nature's resources are raised. The traditional knowledge and skills of people in developing countries have been the source of livelihood for millennia while recent advents in technology have generated economic benefits mainly for outsiders. Technological progress also has often exploited these countries' fragile ecological environment and created serious environmental damage. The first part of the chapter provides some evidence that the exploitation of the biological wealth of the countries on the Indian subcontinent continues. It then explores the village life of Bangladesh and its culture arguing the case for building self-reliant practices for rural sustainability.

## NEW BIOTECHNOLOGY

Despite the formidable work of people such as Shiva (2000, 2005), it is difficult to show direct statistical data of the exploitative use by the West of traditional biosystems knowledge in developing countries. Most of the evidence is qualitative and comes from stories (Posey and Dutfield, 1996) or legal cases where affected individuals or communities, including indigenous people, have fought back to regain ownership of traditional knowledge (Finger and Schuler, 2004). According to Bush and Stabinsky (1996), a major aspect of the legacy of colonial domination is the expropriation of local biosystems resources without due compensation or attribution.

Developed countries have also put systems and mechanisms in place which in many ways legalise these exploitative practices. An example is the patenting system as it fails to recognise intellectual input from traditional and indigenous knowledge sources (Marinova and Raven, 2006).

Patenting in the USA in particular is considered indicative of the development of any new technologies (e.g. Acs and Audretsch, 1989; Archibugi, 1992). The trends in biotechnology patents issued in the USA are presented in figure 1 and 2. Figure 1 shows absolute numbers of biotechnology patents issued in the USA from applications lodged between 1976 and 2005 while figure 2 illustrates their relative shares of all patents for the same period. Since the mid 1990s, the annual number of issued new biotechnology patents has increased drastically reaching a high of 3511 in 2000 (since then it has been gradually decreasing). The year 2000 also represented a significant peak in terms of the relative share of these patents, namely 1.7% of all US patents. Although this share is relatively modest compared to other groups of new technologies (nanotechnologies for example had a relative share of 3.6% in 2000 and it reached 4.8% in 2004), this was a huge improvement on the low numbers during the 1980s.

Interestingly enough the late 1980s and 1990s – the period during which biotechnology patents sharply increased – also witnessed the greatest expansion in the scope of intellectual property rights (Maskus, 2000) as well as a surge in the number of successful US patent applications related to native species (Marinova and Raven, 2006). As a consequence, in the 1990s the UN Commission on Human Rights opened for the first time the debate about the use and recognition of indigenous knowledge and protection of indigenous intellectual property. This debate is yet to be resolved.

The contribution of countries such as India, Bangladesh and Pakistan to the booming market of US biotechnologies has been very modest (154 patents in total during the 1976-2005 period). A larger number of patents however build on traditional plants and other biosystems indigenous to this part of the world (510 patents in total for that same period or 3.3 times more) confirming the concerns raised about exploitation of local knowledge. Table 1 shows the number of biotechnology patents that list India, Bangladesh and Pakistan as part of the patent descriptions (including abstract and patent claims). While 402 US patents were sourced from India only 152 patents had an Indian inventor. In the case of Bangladesh, 30 biotechnology patents were sources from this country's biosystems but only 1 had a Bangladeshi inventor. Similarly for Pakistan, there was only 1 Pakistani inventor for the 72 US biotechnology patents sourced there. Even a smaller number of US biotechnology patents (99 in total) were assigned for further commercial exploitation to representatives of these countries.

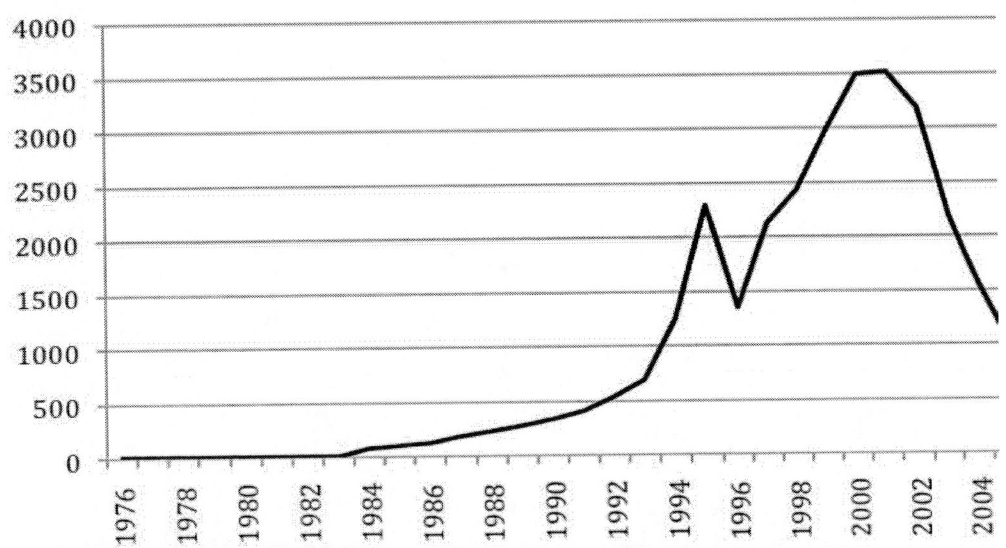


Source: Compiled from US Patent and Trademark Office, as of 24 June 2008.
Note: Data are by date of application for patents issued.

Figure 1. Biotechnology patents in the USA, 1976-2005.

The new biotechnology products, including those produced through exploitation of native biomaterial, are highly likely to hit the soil of the Indian subcontinent in the name of providing more intensive ways of farming or food manufacturing. The sheer numbers of these newly patented biotechnologies imply concerted efforts on the part of their developers to make the best use of the inventiveness that had gone into the sector.

It is difficult to avoid a parallel with the technologies of the Green Revolution of the 1970s which aimed at modernising these developing countries but instead created ecological disasters, health hazards and increased food prices (Charkarborty, 2008). The Green Revolution was based around the application of chemicals and fertilisers synthesised by the West, such as urea or triple superphosphate (TSP). It was also accompanied by particular new high yield crop varieties that were expected to feed better the growing population but required the application of the synthetic fertilisers and encouraged monocropping. This combination caused soil depletion, water poisoning and the need to apply more and more of the chemical substances to achieve the same yield. The Green Revolution also had major negative social consequences contributing to the establishment of cash driven poverty in Bangladesh (Hossain and Marinova, 2005). Are there any guarantees that the case of biotechnologies will be any different?

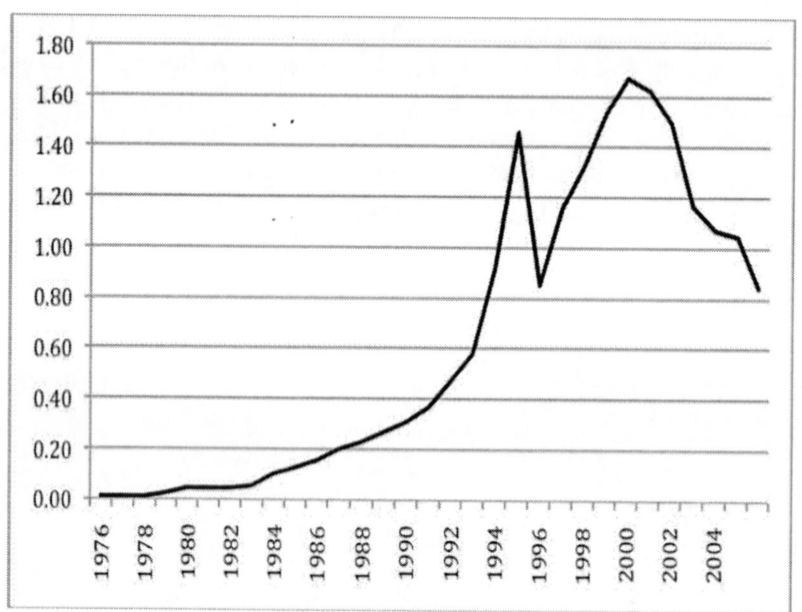


Source: Compiled from US Patent and Trademark Office, as of 24 June 2008.
Note: Data are by date of application for patents issued.

Figure 2. Share of biotechnology patents of total US patents, 1976-2005.

**Table 1. US biotechnology patents, 1976-2005**

| | US patents sourced | US patents with indigenous inventors | US patents assigned to indigenous bodies |
|---|---|---|---|
| India | 402 | 152 | 99 |
| Bangladesh | 30 | 1 | 0 |
| Pakistan | 78 | 1 | 0 |
| TOTAL | 510 | 154 | 99 |

Source: Compiled from US Patent and Trademark Office, as of 24 June 2008.
Note: Data are by date of application for patents issued.

Countries such as India, Bangladesh and Pakistan are under a double threat from the newly developing biotechnology. Firstly, they continue to provide the rich biomaterials without being able to fully (if at all) exploit the economic benefits from this. Secondly, sooner or later they will be targeted as an export market for these new biotechnologies. This can happen under the umbrella of development aid or food provision for their constantly growing populations.

A way to counteract such possible developments is to make sure that the population in these places can be self-reliant. The remainder of the chapter explores the concept of self-reliance within the context of preserving traditional biosystems and biotechnology knowledge in Bangladesh – a country where the Baul tradition understands traditional biotechnology as the technology of nature but also as the technology of the future.

## SELF-RELIANCE

Self-reliance is a state of acquiring of basic needs for enjoyment of a simple lifestyle on one's own accord (Hossain and Marinova, 2003). It does not encourage ill health, famine, illiteracy or inadequate living standards but promotes modesty in consumption, caring for nature and other human beings, and hence supports sustainability (Marinova et al., 2006). Self-reliance can be achieved only if the local biosystems continue to provide basic services for "those who are living, those who are dead, and those who are to be born" (Bowers, 2006: 102).

The Millennium Ecosystem Assessment sees human biology basic needs for food, water, clean air, shelter and relative climatic constancy as unalterable (WHO, 2005). It also identifies nature's cultural services, including spiritual, educational, recreational, aesthetic and inspirational. These biosystems' functions are the foundation for self-reliance. However, from a sustainability point of view the biosystems will also have to provide the economic means for the population to support itself. This includes agriculture, aquaculture, food processing, waste management, water use and energy generation but also underpins the basis for the quality of human life which is linked to economic well-being. The use of technology has been the main means to achieve this and subjugate nature to human needs, but have we gone too far? We have witnessed unprecedented changes in the capacity of nature to provide ecosystems services due to the degradation humans have caused to the natural environment.

Rural Bangladesh is still experiencing the consequences of the Green Revolution but the main challenge it faces today is to learn from them and avoid the insensible deployment of new biotechnologies that have not been developed with respect to its fragile natural environment. Bangladesh is known as the country of the whims of biosystems: regular floods, droughts and river erosion. Outsiders see these natural phenomena as disastrous, while the locals understand this as the renewal or transformational functioning of nature that revitalises and (re)generates the resource base. Self-reliance requires people to continue to live a happy and healthy lifestyle within this environment.

Such a lifestyle is strongly supported by the Bauls of rural Bangladesh where these unique philosophers encourage through their songs and music harmony within nature and society, and where the question of rich and poor is transcendent by way of peace and happiness (Hossain, 2001). The Baul philosophers understand floods, draughts and river erosion as the by-product of nature's prime or fundamental technology of energy. Timely floods wash away dirt and germs, recharge water table and ferilise soil; and timely droughts energise the soils with nitrogen fixation. Untimely floods and droughts can cause temporary inconveniences for affected people, but the timely ones, which occur more frequently, bring benefits that outweigh the losses. River erosion breaks riverbanks and shoals, but also makes new ones that are extremely fertile.

The spiritual credibility of the Bauls (see for example McDaniel, 1989) allows them to also be rural mentors and activists in supporting environmental health and the use of the local biotechnologies that the country's rich biosystems provide. In fact, the Bauls' influence on rural Bangladesh is so important that in 2005 they were included in UNESCO's list of Masterpieces of the Oral and Intangible Heritage of Humanity.

Within the context of biotechnology, the future of rural Bangladesh in many ways would depend as to whose influence will be greater – the pressure from the West or the subtle inspirations coming from the Bauls. The Western model is driven by the forces of globalisation and market economy. The Bauls espouse the view that the rural people of Bangladesh have the capacity and ability to create a remarkably different economy. Such a self-reliant economy can restore degraded ecosystems through the extensive use of biosystems and traditional biotechnology and protect the environment while bringing forth innovation, prosperity, meaningful work and true security. The restorative development processes can unite ecology and economic activities into one sustainable act of production and distribution that mimics and enhances the sustainability of natural processes in biosystems (Natrass and Altomare, 1999). This is the only type of new biotechnologies that can create genuine human benefits.

Some of the technologies developed in the West may play a positive role in achieving such a self-reliant lifestyle. However for self-reliance to be attained, it is important to use the creative capacity and knowledge of local people in rural villages instead of exploiting them for the sake of technologies developed elsewhere. Rural people in Bangladesh understand that their livelihood environment requires constant renewal through the integration of diverse human made technologies as well as by using the natural processes and properties of biosystems. For instance, they are known widely for their ability to apply integrated biosystems into permaculture as the examples below demonstrate (Warburton et al., 2000):

- Simple connections: livestock manure – fertiliser for plant crops;
- Intermediate connections: organic waste – compost or vermiculture – plant crops;
- Closed loops: e.g. livestock – manure - fodder crop – feed – livestock;
- Fuel generation: e.g. organic waste - biodigester – biogas.

Under a self-reliance way of living rural people take the responsibility for the long-term health of the biosystems to which they belong and would not trade economic rewards for environmental and social degradation. Most importantly, self-reliance will stimulate human creativity to develop the methods and technologies that can synergistically achieve both. People will be able to build the pride that comes from stability, control, competence and independence (Kumar, 1986:230).

## Conclusion

The Millennium Ecosystem Assessment (WHO, 2005) warns about the threats to humans from disrupting the functioning of biosystems and people's vulnerability in the face of ecosystem changes. According to Shiva (2000), "(t)he consumption patterns of the rich and the production patterns of the powerful can undermine the consumption patterns of the poor by contributing to the erosion of biodiversity" (Shiva, 2000: 19). An added threat to the poor and underdeveloped areas is the deployment of technologies that do not provide genuine benefits to the local people who have lived there for millennia. The concept of self-reliance which implies a natural simple lifestyle with enough for basic needs, is seen as a possible counterforce for rebuilding a healthy social and natural environment that can sustain generation after generation.

Linking biosystems and self-reliance provides the missing economic connection in the perception of ecological services provided by nature. It encourages holistic thinking where all components, relationships, dependencies and interconnections of the sustainability puzzle come together. It requires rural people as well as policy makers to think systematically about the flows of materials and energy, the use of new and traditional technologies and processes and how they can be integrated. The achievement of a self-reliant lifestyle would only be possible with local people within ecologically healthy biosystems and it will also provide for socially and economically healthy human environments.

## Acknowledgement

The authors acknowledge the financial support of the Australian Research Council.

## References

Acs, Z. J., & Audretsch, D. B. (1989). Patents as a measure of innovative activity. *Kyklos, 42*, 171-180.

Archibugi, D. (1992). Patenting as an indicator of technological innovation: A review. *Science and Public Policy, 19*, 357-368.

Bowers, C.A. (2006). *Revitalising the commons: Cultural and educational sites of resistance and affirmation*. London: Lexington Books.

Bush, S., & Stabinsky, D. (Eds) (1996). *Valuing local knowledge: Indigenous people and intellectual property rights*. Washington, DC: Island Press.

Charkarborty, T. (2008). Green Revolution in Bangladesh and its consequences: Importance of sustainable agriculture practices [E-text type]. supro.files.wordpress.com/2008/10/tapan.doc.

Convention of Biological Diversity (1993). [E-text type]. www.cbd.int/convention/convention.shtml.

Finger, J. M., & Schuler, P. (Eds) (2004). *Poor people's knowledge: Promoting intellectual property in developing countries*. Washington, DC: World Bank and Oxford University Press.

Hossain, A. (2001). *Renewing self-reliance for rural Bangladesh through renewable energy technology system*, PhD thesis, Murdoch University.

Hossain, A., & Marinova, D. (2003). Assessing tools for sustainability: Bangladesh context. In *Proceedings of the Second Meeting of the Academic Forum of Regional Government for Sustainable Development*. Fremantle, Australia, CD ROM.

Hossain, A., & Marinova, D. (2005). Poverty alleviation – a push towards unsustainability in Bangladesh? In Proceedings of the International Conference on Engaging Communities. Brisbane, Queensland: Queensland Department of Main Roads [E-text type]. www.engagingcommunities2005.org/abstracts/Hossain-Amzad-final.pdf.

Kumar, S. (1986). *The Schumacher lectures*: *Vol. 2*. London: Abacus.

Marinova, D., & Raven, M. (2006). Indigenous knowledge and intellectual property: A sustainability agenda. *Journal of economic surveys, 20*, 587-606.

Marinova, D., Hossain, A., & Hossain-Rhaman, P. (2006). Sustaining local lifestyle through self-reliance: Core principles. In S. Wooltorton, & D. Marinova (Eds), *Sharing wisdom for our future: Environmental education in action* (373-380). Melbourne, Victoria: Australian Association for Environmental Education.

Natrass, B., & Altomare, M. (1999). *The natural steps for business: Wealth, ecology and the evolutionary corporation*. Gabiola Islands, Canada: New Society Publishers.

Maskus, K. E. (2000). *Intellectual property rights in the global economy*. Washington, DC: Institute for International Economics.

McDaniel, J. (1989). *The Madness of the saints: Ecstatic religion in Bengal*. Chicago: Chicago University Press.

Posey, D. A., & Dutfield, G. (1996). Beyond intellectual property: Toward traditional resource rights for Indigenous peoples and local communities. Ottawa: IDRC Books.

Shiva, V. (2000). *Tomorrow's biodiversity*. London: Thomas and Hudson.

Shiva, V. (2005). Global trade and intellectual property: Threats to indigenous resources. In S. Krimsky, & P. Shorett (Eds), *Why we need a genetic bill of rights: Rights and liberties in the biotech age* (98-106). Lanham, MD: Rowman & Littlefield Publishers.

Ulber, R., & Sell, D. (Eds), (2007). *White biotechnology*. Berlin: Springer-Verlag.

Warburton, K., Pillai-McGarry, U., & Ramage, D. (2000). *Integrated biosystems for sustainable development*. Proceedings of the InFoRM 2000 National Workshop on Integrated Food Production and Resource Management. RIRDC Publication No 01/174 [E-text]. www.rirdc.gov.au/reports/Ras/01-174sum.html.

World Health Organisation (WHO), (2005). *Ecosystems and human well-being: Health synthesis*, Millennium Ecosystem Assessment. Geneva: WHO.

# INDEX

## A

## B

## C

## F

## G

## H

## I

## M

## N



## O

## P

## Q



## R

## S

## T



## U

## Z